SpringerWienNewYork

T0137929

Dietmar Dietrich, Georg Fodor,
Gerhard Zucker and Dietmar Bruckner (eds.)

Simulating the Mind

A Technical Neuropsychoanalytical Approach

SpringerWienNewYork

Dietmar Dietrich
Institute of Computer Technology, University of Vienna, Austria

Georg Fodor
Vienna Psychoanalytic Society, Austria; University of Cape Town, South Africa

Gerhard Zucker
Institute of Computer Technology, University of Vienna, Austria

Dietmar Bruckner
Institute of Computer Technology, University of Vienna, Austria

© 2010 Springer-Verlag/Wien
Printed in Germany

SpringerWienNewYork is a part of
Springer Science + Business Media
springer.at

ISBN 978-3-211-99869-4   ISBN 978-3-211-09451-8

Printed on acid-free and chlorine-free bleached paper

With 63 figures in black and white.

Preface by a Computer Pioneer

The ENF (2007) forum was about new approaches to information processing using a bionic approach by attempting to copy the principles of human information processing. Clearly, current information processing can already do many things automatically which previously could only be done using human intelligence. However, a lot of human intervention is already required before information technology yields results, not only in providing the machinery on which the processing should run, but also for programming the task in question.

In this regard humanity underestimate the by now omnipresent computer; this tool is not only – as it appears – a typewriter also storing and copying. It is intimately interconnected with its environment, with the institutions in which it operates.

Computers work with models. For humans, the use of models is also a basic feature, a "built-in facility" of human thinking, of our dealing with the world. We do not even have to learn the concept of a model. It evolves automatically, a set of common features is remembered and a name can be attached to memory content. However, what sometimes occurs with models is that the difference is neglected and the model is mistaken for reality. This effect is supported by not giving different names to reality (the real "object") and the model.

We use many kinds of models. For that matter a name for a concept is a model: the content of which we build up through experience and thinking, by reading and contextual observation. It is thus evident that one can model everything and, given a model of something, it can immediately be programmed on a computer (not necessarily in a brief period of time).

In the scientific borderland that the ENF encompassed, the distinction between reality and model is by far more important than in individual fields. Equating reality and can lead to very wrong conclusions. Massive redundancy in communication is by far preferable to misunderstanding and errors.

The question of computer consciousness was also raised. The approach in the past was to look inwardly and mimic human consciousness. It obtains input information and, supported by memory, derives output information which triggers action. That is, of course, precisely the structure of a computer – how could the pioneers have done otherwise? However, what have we achieved? The computer could act like a human[1]. Yet we have no basis for thinking that we have created computer consciousness. Our programs have made computer consciousness unnecessary.

However, I wish the authors good luck in their attempt to implement a first hypothesis of computer consciousness.

Heinz Zemanek

---

[1] and easily pass the Turing test – which I consider an error in thinking by Turing, a logical mistake such an accomplished logician should never have committed.

The Cooperation Between Siemens and the TU Vienna

A good two years ago Professor Dietrich from the Institute of Computer Technology at the Vienna University of Technology approached me with an intriguing idea. Why not apply the findings on the human brain, its method of thinking and learning, to computer models and software solutions for industrial automation engineering?

At the Automation Division at Siemens Austria, we are involved in many forward-looking concepts such as the development of the digital factory. It was therefore only logical that Siemens should explore this idea from the Vienna University of Technology in greater depth.

Automation and automation processes are developed on the basis of a multitude of increasingly intelligent sensors and actuators, integrated and interconnected by means of software. Consequently, all components are networked and communicate acquired "knowledge" in centralized and decentralized control units. This interplay is reflected in manufacturing and processing sectors with ever greater complexity. Processes are determined by increasingly complicated algorithms.

These advances in automation give rise to the questions: can these models be explained or described by the findings that have been made on how the human brain functions or by models of the human brain and could we also derive simplifications from these findings? The question is one that we wrestle with daily and that is what determines our approach in research and development.

As Siemens automation experts, we voiced our clear support for close cooperation in staging the conference week 2007 in Vienna (npsa, ENF, IEEE INDIN, industry day) and participated in it wholeheartedly. Helmut Gierse, the former director of Siemens Automation Engineering worldwide, shared our interest in the interconnection between these fields and responded promptly to the request for support. He demonstrated his foresight, by continually pushing for strategic derivatives to be found in this field for this change of paradigm.

What underlies the challenge to learn? This question was answered in the interplay of time and content at the conference of neuropsychoanalysts and through a partnership with the world's top researchers on the subject of brain functions. I believe this approach lent a valid scientific base and reliability to the structural organization of the arrangement for cooperation with the Vienna University of Technology, the host of the joint forum.

We engineers wondered how we would even understand this "medical" language used by the neuropsychoanalysts. A model of the brain's structure based on the ideas of Sigmund Freud provided the "consoling" answer to that question and allowed us to lay the groundwork for an intelligible discussion for the varying scientific disciplines of the analysts and the engineers.

Siemens agreed to provide financial and organizational support for staging the conference week. The initial budget was straightforward and the number of participants was clearly defined.

The announcement of a conference of psychoanalysts in Freud's home town was well-received internationally due to the overarching theme and became a genuine challenge for Siemens in Vienna. We were likewise aware of helping to bring about something new, something revolutionary with the foundation laid by the Vienna University of Technology. In addition, we definitely wanted to offer the visitors coming in from around the world not only new insights from the lectures and workshops but also a professional event and a memorable stay in Vienna.

To ensure a good conference of practical value, we at Siemens Austria invited excellent and visionary guest speakers on production engineering ("Trendsetting Automation for the Entire Value Chain") and on process engineering ("Trends and Innovations in Process Innovation"). They presented the world's most advanced research findings in these fields and provided food for thought to the participants in the IEEE INDIN conference, all of which was intended to smooth the way from research to application.

We thank the organizers and initiators from the Vienna University of Technology for allowing Siemens to contribute actively to this subject. We took great pleasure in co-organizing this event. The praise voiced by participants from many countries is compelling indeed.

Dir. Ing. Wolfgang Morrenth

Preface by the Editors

The attempt at modeling and finally simulating the human mental apparatus is a sublime goal and a real challenge. The authors aim to attain this goal with the cooperation of engineers and neuropsychoanalysts.

This book contains thoughts and ideas of a lot of people who work in rather different fields. We are happy that we managed to bring them all together and unite them in this book. Psychoanalysis and engineering have a lot to contribute to each other and we hope that this book continues to foster cooperation between these two disciplines. A first milestone was the "Engineering and Neuropsychoanalysis Forum (ENF)", of which we have included the full proceedings. It was a day full of fascinating presentations and fruitful discussions between representatives of psychoanalysis, neuropsychoanalysis, neurology, engineering, and many other fields. However, the ENF also revealed that it is likely that when two persons from two different fields have a conversation, that though they may believe that they understand each other, they in fact do not and also cannot.

We packed all the video recordings of that day onto three DVDs, so that not only the participants have access to it. Another reason is to be able to learn – in retrospect – from the obvious misunderstandings for the next meetings.

A further step was to issue a Call for Participation, which brought forth more than twenty publications, of which ten are included in Part III of this book.

Engineering and psychoanalysis at first sight do not have much in common; they have different ways of working and use a different vocabulary or – even worse – the same vocabulary, but with different meanings. We wanted to build a bridge between these two and we hope that this book will afford researchers a basic understanding of each other's discipline and an idea as to how cooperation may be deepened. If we manage to employ psychoanalytical findings in engineering research and the other way around, then the book was well worth the effort.

We want to thank the HarrisonMcCain Foundation for supporting author's cooperation. We strongly want to thank all authors for their contributions and hard work to make this book successful. We also would like to express our gratitude to the ENF session chairs and all attendees, and all who encouraged us afterwards to continue, or submitted suggestions for further progress. In particular we want to thank Elisabeth Brainin, Tobias Deutsch, Dorothee Dietrich, Harald Hareter, Wolfgang Jantzen, Friedrich Kamm, Roland Lang, Josef Mitterbauer, Brit Müller, Brigitte Palensky, Peter Palensky, David Olds, Marianne Robert, Charlotte Rösener, Mark Solms, Samy Teicher, Anna Tmej, Mihaela Ulieru, Rosemarie Velik and Heimo Zeilinger for their constructive contributions and vital remarks which went far beyond their duties as chairs or authors. We appreciate all insights into scientific disciplines that may appear foreign to other researchers. In fact, putting this book together showed clearly what our research field is all about: working cooperatively on the same task and gaining mutual benefit from the results.

Dietmar Dietrich   Georg Fodor   Gerhard Zucker (né Pratl)   Dietmar Bruckner

How it all began ...

In 1998 I was told that a child in my neighborhood was hurt in an accident with boiling water in the kitchen. The injuries were so severe that they would be visible for her entire life. At this time my area of research, building automation, concerned topics like energy efficiency, security and comfort, but not safety. This event made me aware of the lack of effort put into safety. I took the precise scenario of a child in danger in the kitchen environment as the basis for a ground breaking research project, namely the perception in building automation for recognizing potentially dangerous situations which require adequate protective measures. Nanotech as well as modern web-cams should help to make it possible to install hundreds of smart, distributed sensors in the living environment, for example in kitchens, to allow a system to perceive precarious situations.

One of the major boundary condition of building automation was and still is the necessity of using inexpensive sensory equipment instead of very costly industrial cameras. Sensors should be diverse and complementary, and the system should be able to use sensors which will only arrive on the market in 10 to 15 years. The whole electronic section must be kept simple. Everything must be plug & play, which means that the network must be able to incorporate additional components without configuration requirements. To achieve this level of interoperability, further research in the areas of embedded systems and software compatibility is required.

In embedded systems, when considering a bionic approach, the diversity of sensors and their interconnectivity are highly relevant. Unfortunately, artificial intelligence, cognitive science, and related fields are still hardly able to transfer the principles of intelligence occurring in nature into technical models. As the market also does not require such systems *now*, there is no indication of a shift in the way of thinking; no new developments may be expected in the foreseeable future. In this situation the question arises: which chance do we have for using the intelligence occurring in nature as a model for developing technical solutions?

In 1998, while searching for solutions for principles of intelligence I had a crucial get-together: I met Helmut Reiser, professor in the field of special education in Hannover, Germany. We discussed how humans perceive and recognize their environment. Mr. Reiser explained in detail recent research results from the fields of psychology, pedagogy, and psychoanalysis. He told me that humans associate sensory images; however association is just a small piece of the puzzle. The whole procedure leading to perception is much more complex than was previously thought. In all sensory modalities only characteristic features can be perceived. Via the perceived features, images already stored in memory are associated with each other. With the help of the characteristic features and the associated data from memory the mental representation of the perception is constructed. Therefore, humans can only recognize what they can associate to images which have been internalized previously.

The discussion with Reiser lasted several days until I came to the conclusion that bionics in the area of artificial intelligence has to follow completely new paths. Back in Vienna at the Institute of Computer Technology I found two PhD students – Ms. Clara Tamarit and Mr. Gerhard Russ – who I could convince to assist me exploring new ways for automation. Soon we had the support of "diploma" students.

Being a chip designer, it was clear to me that I would have to utilize a top-down design methodology to have a chance of success. Unfortunately, this approach was incompatible with contemporary results from artificial intelligence as well as psychological and pedagogical sciences. Furthermore, these sciences did not possess a technically implementable unitary model of the mental apparatus at hand, just a collection of psychological views of various schools which mutually reject parts of each other's concepts. Computer engineers need consistent models; ambiguity is not compatible with computer programming. So, how were we to proceed?

My wife Dorothee, a psychoanalyst in training, gave me the decisive hint: she told me of two well known researchers in the field of neuroscience/psychoanalysis/behavioral neurology: Oliver Sacks and Mark Solms. She also put me in touch with a Viennese psychoanalyst, Thomas Aichhorn, with whom I could establish a fruitful relationship. It was him, who then introduced me to Elisabeth Brainin, a neurologist, psychiatrist and training analyst. This connection was the key to a breakthrough. Elisabeth and her husband Samy Teicher – also a psychoanalyst, – became our permanent consultants and reviewers and helped us develop our unitary concept. That was the time of the 2nd generation of PhD students, of which one after another completed their thesis: Gerhard Zucker (né Pratl), Dietmar Bruckner, Wolfgang Burgstaller, Charlotte Rösener, and Brigitte Palensky.

In the early days of this second phase my wife Dorothee introduced Georg Fodor to me who is also a psychoanalyst and a colleague of Mark Solms. Things started to speed up. I was convinced especially by Elisabeth to visit Mark in Cape Town. An opportunity arose when I gave a guest lecture in Pretoria. I will never forget the meeting with Mark. We soon came to the understanding that our views and aims were very similar in that the human mind has to be investigated using scientific methods and the psychoanalytical model.

Subsequently Mark and our group met regularly, which helped us and specifically our project manager at that time, Peter Palensky, to be very successful. He was able to bring the international IEEE[2] conference INDIN[3] to Vienna. At the same time Mark and Georg where also trying to bring the international npsa[4] conference to Vienna close to the timing of the INDIN conference. This enabled us to hold a joint forum called "ENF[5] – Emulating the Mind" on July 23rd, 2007, orga-

---

[2] Institute of Electrical and Electronics Engineers

[3] Industrial Informatics

[4] Neuropsychoanalysis Centre

[5] Engineering and Neuropsychoanalysis Forum

nized by Gerhard Zucker and his team. At this point I have to express special thanks to Wolfgang Morrenth, head of Siemens A&D (Automation and Drives), who believed in the necessity of a paradigm shift in automation and enabled us to realize our ideas for such a forum. With his help the company Siemens sponsored the enormously costly four-part-event for seven days. The npsa conference came first followed by the joint workshop between neuropsychoanalysts and engineers. Then, right after the ENF, the IEEE INDIN conference was held followed finally by the industry workshop with its respective activities. Siemens had surpassed themselves.

The ENF was a great success. One could feel the atmosphere in the hall. It was recorded by several cameras; a set of DVDs was produced. It was completely clear to me that the workshop has to finally lead to the compilation of a book containing more than just the proceedings of the workshop. We tried to include as many discussion results and new insights achieved at the ENF as a result of this forum. The decision to edit this book was the moment the third and current phase of the project was born. Suddenly, many things became clear, but also led to much more complicated questions.

As a consequence we enlarged our team. It now not only consists of engineers and psychoanalysts as consultants, but also of regularly employed psychoanalysts, Brit Müller and Anna Tmej. They are working together with the third generation of PhD students of this project – Tobias Deutsch, Roland Lang, Rosemarie Velik, Heimo Zeilinger, and Tehseen Zia. And we – engineers and psychoanalysts – started understanding each other.

I would like to express my gratitude to those, who contributed to the very difficult first steps of our common path towards a unitary model of the mental apparatus as basis for a new field of research. Reflecting on my experiences with the first steps taken into our new field I found a better understanding of the great achievements of Sigmund Freud and of Mark Solms. Freud knew more than a hundred years ago that the mechanistic way of thinking is not applicable in information theory. This insight has still not reached many IT-engineers (the error will be explained in detail in later chapters). In my point of view Freud's greatest achievement was not to think differently than most of his colleagues, but to stand so firmly and consistently to his convictions. I see Mark Solms' greatest achievement in succeeding to bring together two faculties hostile to each other, neurology and psychoanalysis, and in starting an association with members from both fields, the npsa.

I am curious whether we will succeed in establishing an international cooperation of engineers and neuropsychoanalysts. The ENF workshop raised hopes, but also showed the enormous ditches and reservations and all the diverse sciences' peculiarities.

There is a lot of work still to be done.

Dietmar Dietrich

List of Authors in Alphabetic Order

Etienne	Barnard	Meraka Institute, Pretoria, South Africa
Ariane	Bazan	Université Libre de Bruxelles, Belgium
Elisabeth	Brainin	Vienna Psychoanalytic Society, Austria
Andrzej	Buller	Cama-Soft A.I. Lab, Gdynia, Poland
Dietmar	Bruckner	Vienna University of Technology, Austria
Andrea	Clarici	University of Trieste, Italy
Tobias	Deutsch	Vienna University of Technology, Austria
Dietmar	Dietrich	Vienna University of Technology, Austria
Joseph	Dodds	Institute of the Czech Psychoanalytical Society, Czech Republic; Charles University; University of New York in Prague, Czech Republic
Georg	Fodor	Vienna Psychoanalytic Society, Austria; University of Cape Town, South Africa
Robert	Galatzer-Levy	Institute for Psychoanalysis, Chicago Illinois, United States; University of Chicago, United States
Wolfgang	Jantzen	Universität Bremen, Germany
Wolfgang	Kastner	Vienna University of Technology, Austria
Roland	Lang	Vienna University of Technology, Austria
Brit	Müller	Vienna Circle of Psychoanalysis, Austria; Vienna University of Technology, Austria
David	Olds	Columbia University, New York, United States
Brigitte	Palensky (née Lorenz)	Vienna University of Technology, Austria
Peter	Palensky	University of Pretoria, South Africa; Berkeley National Laboratory, CA, United States
Jaak	Panksepp	Veterinary Medicine Washington State University, United States
Charlotte	Rösener	ACS Technologies, Austria
Matthias J.	Schlemmer	Vienna University of Technology, Austria
Aaron	Sloman	The University Of Birmingham, United Kingdom
Mark	Solms	International Neuropsychoanalysis Centre, London, UK; University of Cape Town, South Africa
Ron	Spielman	Australian Psychoanalytical Society, Australia

Samy	Teicher	Vienna Psychoanalytic Society, Austria
Anna	Tmej	Vienna Psychoanalytic Society, Austria; Vienna University of Technology, Austria
Mika	Turkia	University of Helsinki, Finland
Mihaela	Ulieru	Canada Research Chair; The University Of New Brunswick, Canada
Rosemarie	Velik	Vienna University of Technology, Austria
Markus	Vincze	Vienna University of Technology, Austria
Takahiro	Yakoh	Keio University, Japan
Yoram	Yovell	University of Haifa, Israel
Heimo	Zeilinger	Vienna University of Technology, Austria
Gerhard	Zucker (né Pratl)	Vienna University of Technology, Austria

Organization of the Book

The editors claim that the human brain and mind can be seen as objects which can be investigated with scientific principles. The findings can and must be applied in engineering. One cannot proceed on this path without taking the knowledge of psychoanalysis, the scientists concerned with the human mind, into account. However, their models need to be analyzed from a technological point of view, in cooperation, naturally, with psychoanalysts. And vice versa, the editors are convinced that engineers and computer scientists can pass on knowledge about modeling and synthesis to the science of psychoanalysis.

This book documents the attempt at the "First international Engineering and Neuropsychoanalysis Forum (ENF 2007): *Emulating the Mind*" in working out a unitary view on how to proceed in simulating the mental apparatus.

The book is organized in four parts in order to highlight separate views and to incorporate the contributions of various authors:

Part I constitutes the theoretical base, worked out while strictly following the principles of natural science, to which later contributions will refer. Additionally, major research results from the past five years are presented. They were also presented in abbreviated form during the (ENF 2007) "*Emulating the Mind*", incorporating results from the speakers and their co-authors.

Part II is formed by invited publications that represent the content of the ENF 2007 forum and the summaries of the discussions on that day.

Part III contains strongly reviewed publications collected in a Call for Contributions, representing the reactions to the ENF 2007 forum.

Part IV comprises explanations for engineers and psychoanalysts in a glossary-like fashion. It contains basic explanations of terms to simplify understanding for readers with a different scientific background.

Table of Contents

Part I Theory

List of Authors in Alphabetic Order

Elisabeth	Brainin	Vienna Psychoanalytic Society, Austria
Dietmar	Bruckner	Vienna University of Technology, Austria
Tobias	Deutsch	Vienna University of Technology, Austria
Dietmar	Dietrich	Vienna University of Technology, Austria
Georg	Fodor	Vienna Psychoanalytic Society, Austria; University of Cape Town, South Africa
Roland	Lang	Vienna University of Technology, Austria
Brit	Müller	Vienna Circle of Psychoanalysis, Austria; Vienna University of Technology, Austria
Samy	Teicher	Vienna Psychoanalytic Society, Austria
Anna	Tmej	Vienna Psychoanalytic Society, Austria; Vienna University of Technology, Austria
Rosemarie	Velik	Vienna University of Technology, Austria
Heimo	Zeilinger	Vienna University of Technology, Austria
Gerhard	Zucker (né Pratl)	Vienna University of Technology, Austria

1   The Vision

The approach to developing models described within the following chapters breaks with some of the previously used approaches in Artificial Intelligence. This is the first attempt to use methods from psychoanalysis organized in a strictly top-down design method in order to take an important step towards the creation of intelligent systems. Hence, the vision and the research hypothesis are described in the beginning and will hopefully prove to have sufficient grounds for this approach.

When bringing together two fundamentally different scientific disciplines such as psychoanalysis and engineering, it is of great importance to clearly define the theoretical basis. The first phase of the project revealed very quickly that not just the methods, but also the vocabulary are completely different. Communication was challenging for all partners involved. In order to proceed scientifically it is inevitable that one first defines the building blocks to be used later on to build the model.

The developments in physics, material science, or chemistry are enormous. The packaging density in highly integrated circuits is a very good example. In the 1970s only 10.000 transistors fitted on one chip, in 1990 already 1.000.000 and today we have 100.000.000 transistors on one die, a number far beyond consideration in the 1970s (Khan 2007). Another example, although less spectacular, but of great interest for the market, is home and building automation. Hardly any office building can be found in the western hemisphere that does not utilize any modern building automation communication systems – fieldbusses. This part of automation has grown to a billion dollar market (Loy et al. 2001, Sauter et al. 2001), since the requirements on home and building automation are continuously rising. It may be assumed that 50,000 computer nodes (embedded systems) need to be installed in a large building today. These nodes have to be designed, integrated, and maintained. None of today's tools has sufficient efficiency to deal with this incredible number of network nodes and the information they provide in a corresponding cost efficient way.

Countless other examples could be listed here from the areas of motor vehicles, airplanes, trains, or energy technology, especially where safety and security considerations are of critical concern. Automation technology will eventually be integrated into all areas because "smart systems" are being increasingly integrated into every kind of imaginable object. This implies on the one hand an enormous amount of data that has to be processed, on the other hand that we will be incapable – considering today's technology – to properly process and interpret the accumulating data, let alone efficiently, and comprehensively. Contemporary technical systems are far less intelligent than they are required to be in order to interpret data as well as humans. Just a simple example from the area of recognizing scenarios: Behavior of people can be perceived and recognized via a camera. What problems may be anticipated? The videos from inside and outside of the airport are analyzed by specially trained security staff sitting in front of monitors in the security room. Automating this process is not trivial, but needs to be technically solved. How to implement it is a different question that nobody can yet answer: first of all the trade union blocks the utilization of such systems because they could be used to observe the personnel. That is to say, there is a clear distinction between purely automated surveillance and surveillance by humans[6]. Or another example: During the preparation for the Olympic games in Greece every newspaper featured an article on the huge troubles caused by the software for monitoring the facilities and events , mainly due to the unexpectedly high costs. The reason also lay in the complexity of interpreting scenarios.

Technology needs methods to mathematically describe algorithms, which have not yet been discovered. One way to solve that dilemma is the so-called bionic approach. To give an example of one of the greatest successes of Bionics, the technical modifications which lead to the development of the A320 from regular airplanes like the A31x shall be described: the Airbus A320 was the first plane worldwide to use fly-by-wire control technology, which enabled a reduction of fuel consumption by 27% and more. With this technology it was possible to hold the plane in an instable position – horizontally flat. For several, mainly legal reasons, it took Boeing quite a while to catch up; however both companies now have a roughly equivalent turnover.

The idea of keeping planes in an instable condition was taken from nature. It is a very good example of how a physical phenomenon from nature is investigated, its processes modeled, and then it turns out that purely mechanical control cannot do the same job – in this case for reasons of speed. The process could only be controlled, when a neuron-like communication system was developed – the fly-by-wire system – which was the basis for later developments in the fieldbus area.

The main idea of fly-by-wire systems is communication systems between intelligent nodes, for example between intelligent sensors, actuators, and control units. In comparison to the human body this is exactly the functionality of the peripheral nervous system. However, if we want to copy more than just communication from humans, i.e. their intelligence, we need to work with models of the human psyche, because this is where the major part of information processing takes place. This consideration leads directly to the first problems regarding communication between different scientific cultures or fields. Engineers differentiate clearly between hardware, software, and application. The hardware is the physical part, on which apparently the software runs. Obviously, the software is another description language for certain parts of the hardware. However, two different models are used in order to describe the system. The third model, the application, describes the function – what the thing, the computer actually does for the user. With the help of these models one can describe the whole computer. But what is the psyche in this scenario? The software? The application? We think that this differentiation has not yet been attempted. If we want to model the mind and the functions it performs,

---

[6] Surveillance done by humans clearly interferes with privacy, whereas surveillance by machines does not – if the information is not passed to humans again.

and if we additionally want engineers to understand it to enable them to simulate or emulate something, we need to define the terms. Therefore, in our approach the mental apparatus is seen as the equivalent to software whereas the psyche is defined as the application. We will discuss this in more detail later on.

First it needs to be clarified why we are convinced that psychoanalysis offers the answer. In the previous paragraphs it was stated that bionic communication principles can be used for, e.g. fly-by-wire systems. This was even possible without actually copying the whole physical base, the neurons, because electronics provided another base. The same is true for the mental apparatus. One does not need to model the whole human brain with its billions of neurons and interconnections in order to rebuild the processes of storing and processing information. Modeling the human brain neuron by neuron – the bottom-up approach – is not necessary. If one focuses on the mental apparatus, the top-down approach must be adhered to. If these boundary conditions and considerations are adopted and agreed on for the bionic approach then one only needs to turn to those sciences that are essentially engaged in the study of the mental apparatus for the underlying basis for the model. In our view it is a fundamental error to search for and operate in terms referring to specific isolated psychological functions like emotions, feelings, and consciousness, making various assumptions and creating hypotheses about these terms without actually dealing with the corresponding disciplines. The relevant literature has to be seriously considered. This point will be stressed frequently in later chapters, since it is not a scientific approach to simply reject or ignore existing research results, and not to incorporate the state of the art of the respective field into one's own scientific work.

Coming back to the sciences concerned with the psyche. Which one of them could be utilized for the first step of development if the task is to design a unitary model of the mental apparatus following the top-down approach? Neurology is by definition not applicable, because neurologists primarily work with physical, chemical, and physiological processes. A model for higher cognitive functional units is not available. Another science, pedagogy, focuses its interest on learning and education. Psychology on the other hand investigates only very specific aspects of psychological functioning. Such results cannot be used for synthesizing holistic models (a fact that will be expanded on later), mainly because the relevant aspects of control theory – interconnectedness and mutual dependencies through feedback and regulatory circuits – are insufficiently considered. Another candidate, psychiatry, is exclusively concerned with pathological phenomena. Hence, the only science remaining to be considered is psychoanalysis. It has – as the only candidate – an approximation of what engineers would regard to be a unitary model, which gives engineers a very good starting point. If a system is designed from scratch it is not possible to integrate every function in detail. Just think of the first computer developed by Zuse compared to contemporary machines. Every scientific pocket calculator today is capable of performing more functions. Other examples are software tools for architects from the 1980's. Compared to modern ones they were just programs to draw lines.

Simple outline models will have to be used as a basis, which – following a top-down approach – can only represent simple functions. Once this first basic concept is available, its functionality can be refined and enhanced.

But what is the psychoanalytical model – even in its most simple form? It is obviously not exactly what Freud defined. As a natural scientist Freud frequently refined his previous findings and revised his assumptions and hypotheses in cooperation with others. The important set of knowledge for our endeavor is the model available today. However, this immediately poses the first major obstacle. The training of psychoanalysts lasts between 7 and 10 years and focuses on clinical application and not on research[7]. So how can an engineer acquire this knowledge? We engineers will have to accept that this will not be possible.

Every electrical engineer would immediately understand and agree that it would take a psychoanalyst an unbelievably long time to be able to understand and apply Maxwell's equations, which form the foundation of electrical engineering, at a scientific level. Conversely the question arises how any scientist in Artificial Intelligence or Cognitive Science, who grew up in the engineering world, dares to claim that they can understand the psyche in such detail and such depth that they can implement it in a robot. Obviously this is a gross misperception of realities. Understanding this however leaves us with a grave dilemma: How can we engineers acquire the necessary expertise to achieve our ambitious goal of creating the aforementioned unitary model? The only viable way out of this dilemma is – in our view – to incorporate "sources" of such knowledge, namely psychoanalysts, into our project team and work very closely with them. We see no other way.

In this way the problem of communication between cultures, which was mentioned above as a critical issue, can be overcome. If such different cultures collide – as happened between the npsa conference and the INDIN conference in July, 2007 (see (ENF 07) and respective DVDs) – it is very interesting to observe that it takes the partners concerned quite a long time to recognize that they talk at cross purposes. Overcoming this obstacle is essential and will be very beneficial for the envisaged joint model.

Another important phenomenon has to be mentioned: We have to re-question again well-loved terms, which we have become used to as technical terms.

Let us first take up Freud's insights from his effort to describe cognitive processes based on neurological investigation. His failure to do so allowed him to understand – and that has to be seen as an outstanding achievement – that mechanistic thinking would never allow a deeper understanding of an information processing apparatus. Let us recapitulate the fly-by-wire example. The great success of the Airbus was the replacement of traditional mechanical control devices (which combine information and power transmission) with electronic ones. The control system consists of sensors capturing the input signals, a centralized control unit computing and sending the necessary actuator signals and the actuators, that

---

[7] Wiener Arbeitskreis für Psychoanalyse; http://www.psychoanalyse.org/

translate the signals into mechanical movement. All components use decentralized power supplies connected via communication systems called fieldbus systems. A similar development is currently taking place in the areas of vehicle technology and railway technology, and one can already predict today that this will have to happen in all areas where automation technology will be used to control the process. This has been a massive trend for the last 15 to 20 years: to strictly separate energy flow and information flow within a system in order to allow a computer system to process the data.

Looking at Darwin's theory of evolution, nature serves as an example for this approach (Dietrich and Sauter 2000). In the amoeba the two flows, energy and information, are connected while in creatures of a higher order like insects or bugs communication and information processing units have already developed: neurons. The reason for this development is easily explained: If information flow and energy flow are combined in the same process, compromises have to be made. Every mechanical transmission in a vehicle is based on such compromises between different requirements, for example the requirement for the car to be moving along a curve or the requirement of driving straight on a highway. Processes can only show some intelligence in terms of flexibility if they incorporate information processing units (e.g. as in the bug).

These considerations lead to the next aspect, why it is not just adequate, but essential to take a deeper look at control systems like the mental apparatus. The answer is that we want automation systems – machines – to solve problems which require far more intelligence than a bug offers. Hence, there are two single possibilities: Either we try to enhance previous findings from natural science research, or we attempt to follow the path of the Airbus and try to find out how nature constructed the mental apparatus. In my personal view the latter strategy is the more promising as it appears to be the shorter and easier way, considering Darwin's hypothesis that in nature the optimal solution to a problem will always succeed. Therefore, it is always promising to look for solutions in the nature surrounding us.

What are the key features that define the mental apparatus? The first insight has already been mentioned: information is processed in a dedicated apparatus. The inputs for that apparatus are the sensors and the outputs are the actuators. But some phenomena even increase complexity: One example would be feedback loops in hormonal systems. Hormonal output is not just activated by actuators and sensed by sensors; hormones also have the potential to change the behavior of the physiology (hardware). However in this first approach it is not possible to elaborate on this particular aspect in more detail.

The second important insight was to understand how humans recognize images and scenarios (see Section II.2). We do not perceive our surroundings like a camera, pixel by pixel. Rather, human perception and recognition turns out to be a highly complex sequence of processes. The first step is to depict characteristic parameters like lines and curves. These characteristic features are then associated with images which have been stored previously in our "big" database – our memory. All this data is then processed to what we consciously and unconsciously see, smell, taste. Following this picture, the mental apparatus is just an enormous database that mainly processes images and scenarios. This model of perception and recognition provides us with a theoretical basis for a description of the functionality of the psychic apparatus which in due course can be used for modeling.

The third insight, which Freud also arrived at early on, is: the psyche consists of dynamic processes of inhibition and excitation. The ideal world of stable tranquil harmony is a product of wishful thinking and does simply not exist in the real world. Inside the mental apparatus antagonistic forces collide continually and have to be constantly counterbalanced by the mental apparatus. This means that the process or processes of the mental apparatus are constantly in danger of slipping, getting out of balance.

Let us go back to the example of the Airbus 320. It has a physical stability problem: as long as the plane is in a stable position, it needs more fuel. Therefore, it flies in an unstable position and has to be kept in balance between climbing and sinking at a frequency of 10 Hz. It seems that nature has found large advantages in the instability of dynamic processes. This is the direction we need to direct our investigations. The only thing we can say at this point is that there must be reasons why such systems prevail. Let us adopt the principle and analyze it.

All abovementioned considerations lead in one direction: How can an engineer utilize the findings of psychoanalysis? It has been mentioned only once that the mutual communication between both disciplines has advantageous consequences for both. Being engineers, we can only talk about a few psychoanalytical expectations, because we cannot speak for psychoanalysts here. However, we can decide together what is possible and what is not. Though there is one thing we have learned in these last nine years of hard, collective research work: the way that psychoanalysts think will sometimes differ to our opinion as engineers, especially where developing a model is concerned. Our strength as engineers is developing systems. That is what we have to offer.

2   Basics

In this chapter we wish to build a common understanding of the scientific disciplines of engineering and psychoanalysis. Since they not only differ in vocabulary, but also in methodology and way of working it is not easy to build this bridge, and it will require at least two textbooks to do so. However, as this is not feasible, we have included short introductions to the main scientific fields, followed by discussions about their position in science and the relations between them.

2.1   Introduction to Automation

The introduction of the term automation has been awarded to the Ford manager D. S. Harder in 1936. Originally, it defined the automatic transport of a mechanical object from some point A to point B. Later the term comprised a complete manufacturing process. Nowadays automation refers principally to any kind of feed-forward or feedback control of processes.

While early automation was only concerned with purely mechanical processes, it was realized quickly that electrical engineering had the potential to not only transfer control impulses, but also to generate information from diverse sensory sources. In this way it was possible to design devices capable of starting engines or operating valves dependent on particular measurable conditions like temperature or pressure. Due to these historical roots, the whole idea of automation was originally based on a mechanistic way of thinking. The first transformation in this respect started with the presentation of the first programmable computers: the *Z3* in 1941 by Zuse in Germany and the *Mark 1* in 1944 by Aiken in the US. This development led to fundamental reconsiderations in automation: The new devices offered the possibility to work with information about the observed process, separated from the process itself. In other words: to replace the built-in mechanical control systems by much more flexible and precise electronic automation systems which operated remotely from the work process proper.

To summarize, the development of automation up to its current level may be subdivided into four steps: The very first automated processes were highly sophisticated mechanical processes (like wind mills, steam engines, or mechanical watches). In the next step, electrical communication was used to transfer the information about events in automated systems. In a following step, more information about the process was collected via sensors and translated into electric signals. These signals were then used for computation in fixed electronic circuits in order to better control the process. The fourth step to modern automation was the introduction of computer systems instead of the electronic circuits, which allowed process control using free and flexible computation[8]. The whole automation system thus consists of the following components:

a) Sensors to observe various physical parameters of the process and translate them into electrical signals.

b) The control unit, which is usually a computer, collects data, stores it and computes signals for the actuators. Those signals are based on target-performance comparisons between collected data and a programmed abstract model of the process, and

c) The actuators that translate the computed output signals of the control unit back into physical actions – which consecutively affect the process such that it attains its desired physical condition.

This structure is typical for a so-called fieldbus system, which will be referred to later.

---

[8] One example to illustrate these developments is light: The first lights after torches were gas lamps (totally mechanical), later replaced by electric lamps controlled by switches (simple electrical communication). The next generation including electronic control was light controlled by a dimmer switch. Finally, modern illumination systems are controlled by fieldbus communication systems.

This functional distinction between the four areas: sensors, controllers, actuators, and communication systems led to the current sub specializations in automation.

There is sensor development on the one side – which is basically a physical or mechanical topic. It has specialized towards micro technology or, later, nanotechnology. Recent research results from this field predict a considerable increase in low-priced, but highly capable sensors.

The second field, communication technology, was accepted in automation since the late 80ies with the rise of fieldbusses. A fieldbus is defined as a communication system connecting sensors, actuators, and controllers in order to exchange data. Fieldbusses are bidirectional, which means each device can send and receive information as required. Sensors and actuators received their own control and communication modules to make them "intelligent" (or "smart"). Without this additional feature it was just possible to read a sensor's value or to send a control signal to an actuator. Bidirectional communication e.g. allows the control unit to (re)parameterize sensors (program them with new parameters like exposure time for cameras) or actuators, if necessary, to communicate their range of abilities to the control unit during operation.

With the capabilities of the various devices enhanced in such a way, the separation between sensor, actuator, and control unit becomes less and less important, since devices originating from the three fields of specialization can nowadays perform almost all functions involved in the entire automation system. Therefore, today the function type is distinguished, not the device type, since a smart sensor possesses fully integrated controllers and communication systems.

The third focus in automation is the discipline concerned with the unit between sensors and actuators – the controller, or control unit. A controller processes the essence of the automation system: the information about the process. This research field is divided into several parts, mainly dependent on the area of application. Two of them are of greater concern for us, namely computer engineering and electrical engineering. Computer engineering is only concerned with software. Automation in computer engineering deals with mathematical descriptions and algorithms of and for processes creating abstract models of the real process implemented in software. Electrical engineering additionally deals with the computer hardware. Special purpose chips and devices – embedded systems[9] – are an emerging topic. They are specifically designed for each automation system in order to provide the respective controllers with exactly the functionality they need for their specific performance. In recent times the interconnection of such devices has become increasingly important again.

The forth area is the area of propulsion technology. In earlier times this was an application-dependent, highly specialized field, because e.g. very different technologies are involved in building a motor, depending on the power that the motor

---

[9] The first step in automation towards computer controllers was to use so-called PLCs (Programmable Logic Controller), central processing units.

must provide. Nowadays, the separation between sensor, propulsion, communication, and control technology softens, because computer technology unites the technologies and allows systems to be designed that integrate all of the above technologies in one device.

Nowadays, fieldbus and related technologies can be seen in virtually all areas of technology, starting from air and space technology (where it originated) to industrial automation, building automation, and forest or railroad technologies.

Hot topics in current automation research are the following:

- Some applications demand reaction times within specific time constraints, so that the process can be kept on track – so-called *real-time systems*. How fast this reaction is in terms of time units, depends on the application: it can be minutes for controlling the temperature in a room, or small fractions of a second for controlling the braking system of a car.

- Another new direction of research is targeted at *decentralized systems*, where the computational power and the intelligence is distributed over a network. This is in contrast to traditional centralized systems where a dedicated (controller) unit is responsible for all computations and control.

- The latest areas of interest are: embedded systems, ad-hoc sensor networks, and smart dust. These terms refer to various stages of development of basically the same idea – having large numbers of small, wirelessly interconnected sensing devices embedded into physical objects. These networks of sensors are supposed to measure any desired physical condition in any desired location.

  The vision of an *embedded system* is to integrate the whole electronic part into one single chip or small device.

  The specific criterion of *ad-hoc sensor networks* is that not all units within the network are directly connected. Direct communication depends on the distance between devices.

  Smart dust, today being merely a utopia, is the term for having one single unit containing sensor, controller and communication, that is so small that it virtually requires no energy source and can be placed within wallpaper.

- The area of safety deals with ensuring reliability and availability. The problem with electronic devices is they may fail from the very beginning on (in contrast to mechanical devices). This probability for failure remains approximately constant over their lifetime (while mechanical systems fail with higher probability as they get older due to frictional loss and material fatigue).

- The next direction targets the costs of the entire life-cycle of electronic devices considering design, fabrication, and maintenance. This is particularly important in Europe, since personnel costs here are decidedly high, and natural resources are only available in a few countries like Norway or Great Britain. It is therefore no wonder that in the area of automation Germany and France lead the world market, both, in industrial and building automation.

Finally, a general trend in automation needs to be highlighted in particular, being the paradigm shift away from mechanistic thinking. As mentioned in the first chapter, Freud claimed 100 years ago, that, in order to understand the mental apparatus, mechanistic ways of thinking have to be relinquished. The same is true for automation. In the current situation we engineers learn (and teach) modeling physical processes via abstraction. We learn to linearize (to describe complicated correlations with simplified, linear correlations), to minimize the number of observed parameters, to ignore side effects, and to split complex processes into sub processes and thus to create models with strongly reduced complexity. In addition to the obvious reasons, to promote easier understanding and handling of correlations, another practical reason is that today's sensors are expensive to purchase, install, and maintain as well as often being of low quality. Hence, the process cannot be observed in too much detail. We can therefore afford to have not very differentiated models.

However, our goal in process control must be the opposite! The whole process should be observed in full detail including all influences coming from outside the process ("disturbances"). And if it is necessary to split the model of the process into smaller functional units, the mutual interferences of those functional units must be described ("interfaces").

Such requirements imply mathematical and computational challenges. The incorporation of all – presumably non-linear – influences will not allow for simple solutions. However, taking ever-increasing computational resources into account, it is no longer necessary to limit the model's description to simple mathematical specifications. Complex problems can be simulated in split seconds on standard desktop PCs. Even the computational power of most embedded systems is sufficient to handle such problems. Additionally, if units with less capacity are involved in distributed systems, they can outsource computational tasks to more potent units in the network.

The level of available resources today is favorable; only one thing needs to be considered: There must be enough sensors to observe the process. By using a large number of sensors automation systems imitate a rule of nature. Creatures at a higher level of development tend to have more sensors to perceive their environment with. So if we want automation to improve its abilities, we need to provide a sufficient amount and variety of sensors.

As stated above, a sensor translates a physical property into an electric unit. Modern (smart) sensors do not send analog values, but digitize them and send them via a communication system (e.g. fieldbus) to the controller. Aside from the great advantages of flexible computation, digitizing information close to the process allows the replacement of electronic circuit elements. This is desirable because electronic circuit elements always obstruct complicated, higher-order circuits due to their inaccuracy. Additionally, due to their size they are influenced by electro-magnetic waves. All these side effects can be avoided by processing the data in a computer.

In the areas of digital technology, the principles mentioned are already state-of-the-art. Two impressive examples are software-radios and vehicles. A software-radio is a device that digitizes radio waves – which a normal radio also receives – i.e. it turns the radio waves into digital data. All processing, filtering, amplifying, etc. is done digitally. Non-linearity or inaccuracy of components no longer constitutes implementation problems. All circuit related electrical engineering problems are easily solved in the computer. Finally, only the computer-generated signal for the speaker is converted back to an analog signal. The only significant effort remaining is the mathematical description of the process itself.

Secondly, the current trend with vehicles is to replace mechanical and especially moving parts by electronics. In particular that means the engines in future will be directly integrated into the wheels in order to eliminate the steering wheel and the brakes. Friction will be dramatically reduced, while completely new ways will open up for the design of the passenger compartment in a car. Such cars will have to have a large number of integrated sensors.

To summarize, the future trends in automation are: Processes will be observed by an increasing number of sensors. Sensors as well as actuators will become more intelligent. All units in automation systems will become interconnected and transfer their knowledge to control units which perform the necessary mathematical computations. These control units will either be organized centrally or distributed over networks. Finally, we claim that these computations will also incorporate principles of the human mental apparatus – the highest developed control unit known in nature. In this respect, Darwin will again be proven right.

2.2 Introduction to Psychoanalysis

Psychoanalysis is a discipline which incorporates subjectivity into natural science. Sigmund Freud (1856-1939), its founder, was the first to conceive and formulate it. Based on Freud's books and lectures, psychoanalysis has since developed into an intellectual movement with a lasting influence on Western culture. Psychoanalysis significantly determines the present concept of psychotherapy and in a wider sense our idea of the human condition as a whole. It has opened up new perspectives for neuroscience, psychiatry and psychology and is a source of inspiration for many fields in the humanities and for a great variety of arts. Against the background of an understanding of unconscious processes, the assessment of civilizing, societal, cultural and artistic processes has been substantially changed and enhanced. The description of psychoanalytic theory below is closely associated with Sigmund Freud as a person. This should not be misunderstood as an effort to place the person of the founding father of psychoanalysis centre stage; in the first decades of psychoanalytic theory, however, its development was strongly linked to Sigmund Freud's research, his development and later of course also his rejection of his theories.

What remains central, however, is that psychoanalysis represents a theory of the development and functioning of mental processes based on clinical observations made in a specific relationship situation which can be described as the psychoanalytical setting.

Freud describes psychoanalysis as follows: *"Psycho-analysis is the name (1) of a procedure for the investigation of mental processes which are almost inaccessible in any other way, (2) of a method (based upon that investigation) for the treatment of neurotic disorders and (3) of a collection of psychological information obtained along those lines, which is gradually being accumulated into a new scientific discipline"* (Freud 1923a, p. 235). He goes on to differentiate: *"The assumption that there are unconscious mental processes, the recognition of the theory of resistance and repression, the appreciation of the importance of sexuality and the Oedipus complex – these constitute the principal subject-matter of psycho-analysis and the foundations of its theory"* (Freud 1923a, p. 250).

Accordingly, psychoanalysis is:

- A psychological theory of the mental life and experience, particularly of their unconscious parts;
- A procedure for the investigation of unconscious mental processes;
- A method for the treatment of mental disorders;
- It is a process to understand the unconscious psychic reality of a person and its method is primarily that of observation and interpretation.

Since its conception more than 100 years ago, psychoanalysis has developed further and gone in different directions resulting in different methods of psychoanalytic work and thinking. Freud's writings, however, continue to be the basic introduction to psychoanalysis and reading his texts remains indispensable for any in-depth study of this particular science.

Born in 1856, Freud came from an already widely assimilated and liberal Jewish family. In the Freud biography his markedly enlightened and rationalist attitude is related among other factors to this background.

The main publications written in the initial years of his work concentrate on neurological topics – for instance his essays on aphasia (Freud 1891) and on infantile cerebral palsy (1897a). In his essays on aphasia, Freud argued decidedly against a general localization of brain functions and thus explaining, among other things, brain activity. Additionally he describes that language would not be possible without consciousness; for the engineering sciences this must mean that a machine will never be able to understand human language if it does not possess a technical equivalent of consciousness. A first publication on hypnotic suggestive therapy (Freud 1892-93a) was followed by others on psychological-psycho-therapeutic topics, notably "The Neuro-Psychoses of Defense" (Freud 1894a) and "Studies on Hysteria" (Freud 1895d) co-written with Josef Breuer. The "repression theory" set forth in these writings which explores the mechanism of neuroses is later replaced by the "theory of seduction" with the aim of defining the cause of neuroses. This theory is developed further in "Further Remarks on the Neuro-Psychoses of Defense" and in "The Etiology of Hysteria" published in 1896.

In 1900 Freud published "The Interpretation of Dreams", which is considered to be his most important work; it is based on his insights from his self-analysis that started in 1897. The assumptions necessary to explain the events taking place in our dreams, i.e. the existence of two mental agencies, the system "conscious-preconscious" on the one hand and the system "unconscious" on the other with a "censor" localized in between, introduce the fundamentals of "metapsychology" and include the basic psychoanalytical assumptions about mental processes.

In "Three Essays on the Theory of Sexuality" (Freud 1905d) Freud undertakes to substantially expand the general understanding of sexuality. In the time span of only a few years he proceeds from abandoning the theory of seduction in 1897, thus bringing the "prehistory" of psychoanalysis to an end, to the development of its essential contents – the theory of the unconscious, of infantile sexuality and the conception of neuroses being based on unresolved conflicts from childhood.

Later publications include Freud's writings on the general psychological and biological fundamentals of psychoanalysis (theories of the unconscious, the drives and sexuality) clinical papers (case studies, essays on the general theory of neurosis and psychotherapy) as well as texts on the application of psychoanalytic findings to non-clinical contexts and certain phenomena (parapraxis, jokes, literature, religion, culture and society).

According to the above assertion that apart from being a method for investigation and a psychological theory, psychoanalysis also represents a method for conflict resolution, Freud's writings also provide meta-psychological observations as well as reflections on technical and methodological issues in treatment. With regard to the latter he develops concepts important for psychotherapy such as "free association", "poised attention", "resistance", "transference", "abstinence", "importance of the setting" and others.

Attacks on psychoanalysis are part of the history of psychoanalysis. From its inception in turn-of-the-century Vienna, psychoanalysis has inspired strong feelings. Early reviews of Freud's account of his novel treatment of hysteria emphasized the humanity of the talking cure, the methods it offered for exploring the inner world of human emotional life. At the same time, the negative reviews were hostile to the point of dismissal, attacking the subjective, "unverifiable" nature of the analyst's report as well as ridiculing the emphasis that Freud placed on the sexual origins of mental distress.

The famous splits in psychoanalysis in the years 1911-13 are associated with the names of Sigmund Freud, Alfred Adler and Carl Gustav Jung. Freud and Adler were divided about the fundamental question of human action: are we masters of our fate or do we act out of instinctual conflicts of which we are largely unaware? Adler, inspired by his personal and clinical experience, felt certain that human beings do create their own world. In his view, we do things not *because of* but *in order to*. Freud could not accept Adler's view of the centrality of human agency in light of his experience of the unconscious motivation in human affairs (Handlbauer 1990).

The fundamental disagreement between Freud and Jung was the conflict between modern Western science and traditional Western spirituality (Freud 1974). Jung took a classic position in opposition to the materialism of Western science by insisting that nothing could exist unless it was perceived. Jung created a psychology by evocation of symbol and myth and the intriguing question of the collective unconscious. It is the idea of a collective psychology, operating at a very deep level in the human psyche (Jung 1976).

A paradigm shift of psychoanalysis began to emerge in the 1920s with the psychoanalysis of children associated with Melanie Klein's work. The fundamental theoretical problem posed by child analysis was to understand the origin of childhood anxieties. Klein pioneered the technique of play analysis – the use of a set of small toys as a substitute for the technique of free association. Through observation of the play, the depths of a child's inner world could be reached by making immediate interpretations about the child's earliest feelings and unconscious fantasies, many of which have made an important contribution to our understanding of early separation anxiety. Within the framework of the existing metapsychology, Klein added on and redefined highly relevant aspects of the early development of the psyche and to the technical repertoire of psychoanalysis (Klein 1935, 1946).

Now psychoanalysis is still widening its scope. With the most recent advances of the neuropsychoanalytic method psychoanalysis once more is enlarging its scientific possibilities. There is an approximation between psychoanalysis and neuroscience, where neuroscience could provide a more concrete empirical and conceptual basis for correlations with psychoanalytical concepts. Since the 1990s, Mark Solms, president of the International Neuropsychoanalysis Society (npsa has worked on presenting a new method of research in neuroscience (Kaplan-Solms & Solms 2000) adapting the traditional neuropsychological method developed by (Luria 1973) for the study of cognitive functions extended to emotional phenomena, which psychoanalysis has been investigating for over a century.

Essential for the topic of this book, however, is Freud's metapsychology as it forms the psychoanalytic frame of reference of the ARS-PA model described later on.

In 1915 Freud defines the concept of a "metapsychology" more specifically as a system informing about the localizations of mental processes in the postulated parts of the psychic apparatus (topography), the forces responsible for their occurrence (dynamics) and the amounts of energy (economics) thus relocated: "I propose that when we have succeeded in describing a psychical process in its dynamic, topographical and economic aspects, we should speak of it as a meta-psychological presentation" (Freud 1915e, p. 181).

As the essential characteristics of topography, dynamics and economics are the defining factors of Freud's metapsychological theory and as they are repeatedly applied to different topics, these three perspectives shall be summarized here:

The topographical perspective: *"The crucial determinants of behavior are unconscious"* (Rapaport 1960, p. 46).

Psychoanalysis names and conceptualizes what is not perceived or not perceivable and does so exclusively in psychological concepts such as motivation, affects or thoughts. Therefore Freud developed the topographical models with their locations of unconscious, preconscious and conscious contents (Freud 1915e), as well as of the Ego, the Id and the Super-Ego (Freud 1923b). It thus dissociates itself from other psychological systems using non-psychological concepts for what is not perceivable (e.g. brain fields, neuronal connections, etc.). Differences between determinants defining the conscious and unconscious are expressed in the concepts of the primary and secondary process. In his early work Freud described the "preconscious/conscious" and the "unconscious" as the basic systems. Later they are subordinated in the structural concepts of Ego, Id and Super-Ego.

The dynamic perspective: *"The ultimate determinants of all behavior are the drives"* (Rapaport 1960, p. 47).

The central role Freud confers to the drives is the result of two of his basic assumptions: Behavior is determined by drives and there is an infantile sexuality. Freud revises his drive theory on several occasions. Initially he assumed a duality between sexual drive and self-preservation drive and Ego drive. In his third and last drive theory he distinguishes between the libidinal sexual drive with its partial drives and the aggressive death drive.

And last, but not least, the description of the third perspective, the economic perspective: *"All behavior disposes of and is regulated by psychological energy"* (Rapaport 1960, p. 50).

It is important to note that Freud does not identify psychic energies with any kind of biochemical energy: *"From the point of view of the energy economy of the organism, the exchanges of psychological energy may be considered as the work of an information engineering network which controls the biochemical energy output of overt behavior"* (Rapaport 1960, p. 52). The primary process of the unconscious is determined by drive energies. It is regulated by the tendency to ease the drive (pleasure principle). The aim is to instantly discharge agglomerations of energy by means of the mechanisms of displacement, condensation, symbolization, etc.

The secondary process accords to objective reality and reaches the object in reality through postponement and trial action by way of thoughts thus postponing the discharge of drive energy until the object has been found.

As the ARS-PA model, which will be explored in Chapter 3, refers to Freud's second topographical model, it will need to be described in more detail.

While in the first topographical model a distinction between the agencies of unconscious and preconscious/conscious was made, in 1923 – against the background of his theory of the unconscious conflict – Freud conceived the second topography which is also called the structural model.

Id, Ego and Super-Ego are now the three agencies of the personality: The Id involves everything representing the drives mentally. The characteristics of the Id are identical to those Freud attributed to the system unconscious in the first topography.

The concept of the Ego comprehends all organizing, integrative and synthetic functions of the mind.

The Freudian concept of the Super-Ego consists of the prohibition/imperative and ideal functions. Freud first introduced it in "The Ego and the Id" (1923b) explaining that the critical function called Super-Ego represents an agency which has become separate from the Ego and seems to dominate it.

2.3   Psychoanalysis, a Natural Science?

Freud's notion that psychoanalysis should be classified as a natural science was fiercely contradicted from the very beginning. This question was however reduced to the *verification* of hypotheses, thus, the apparent problem has been the natural scientific verification of psychoanalysis ever since it was first introduced. Freud himself unequivocally understood his new science as pertaining to the natural sciences. However, taking a look at the various areas of the natural sciences today, such as mathematics, modern astrophysics, nuclear physics, theoretical computer science, etc., it becomes apparent rather quickly that a definition of the contents of the natural sciences as being verifiable cannot and must not be sufficient any longer. Engineers have therefore adopted the following easy but very agreeable principle: The physical world can *only* be described via the employment of models. These models can at first be hypothetically accepted to be subsequently critically challenged by experimental observations. Hence, these models pass as "true" as long as they fit, stand the tests and are therefore not *falsified*. According to Popper (1971) this scientific approach is correct, because:

(a) Statements are never definitely verifiable, because in principle, "the truth" can never be found.

(b) But they are definitely falsifiable.

Hence the following ensues: Only those statements declare something about the reality of experience that can fail, i.e. those which can be submitted to a methodical investigation of this kind. Engineering as well as psychoanalytical models are such kinds of statements, describing the reality of experience. Both types of models bear up against the criterion of falsifiability. But in psychoanalysis the consideration method is a humanistic one (e.g. hermeneutics) besides a natural scientific approach, which nowadays increases with the development of the neuropsychoanalytic research. Its methods and results of the latest research approximate psychoanalysis to natural science, where psychoanalysis originally comes from.

The question must however be asked why in the realms of psychoanalysis and especially regarding the subject of "consciousness", so many people try to declare this approach to be non-scientific – without any scientific acceptable arguments.

In keeping with Waelder (1983), it is possible to say that the older sciences have acquired a certain prestige and are characterized by relative consent among their exponents which in turn resulted in the general assumption that they were universally verifiable. In contrast to the long-established natural sciences, the more recent behavioral sciences are not granted this kind of renown.

Freud identifies the observation as the foundation of his science. Yet, he adds *"[...] so psychoanalysis warns us not to equate perceptions by means of consciousness with the unconscious mental processes which are their object. Like the physical, the psychical is not necessarily in reality what it appears to be" (Freud*

1915e, p. 171). And he goes on to say, "[...] that we have no other aim but that of translating into theory the results of observation" (ibid., p. 190).

A reflection of the scientific qualities of psychoanalysis inevitably raises the ancient philosophical issue of the "mind-body" dichotomy. The old question is based on the idea that mental processes are opposed to bodily states while the psychoanalytical theory proceeds from the assumption that all manifestations of mental processes (it is no coincidence that Freud speaks of the psychical apparatus) are caused by the body and its demands upon the mental apparatus.

Today we find various scientific schools of thought in fields such as biology, genetics, ethnology concerned with the psyche, the instincts and other unconscious mental states such as the dream. It was not until the development of modern imaging techniques[10] and the investigation of hormonal cellular substances effective within the central nervous system that it became possible to turn to these research fields in a manner different from the one Freud applied. In modern neurosciences Freud's struggle for a scientific perception of the psyche as opposed to the metaphysical comparison of body and mind is no longer questioned and has resulted in an approximation between these disciplines and psychoanalysis. The outcome is the emergence of "neuropsychoanalysis".

Returning to Freud, the concept of the unconscious and the concept of the instincts have prompted repeated attacks on psychoanalysis. Freud defined instincts as follows, *"If now we apply ourselves to considering mental life from a biological point of view, an 'instinct' appears to us as a concept on the frontier between the mental and the somatic, as the psychical representative of the stimuli originating from within the organism and reaching the mind as a measure of the demand made upon the mind for work in consequence of its connection with the body"* (Freud 1915c, p. 121-122). The erotic instincts and the destructive instincts are psychological concepts whereas the life and death instincts are biological notions further removed from psychological observations. They correspond to biological forces.

Waelder (1983) provides a phylogenetic definition of instincts and intelligence determining them as differentiation products of the animal instincts.

Freud (1940a) defines the Ego as follows:

"Here are the principal characteristics of the Ego. In consequence of the pre-established connection between sense perception and muscular action, the Ego has voluntary movement at its command. It has the task of self-preservation. As regards external events, it performs that task by becoming aware of stimuli, by storing up experiences about them (in the memory), by avoiding excessively strong stimuli (through flight), by dealing with moderate stimuli (through adaptation)

---

[10] In this context it must however be pointed out that imaging techniques have to be critically questioned. In Luria's model (1973), in which three hierarchical layers are defined, the lowest layer and maybe also the second lower layer can be "scanned" to a limited extent, while the third layer is processed distributed and therefore cannot.

and finally by learning to bring about expedient changes in the external world to its own advantage (through activity). As regards internal events, in relation to the Id, it performs that task by gaining control over the demands of the instincts, by deciding whether they are to be allowed satisfaction by postponing that satisfaction to times and circumstances favorable in the external world or by suppressing their excitations entirely" (Ibid., p. 143-144).

He writes about the basic assumption of psychoanalysis, that

"[...] we know two kinds of things about what we call our psyche (or mental life): firstly, its bodily organs and scene of action, the brain (or nervous system) and, on the other hand, our acts of consciousness, which are immediate data and cannot be further explained by any sort of description [...] We assume that mental life is the functioning of an apparatus to which we ascribe the characteristics of being extended in space and of being made up of several portions – which we imagine, that is, as resembling a telescope or microscope or something of the kind. Notwithstanding some earlier attempts in the same direction, the consistent working-out of a conception such as this is a scientific novelty" (Ibid., p. 144-145).

He continues that the view

"[...] which held that the psychical is unconscious in itself, enabled psychology to take its place as a natural science like any other. The processes with which it is concerned are in themselves just as unknowable as those dealt with by other sciences, by chemistry or physics, for example; but it is possible to establish the laws which they obey [...] in short to arrive at what is described as an 'understanding' of the field of natural phenomena in question. This cannot be effected without framing fresh hypotheses and creating fresh concepts; [...] but these are [...] to be appreciated as an enrichment of science. They can lay claim to the same value of approximations that belongs to the corresponding intellectual scaffolding found in other natural sciences, [...] so, it too will be entirely in accordance with our expectations if the basic concepts and principles of the new science (instinct, nervous energy, etc.) remain for a considerable time no less indeterminate than those of the older sciences (force, mass, attraction, etc.).
Every science is based on observations and experiences arrived at through the medium of our psychical apparatus. But since our science has as its subject that apparatus itself, the analogy ends here" (Ibid., p. 158-159).

And finally,

> *"[...] the hypothesis we have adopted of a psychical apparatus [...] has put us in a position to establish psychology on foundations similar to those of any other science, such, for instance, as physics. In our sciences as in the others the problem is the same: behind the attributes (qualities) of the object under examination which are presented directly to our perception, we have to discover something else which is more independent of the particular receptive capacity of our sense organs and which approximates more closely to what may be supposed to be the real state of affairs. [...] it is as though we were to say in physics: 'If we could see clearly enough we should find that what appears to be a solid body is made up of particles of such and such a shape and size and occupying such and such relative positions.' In the meantime we try to increase the efficiency of our sense organs to the furthest possible extent by artificial aids; but it may be expected that all such efforts will fail to affect the ultimate outcome. Reality will always remain 'unknowable' "* (Ibid., p. 196).

Karen and Mark Solms (2000) underline Freud's conceptual leap by which psychoanalysis became a science, that is, a natural science. Freud treated processes that were not conscious as if they were conscious, thereby translating them into the language of perception.

Since Freud did not concentrate on the neurological manifestations of the psychic apparatus, psychoanalysis departed from the physical world. It could not benefit from technological progress which significantly increased the efficiency of our sense organs. Today we are able to proceed from a psychological understanding of the unconscious, from a neurological understanding of the mental apparatus and from a re-conceptualization of the classic theories of function and localization.

Kaplan-Solms and Solms refer primarily to these three developments which enable us today to correlate the discoveries of psychoanalysis with those of neurosciences: *"Rather, our aim is to supplement the traditional viewpoints of metapsychology with a new, 'physical' point of view"* (Kaplan-Solms & Solms 2002, p. 251). They show that psychological structures such as Ego, Id and Super-Ego can be ascribed to certain functional systems and regions inside the brain. The physiological view of the psychic apparatus leads them to the thesis: *"If we extend our 'physical point of view' of the mental apparatus to its logical extremes, we would have to say that whereas the sensory and motor organs at the periphery of the body are under the purview of the Ego, the vital organs in the interior of the body fall within the domain of the Id. [...] the Id, although it seems to reside in the deepest recesses of both the mind and the body, is in direct contact with the outside world in three important places. The viscera emerge from under the skin in the mucous orifices of the mouth, the anus, and the genitals"* (Ibid., p. 283). They refer to these apertures as the "senso-motoric" organs of the Id.

The approach of Kaplan-Solms and Solms shows the need for a collaboration of the most diverse scientific fields in order to arrive at an approximation of the complex problems of neurosciences and psychoanalysis.

In the year 1948, Warren Weaver, a US mathematician and information theorist, wrote the following about "mixed teams" of sciences from the most diverse fields:

"The second of the wartime advances is the 'mixed-team' approach of operations analysis. These terms require explanation, although they are very familiar to those who were concerned with the application of mathematical methods to military affairs.

These operations analysis groups were, moreover, what may be called mixed teams. Although mathematicians, physicists, and engineers were essential, the best of the groups also contained physiologists, biochemists, psychologists, and a variety of representatives of other fields of the biochemical and social sciences. Among the outstanding members of English mixed teams, for example, were an endocrinologist and an X-ray crystallographer. Under the pressure of war, these mixed teams pooled their resources and focused all their different insights on the common problems. It was found, in spite of the modern tendencies toward intense scientific specialization, that members of such diverse groups could work together and could form a unit which was much greater than the mere sum of its parts. It was shown that these groups could tackle certain problems of organized complexity, and get useful answers" (Weaver 1948).

Will we again have to wait until we will be pressured enough so that a synergy between the two worlds of psychoanalysis and engineering will simply unfold, or can we already start to think about its reasonableness? Should we not already expand bionics by psychoanalytic principles now?

These deliberations are essential for the notions presented in this book. They provide the basis for natural scientists, engineers and psychoanalysts to be able to co-operate on a model of the mental apparatus which would undoubtedly meet all the terms set forth by a problem of "organized complexity" – a subject which is ranked among the most important topics in science nowadays.

2.4   Neuropsychoanalysis

While Sigmund Freud's scientific reputation has been under serious attack for decades especially from the neurosciences, the originality of his ideas was far less doubted. It is therefore even more surprising to learn that current neuroscientific developments seem to support a significant number of the basic assumptions that are fundamental to the psychoanalytic model of the development and workings of the mind, which Freud named the "psychic apparatus".

A man of no small reputation, Eric Kandel, the second psychiatrist in history to be awarded with a Nobel Prize in medicine, only recently spoke of the psychoanalytic model as being "still the most coherent and intellectually satisfying view of the mind" (Kandel 1999, p. 505) and expects it to play a decisive role in his "new intellectual framework for psychiatry".

If so, Sigmund Freud and his model may after all not only have been original, but also truly scientific. For those who think that the psychoanalytic model of the mind is a highly original work of neuroscience, namely the neuroscientific field covering the natural phenomenon subjectivity, an exciting new opportunity evolves. Emotionality and motivation, areas of nature which until not very long ago were placed by the neurosciences somewhere between neglect, dread and contempt, have become respectable fields of neuroscientific investigation. Moreover, neuroscientific research has produced a solid bulk of results about the functioning of the brain, which are surprisingly similar to psychoanalytic hypotheses.

2.4.1 Psychoanalysis and Neuroscience: The Time has Come to Correlate

Solms (2004) summarized basic psychoanalytic concepts which find support in recent neuroscientific concepts.

While Freud's contemporaries rejected his central notion that most mental processes occur unconsciously, this view is hardly disputed anymore in today's neuroscience. This does not prove though that we actively repress conflictual thoughts as Freud suggested when he spoke of the dynamic unconscious. But even this concept finds growing support in case studies but also in laboratory research reported about by neuroscientists.

Neuroscientists have identified different memory systems, some of which are unconscious, which mediate emotional learning. The hippocampus, which lays down memories that are consciously accessible, are not involved in such processes. Therefore, no conscious memories are available and current events can "only" trigger remembrances of emotionally important memories. This causes conscious feelings, while the memory of the past event remains unconscious. In other words, one consciously experiences feelings but without conscious access to the event that triggers these feelings.

The major brain structures for forming conscious memories do not function in the first two years of life. Speaking of infantile amnesia Freud surmised, that it is not that we forget, but rather that we cannot recall these early childhood memories to consciousness. But this inability does not preclude them from affecting adult feelings and behavior. Developmental neurobiologists largely agree that early experiences, especially between infant and mother fundamentally shape our future personality and mental health. Yet none of these experiences can consciously be remembered. It becomes increasingly clear, that a great deal of our mental activity is unconsciously motivated.

The repressed part of the unconscious mind was described by Freud (1911b) as not operating according to the reality principle like the conscious mind, but rather in a wishful manner, blithely disregarding the rules of logic and the arrow of time. Indeed this has been shown in patients with damage to inhibitory structures of the brain (frontal limbic region) who developed a syndrome called Korsakow's psychosis. These patients are amnesic but unaware of this fact and replace the gaps of memory with fabricated stories called confabulations. Several studies showed the wishful quality of these confabulations but also of the entire process of perception and judgment.

Freud claimed that the pleasure principle gave expression to primitive animal drives placing the human being in very close vicinity to the animal world. Neuroscientists of our time now believe that the instinctual mechanisms governing human motivation are based on emotional control systems which we share with our primate relatives and with all mammals. Even if more basic mammalian instinctual circuits are classified today than the two drives hypothesized by Freud – sexuality and aggression – it is remarkable to find the "seeking system" (Panksepp 1998) regulated by the neurotransmitter dopamine bearing such remarkable resemblance to the "libido" which Freud has described as a pleasure seeking system that energizes most of our goal directed interactions with the world.

The "seeking system" is also of particular interest in dream research, where Freud's ideas regarding dreams – being instigated by the drives and expressing unconscious wishes – were discredited when rapid-eye-movement sleep and its strong correlations with dreaming were discovered. REM sleep occurred automatically and was driven by acetylcholine produced in a "mindless" part of the brain stem, which has nothing to do with emotion or motivation. Dreams now were regarded as meaningless, simple stories concocted by the brain under the influence of random activity caused by the brainstem. Yet recent work has revealed that dreaming and REM sleep are dissociable while dreaming seems to be generated by a network of structures centered in the forebrains instinctual-motivational circuits. These more recent views are strongly reminiscent of Freud's dream theory.

These developments and the advent of methods and technologies (e.g.: neuroimaging) unimaginable 100 years ago call for a scientific endeavor to correlate the psychoanalytic concepts, derived from observation and interpretation of subjective experiences, with the observations and interpretations of objective aspects of the mind studied by the classical neurosciences. The field of scientific research on these correlations bridging the two fields has been named neuropsychoanalysis. Its aim and raison d'être is to restart a direction of scientific inquiry that Freud and his psychoanalytic contemporaries postponed – for good scientific reasons, as we believe. There is a unique chance now to "finish the job" as eminent neuroscientist Jaak Panksepp puts it (Solms 2004, p. 62), to arrive at a scientific model of the mind where correlations between the subjective and the objective are solidly established.

2.4.2   Freud and Neurology

Freud was trained by outstanding authorities of the Medical School of Vienna, an institution highly renowned in his time, and was well versed in both neurological research and clinical diagnosis and treatment. It may be of interest specifically in this book to look into the questions of just why did Freud when it came to the understanding of the workings of the mind – the central steering system of the human individual - seemingly break with those paradigms in his field, which were dominant in his time. How did it come about that psychoanalysis has a functional model of the mind, and why do we think today that it lends itself for a top-down engineering procedure for automation systems?

In a few brushstrokes the milestones of a scientific development shall be sketched out in order to illustrate the seemingly abrupt turn (which is in fact – as we believe – a natural development) from mainstream neurology to the then new school of psychoanalysis, the culmination of Freud's scientific endeavors.

In his scientific career Freud started from the study of the elementary morphology of individual nerve cells, and of simple spinal structures of some of the lowest vertebrates, gradually moving upwards through the anatomy of human fetal cranial nerve nuclei and other brainstem structures, until finally reaching the human adult cerebrum, where he tackled some of the most complex clinical and theoretical problems of human neurology and neuropsychology. During two decades (1877-1900) of painstaking neuroscientific research Freud had published more than 200 books, articles and reviews including numerous significant contributions some of which established Freud as a leading international authority in two groups of neurological disorders involving language and movement, aphasia and cerebral palsy. (Accardo 1982, Amacher 1965, Jeliffe 1937, Jones 1953, Meyer-Palmedo & Fichtner 1982, Solms & Saling 1986, Sulloway1979, Vogel 1955, 1956, Triarhou & Del Cerro 1985)

It is obviously in relation to these latter problems that he began to confront the questions that concern us here and, specifically, the questions arising from the metaphysical problem of the relationship between brain and mind.

When Freud first tackled these questions, they had already been subject of intense scientific interest in European neurology for many years. Internal medicine, with neurology being one of its subspecialties slowly developing into a medical discipline in its own right, concerned itself with the diagnosis and treatment of diseases attacking the interior of the body – diseases which could for that reason not be apprehended directly in the living patient, but rather had to be inferred from their indirect manifestations, external symptoms and signs. The physician had to wait for the death of the patient, and the pathologist's report, before he could determine conclusively whether the physician's diagnosis had been correct or not. With the accumulation of experience over generations – regarding the sort of clinical presentation during life that tended to correlate with particular pathological-anatomical findings at post-mortem – it gradually became possible for the internal physician to recognize typical constellations of symptoms and signs, and thereby predict from the clinical presentation (with reasonable accuracy) what and where the underlying disease process was. This method of clinico-anatomical correlation was very successfully used in medicine.

Due to its convincing results it was also applied in the field of phrenology to serve a new and radically different purpose. This is because diseases of the brain, unlike those of any other organ, have immediate and dramatic effects upon the mind. Brain damage can change the patient as a person (Harlow 1868). The classical notion that the "brain is the organ of the mind" was therefore from the very start based on a prototype of the method of clinico-anatomical correlation. Its underlying hypothesis was that human mental faculties could, be "localized" in distinct areas of the cortex, like other functions of the body can be localized in specific organs or groups of cells (e.g. production of insulin in the ß-cells of the pancreas).

The clinical case studies reported by Paul Broca in 1861 and 1865 seemed to strongly support this hypothesis. Broca observed that damage to a circumscribed region of the brain (the third left frontal convolution) resulted in a loss of the power of speech. On this basis, deducting the localization of the function from the localization of the symptom, Broca localized the psychological faculty of language to a specific anatomical structure (subsequently known as Broca's area). From 1861 onwards this method began to be used in a formal, systematic way to map the different faculties of the human mind onto the multifarious convolutions of the human cerebral cortex. In 1874, Carl Wernicke, who – like Freud – was a pupil of the great Austrian neuroanatomist Theodor Meynert, discovered that damage to another part of the left cerebral hemisphere (the first temporal convolution – Wernicke's area) resulted in a loss of speech comprehension. Wernicke concluded that the cells of these two cortical areas – Broca's and Wernicke's centers – the centers for expressive and receptive language, respectively – were the repositories of memory images for the movements and sounds corresponding to spoken and heard words. These images were deposited in these areas by virtue of their relatively direct anatomical connection – via Meynert's "projection" fibers – with the relevant peripheral sensory-motor organs. Wernicke speculated further that the two centers in question were connected with one another by means of Meynert's "association" fibers, which passed through (or under) the intermediary region of cortex known as the insula. Eleven years later, Ludwig Lichtheim (1885) – also under the influence of Meynert's teachings – further elaborated this scheme, and postulated additional centers (one each for writing, reading and "concepts", all of which were likewise connected with one another[11].

Proceeding in this manner, the whole complex faculty of language was gradually localized in specific tissues of the brain. Subsequently – still under the influence of Meynert's teachings – numerous other complex psychological faculties, such as visual recognition (Munk, Lissauer) and voluntary action (Liepmann), were similarly localized in discrete cortical areas. (Riese, 1959)

By 1934, a few years before Freud's death, this approach to clinico-anatomical correlation had produced detailed maps of the psychological functions of the various cortical convolutions, including such complex faculties as the body scheme, emotional sensations, and the social ego (Kleist 1934). This research program resulted in the gradual development of a further sub-specialty within the new field of neurology, known as *behavioral neurology*.

---

[11] Lichtheim's model of the aphasias is being used in neuropsychological assessments to this day.

We know that Freud was thoroughly versed in the art of rational diagnosis and treatment of neurological diseases through the recognition of syndromes, based on knowledge obtained by the method of clinico-anatomical correlation. In fact we know that Freud was a particularly gifted practitioner of this art.

He published a series of articles at the time, attesting to his mastery of syndrome diagnosis and the clinico-anatomical method – articles which were later described as models of good neurological deduction. Clearly then, Freud was aware, shortly before he made the breakthrough into psychoanalysis, that there was a well-established method available to the clinical neurologist by means of which it was possible to correlate mental faculties on a rational basis with the physiology and anatomy of particular tissues in the brain. But if that was so, it raises a fundamental question: *why did Freud not use this method to identify the neurological correlates of the psychological processes that he later discovered?* And why don't *we* use it today?

In summary the answer to this question, as will be shown in greater detail, was that Freud believed that the clinico-anatomical method as used by his colleagues and teachers could not accommodate for:

(1) The *functional* nature of neurotic pathology.
The classical syndrome method was only useful clinically for the diagnosis of those illnesses which could be traced back to structural lesions of the nervous system, not for functional disorders.

(2) The *dynamic* nature of normal mental processes.
The clinico-anatomical method was only useful scientifically for identifying the physical correlates of *elementary components* of the mental apparatus, but not for representing the *dynamics* of the apparatus as a whole. And it was the functional dynamics that mattered most when it came to the understanding of complex mental processes.

(3) The complex inner workings of psychological processes.
A priori anatomical and physiological speculation was useless as a tool of scientific investigation of psychological processes, which had to be studied in their own right independently of brain anatomy and physiology. Freud's localizationist teachers and colleagues didn't conceive of mental processes as in any detail independent of physical ones. A problem in these works was that descriptions that were previously used for the mind were now redefined and used to describe the physics of the brain. This attitude was based above all on Theodor Meynert's view on consciousness: he understood consciousness and voluntary action as being the middle links in a chain of cause and effect in which the end links were the transmission of excitation in afferent and efferent nerves.

But let us look into these issues in more detail.

2.4.3   Neurosis in the German and the French School of Neurology

There was a group of diseases, the *neuroses*, hysteria and neurasthenia in particular, where no demonstrable lesion of the nervous system could be found at autopsy to account for the clinical symptomatology observed during the life of the patient.

In the German speaking world this absence of demonstrable lesion posed an insoluble problem for the clinico-anatomical approach. How was one to explain in anatomical terms the mechanism of clinical syndromes which had no pathological-anatomical basis? Some neurologists, Freud's teachers and colleagues in Vienna among them, developed elaborate speculative theories of the mechanisms of neurosis, based on a priori anatomical and physiological assumptions, while others simply declared that the neurosis were not fit subjects for serious scientific attention: if there was no lesion, there is no clinical syndrome.

While both, the German and the French school of neurology used the method of clinico-anatomical correlation, they differed in one instance which in regard to neurosis was decisive, namely whether one *started* from the clinical observations (like Charcot and the French school) or from the assumptions of physiological and anatomical theory (like Meynert and the German school). No demonstrable lesion was far less a problem for the French school. Clinical reality was primary. Charcot's goal therefore was not so much to explain the various clinical pictures but rather to identify, classify and describe the clinical syndromes of hysteria and neurasthenia as a *necessary prerequisite for physiological and anatomical explanation*, on the assumption that their pathological-anatomical correlates would eventually yield to advances in laboratory techniques.

In adherence to the French school Freud assumed that neuroses were based on "dynamic" lesions, disorders of physiological function as opposed to anatomical structure. Here the clinico-anatomical method had nothing to offer. How does one set about localizing the "seat" of a dynamic, functional disorder which could only be described as a "mode of reaction" of the nervous system as a whole, or expressed in a formula which was based on the "conditions of excitability" of the nervous system (Freud 1893f, 1956, Charcot 1889, Goetz, Bonduelle, Gelfand 1995)

Like Charcot and many others Freud believed that the clinico-anatomical knowledge of the major syndromes of physical neurology was complete by the end of the nineteenth century. With the French school Freud was moving away from an anatomical localizationist approach towards a dynamic physiological one. What he was looking for now was a formula which expressed the physiological laws of the *functional* apparatus that was disordered in these conditions.

It is notable to point out here that Freud in a monograph in 1893 (Freud, 1893b) made similar remarks about the need to conceptualize the *movement* disorders of children – cerebral palsy, an undeniably physical disorder – in dynamic, functional terms rather than static, anatomical terms. Freud suggested that these disorders, no less than the neuroses, are best understood as reflecting the laws of a complex functional system, with a long developmental history, rather than in localizationist, anatomical terms. This was an approach fundamental to the *dynamic* school of neurology.

But how does one set about exploring the functional laws underlying such disorders? There were simply no physiological methods available. This is where Freud shifted from the theoretical approach of his teachers to the procedure of the French school of neurology with its reliance on a clinical, bedside approach. Freud set about exploring the neuroses by means of purely clinical methods, investigating from effect to cause, commencing with a study of the disease at the bedside, as distinguished from the converse method of *a priori* reasoning, with the teachings of anatomy and physiology for its basis.

Yet in contrast to Charcot, who believed that these dynamic, functional lesions would be localizable one day with sufficiently advanced techniques, Freud's clinical observations led him to conclude that these lesions were purely psychological and would therefore never be anatomically localizable.

2.4.4 *Physiology and Psychology, two Distinct Fields*

The untangling of psychology and cortical physiology was the next step in Freud's transition from neuropathology to psychoanalysis. In this respect the theoretical influence of the English neurologist John Hughlings Jackson (1884) was of great importance. Here again the development occurred not primarily in relation to hysteria or any other neurosis, but rather in relation to aphasia, a fully neurological condition and, moreover, the very condition upon which the clinico-anatomical localization of mental functions was first attempted 25 years before.

In his 1891 monograph on aphasia, Freud exposed the major fallacy upon which the entire localizationist enterprise rested, namely its conflation of psychological and physiological concepts with shifting from descriptions in terms of the mind to descriptions in physical terms, leading to the assumption that individual ideas (German: *Vorstellungen*, "presentations") could be localized in individual nerve cells.

Basically the idea of the neurological function of language of the Meynert school was this: Encoded sonic stimuli travelled up the acoustic pathway in a purely physiological form until they reached the cells in the primary auditory cortex, at which point they were transformed into conscious sensations of sounds, which in turn excited cells in the Wernicke's center which transformed the sounds into memories of words, which were then recognized as such when they were transmitted to the centre of concepts. From either of these two centers sensory word images could in turn excite motor word images, which were located in cell groups in Broca's center, which then stimulated the motor cells in the precentral gyrus, which controlled the glossopharyngeal organs of speech. In this way the conscious and volitional aspects of language were transformed back into physiological processes.

Here is Freud's critique and his alternative view in his own words:

"Is it justifiable to immerse a nerve fiber, which over the whole length of its course has been only a physiological, subject to physiological modifications, with its end in the psyche and to furnish this end with an idea or a memory?" (Freud 1891b, p. 5)

Instead Freud postulated, that *"[...] the relationship between the chain of physiological events in the nervous system and the mental processes is probably not one of cause and effect. The former do not cease when the latter set in; they tend to continue, but, from a certain moment, a mental phenomenon corresponds to a part of the physiological chain, or to several parts. The psychic is a process parallel to the physiological"* (Freud, 1891b, p. 55).

In other words Freud hypothesized, that certain physiological processes occurring at specific points in the causal chain are experienced consciously as meaningful words or objects, as opposed to the idea, that these conscious experiences occur instead of physiological processes, and to the idea, that words or objects can actually be found inside the tissues corresponding to those links in the chain.

He conceived of conscious experiences and physiological processes to be two fundamentally different things. Words and objects are perceived in the mind, in parallel with certain physiological modifications which occur in the brain:

"In psychology the simple idea is to us something elementary which we can clearly differentiate from its connection with other ideas. This is why we are tempted to assume that its physiological correlate, i.e. the modification of the nerve cells which originates from the stimulation of the nerve fibers, be also something simple and localizable. Such an inference is, of course, entirely unwarranted; the qualities of this modification have to be established for themselves and independently of their psychological concomitants" (Freud 1891b, pp. 55-56). Henceforth, ideas would be studied in their own right and in their own terms:

"What then is the physiological correlate of the simple idea emerging and re-emerging? Obviously nothing static, but something in the nature of a process [...] it starts at a specific point in the cortex and from there spreads over the whole cortex and along certain pathways" (Ibid., p. 56).

The relationship between psychology and cortical physiology appears now immeasurably complexified:

"I shall carefully avoid the temptation to determine psychical locality in any anatomical fashion. I shall remain upon psychological ground, and I propose simply to follow the suggestion that we should picture the instrument that carries out our mental functions as resembling a compound microscope or a photographic apparatus, or something of the kind. On that basis, psychical locality will correspond to a point inside the apparatus at which one of the preliminary stages of the image comes into being. In the microscope and telescope, as we know, these occur in part at ideal points, regions in which no tangible component of the apparatus is situated" (Freud 1900a, p. 536).

2.4.5   The Psychic Apparatus – a Functional Model

In addition to conceptualizing mental processes in dynamic, functional terms (as opposed to static, anatomical ones) Freud would conceive of mental functions in purely *psychological* terms (as opposed to physiological ones); and, accordingly, he would recognize them as being unlocatable in the elements of the mental organ. Rather he would see them as being *distributed*; that is, as being located not *within* but rather *between* those elements.

"It will soon be clear what the mental apparatus is; but I must beg you not to ask what it is constructed of. That is not a subject of psychological interest. Psychology can be as indifferent to it as, for instance, optics can be to the question of whether the walls of a telescope are made of metal or cardboard. We shall leave entirely on one side the material line of approach, but not so the spatial one" (Freud 1926e, p. 194).

To summarize here, Freud's neurological views between 1885 and 1893 reveal that, first, he shifted away from an explanatory method based on anatomical and physiological speculation, in favor of a descriptive method, based on clinical observation. Next, the inferences that Freud drew from his clinical observations became increasingly functional and dynamic, that is to say, physiological as opposed to anatomical. Finally, Freud recognized that the functional apparatus he was investigating was constructed according to the laws of psychology rather than those of physiology; in other words that the relationship between psychology and physiology was a complex one, involving relationships between multiple elements rather than simple one-to-one correspondences.

He conceived of the "psychical apparatus" (1940a) as a hierarchical functional system, sub serving psychological processes, with a complex genesis and structure, which occupied the space *between* the basic sensory-motor centers. While the latter – the "cornerstones" of the system in his view could be localized to a certain degree in the conventional sense of the word, a functional system of this sort could not. Instead of conceptions of "centers" for complex psychic processes there arise the concepts of *dynamic structures or constellations of cerebral zones*, each of which comprise parts of the cortical portion of a given functional entity and preserve its specific function, while participating in its own way in the organization of one or another form of activity. Hence it was essential to gain a comprehensive understanding of the *internal psychological structure* of any mental process, before it would be possible – or even useful – to localize it.

A. R. Luria and Dynamic Neuropsychology

The work of the famous Russian neuroscientist A. R. Luria is of pivotal importance for the task in neuropsychoanalysis of correlating psychoanalytic with neuroscientific concepts, but also for the project dealt with in this book.

Luria was strongly committed to the clinical method. His scientific work is based on fundamental concepts and methods compatible with those of Freud, whose scientific writings he knew very well. (Van der Veer & Valsiner 1991, Kozulin 1984) He began with his "Dynamic Neuropsychology" where Freud had left off half a century before when he published his monograph "Traumatic Aphasia" in 1947.

Luria argued that the idea of localizing complex brain functions in discrete centers was untenable and distinguished between the localization of symptoms and the localization of functions. He described the kind of functions he referred to in this context with the following words:

"Most investigators who have examined the problem of cortical localization have understood the term function to mean the 'function of a particular tissue' [...] It is perfectly natural to consider that the secretion of bile is a function of the liver and the secretion of insulin is a function of the pancreas. It is equally logical to regard the perception of light as a function of the photosensitive elements of the retina and the highly specialized neurons of the visual cortex connected with them. However, this definition does not meet every use of the term function. When we speak of the 'function of respiration' this clearly cannot be understood as the function of a particular tissue. The ultimate object of respiration is to supply oxygen to the alveoli of the lungs to diffuse it through the walls of the alveoli into the blood. The whole process is carried out, not as a simple function of a particular tissue, but rather as a complete functional system, embodying many components belonging to different levels of the secretory, motor, and nervous apparatus. Such a 'functional system' [...] differs not only in the complexity of its structure but also in the mobility of its component parts" (Luria 1979, pp. 123-124).

In Luria's view focal brain lesions did not lead to "loss" of psychological functions but rather to complex, dynamic disorganization, and expected lesions affecting different component zones to lead to characteristic syndromes. In the conceptualization of these syndromes the secondary effects of the lesions *upon the functional system as a whole* also had to be considered.

Symptoms evoked by disturbances of different factors would have complicated structures and could have different causes. For this reason, symptoms had to be carefully analyzed and "qualified". The "qualification of the symptom" depended on a careful analysis of the patient's defects. The singling out of a symptom therefore was not the end but rather the beginning of an investigation, *aiming to obtain a detailed picture of its internal psychological structure in order to elucidate its psychological mechanism as well as to find out which other functional systems are disturbed by the same lesion and what factors underlie those other disturbances* (syndrome analysis). This allows identifying the *single, basic factor which underlies all of the symptoms produced by the one particular lesion.* The common underlying factor, in turn, points to the *basic function of that particular part of the brain.* In a next step the different ways in which each functional system is disturbed by lesions to *different parts of the brain* could be studied. Lesions in different parts of the brain will disturb the functional system in different ways. (Luria 1973, 1980, 1987)

Freud always believed that a correlative study of this sort was feasible in principle (although it was impossible for him in practice, given the limitations of the methods available at that time). He was well aware that every mental process must somehow be represented as a physiological process which occurs in the tissues of the brain; but he also held to the view that it was an error to *localize* complex mental faculties within circumscribed neurological "centers". His essential objection was that human mental processes are dynamic (i.e. "virtual") entities, which cannot be correlated isomorphically with the concrete, static structures of cerebral anatomy. He believed, moreover, that, "even if a localization of conscious processes was possible, it would still give no help towards *understanding* them" (Freud 1940a, pp. 144-5). He concluded that it would not be possible to understand conscious phenomena in neurological terms until their unconscious psychological substructure of the conscious phenomena had been laid bare. To do otherwise would be akin to trying to explain how a computer displays an image by studying its circuitry, but ignoring the program that uses this circuitry to create the image. Freud therefore devoted his scientific energies to the latter (purely functional) task, and deferred the correlative (neuro-psychological) task to future investigators, anticipating methodological advances.

Luria's modifications of the clinico-anatomical method (the method that Freud originally rejected) have made the correlation of dynamic mental functions with their neuroanatomical substrata possible, in a manner which accommodates both of Freud's major methodological objections. Luria's method of "dynamic localization" recognizes that the underlying *psychological* structure of a mental process needs to be clarified before it can be localized, and that these complex, *dynamic* processes cannot be correlated isomorphically with static anatomical structures.

2.4.6   Neuropsychoanalysis and Automation

One could describe the various attempts of using biological models for automation up to this point as being based on "simple localizationist" neurological models. We do not believe that this approach will push our research forward. Instead, we follow a new approach that is comparable to a dynamic neuropsychological model in its endeavor to accommodate the complexity of the central steering system of the human being – the mind.

The neuropsychoanalysts job is in one way surprisingly similar to the cooperative work with neuroscience. In both cases it is the functional concepts and models of the "psychic apparatus" that have to be contributed by neuropsychoanalysts for the search for possible correlation. However while neuroscientists obtained their concepts and models by researching a preexisting part of nature, the brain, we now use the "psychic apparatus" model for the construction of a technical system in a top down engineering process. It is a very new and amazing challenge to have neuro-psychoanalysts cooperating with engineers instead of biological scientists promising to give insights and raise questions from a very different viewpoint.

2.5   Realizing Psychic Functions in a Machine

To meet the challenges of modern automation as they were outlined in Section 2.1, we have to build a system with capabilities that are more human-like than today. Artificial Intelligence has brought forth many proposals and methods to solve partial problems. A machine today can find the shortest path between two locations on a map, it can beat humans in the game of chess, and it has the ability to recognize human handwriting and even has limited understanding of spoken language. All these abilities, which have been developed in the last five decades, are well suited for making machines better and more useful for human users. However Artificial Intelligence seems to have reached a dead end, since many of the promises that have been made could not be kept. It always seems just a small step ahead – from using spoken language to controlling the navigation system in a car to a machine that is able to hold a conversation about everyday topics – but in the end we actually never got there. Technological development naturally happens at the border of existing solutions. Pushing forward a technology like language recognition will yield better algorithms for improved recognition and would thus raise the expectation that in a time not far away a machine will be able to understand everything it is told. It only requires a bit more research, some tweaking and tuning of the algorithms and – the deus ex machina for all information technology problems – more computing power. However, big companies have spent considerable amounts of their research budget to build machines that would listen and *understand*, but the breakthrough is still to come.

Apparently the commitment to continue existing development until the desired human-like capabilities are achieved is fading. Working the way up from existing solutions, refining them and combining them with other solutions has its limits. It is necessary to look at the requirements from a different perspective, let go of the urge to build a working product within the next technological development cycle and examine the prerequisites for a breakthrough in innovation. We want to build machines with human-like abilities; machines that can better understand the real world and take over duties which are too tedious or dangerous for human beings or which improve our level of comfort.

To do so requires an analysis of these abilities together with the prerequisites that enable a human being to have such abilities. Humans are the most evolved species on the planet; our cognitive abilities rely on a vast amount of complex interrelated mechanisms that are based on millions of years of evolution. Obviously it is not feasible to examine all areas at the same time, studying motor abilities, the structure of sensors, blood and hormonal cycles together with all embedded control loops. What appears most interesting and is also the focus of this book is the flow of information, the mechanisms for processing information from different sources, the separation of important and unimportant information and the selection of an appropriate course of action based on the processed information. This is the task of the mind, the human psychic apparatus. It may appear quite ambitious to take the human psychic apparatus as an archetype for a model of a machine. Much research that was conducted in the last decades avoided this challenge and attempted to solve partial problems – with considerable success. However, if we want to achieve human-like abilities, we have to accept that we have to look at the problem as a whole with all its complexities. Only if we consider the whole human psychic apparatus – and not parts of it – and take it as the base for our research, we can succeed in understanding what it takes to make a machine understand.

Facing the challenge to describe and understand something as complicated as the human psychic apparatus we have to look at ways of accessing knowledge about it; we cannot start from scratch, since this would take too long. We have to find an existing discipline that has the required knowledge available. As long as we are looking at human-like abilities, many schools appear to be candidates as knowledge sources. Behaviorist approaches, for example, uses observational correlates as the base for building theories about the mind, they study behavior of individuals and deduce the causes of behavior. Other models like Believe-Desire-Intention (BDI) in (Georgeff & Ingrand 1989) use simplified models as synthetic agents allowing them to interact with their environment and with other agents and act autonomously. This is not satisfactory for our goal. To build such a machine we need a description of the human psychic apparatus that does not describe behaviors of the system as a whole, but rather allows us to break it down into separate functions. We need a functional description, which allows us to examine and describe each function and its interaction with other functions. And we need a complete description, which does not leave out any aspect of the human psyche.

The discipline which best suits these requirement is psychoanalysis (see Section 2.2) and neuropsychoanalysis (see Section 2.4). The psychoanalytic model is a functional model of the human psychic apparatus and as such it allows engineers to transfer it to a functional model for machines with human-like abilities. Since the problem at hand cannot be solved in one piece, it has to be broken down into partial functions, which can then be examined closely. Sigmund Freud and the researchers who continued his work until today provide this functional model. It allows us to describe the psychic apparatus in terms of functions, which have inputs and outputs by which they interact with other functions. Such functions are grouped in modules; they have properties and create behavior. Using this approach proved very fruitful and yielded a thorough technical description of the human mind that will enable engineers to consider an actual implementation of human-like abilities.

As opposed to the approach of working *bottom-up*, which means that an existing solution is improved by refining its mechanisms and combining it with other solutions, the use of psychoanalysis in engineering requires us to work in a different way. Psychoanalysis knows about psychic functions, but it does not always have the answer to how these functions are "built". The correlate, which is responsible for making a function work the way it does is not the main concern of psychoanalysis. To follow the psychoanalytic model we therefore have to take a different approach: we work *top-down*, meaning that we first describe the psychic apparatus on the topmost level, where we identify all modules and the interfaces between them. Then we examine each module and break its functionality down into sub modules, where we again define the interfaces to other sub modules. Using this principle we can finally look at a model that consists of several levels of modules, where each level relies on the functions of the modules below. This is the big advantage of psychoanalysis compared to other disciplines: functional descriptions are well-suited for being translated into technical implementations. This promising starting point is expected to yield the best results for a technical implementation.

2.6 Automation as the Challenge for Psychoanalysis

The ARS-PA project, as will be seen, intends to describe Freud's second topographical model with its agencies Ego, Id and Super-Ego and their interrelations following the top-down approach in accordance with the scientific modeling applied in engineering.

This involves systematizing partial areas of Freud's metapsychology in line with the rules of this particular approach. Freud's writings do not provide a systematic presentation of psychoanalytical theory. They offer insight into Freud's decades of developing his theories, his reasons for later ruling out certain hypotheses again or for holding on to them and conceiving them further.

Transfer of the second topographical model and its implications to a model designed in line with the top-down approach presents the opportunity to systematize psychoanalytical metapsychology under the following premises:

(1) Freud's writings and more generally all psychoanalytic theories that do not contradict original Freudian contents constitute the exclusive frame of reference for a systematization according to the top-down approach;

(2) The attempt at systematization should adhere to the wording of the original source literature as closely as possible; and

(3) The systematization must preserve the contents of Freudian concepts.

Against this background a significant challenge will be to consult Freudian terms in a new manner, namely a scientific engineering way in order to correctly place them within the model. Meeting the exigencies of the model while at the same time preserving the implicit contents of psychoanalytic metapsychology will require subjecting each of its terms to an interpretation.

This interpretation should allow for an assessment as to whether (1) a term may be transferred literally from the source literature to the model and (2) where to place it in accordance with the top-down approach. In such instances where terms are changed and/or newly assembled in order to meet the requirements of the model, the challenge will be to thoroughly examine whether the change still complies with Freudian deliberations or not.

From a psychoanalytical point of view the top-down approach is clearly immanent in metapsychological concepts. The top-down approach involves describing levels of lower complexity to underlying levels of increasingly higher complexity and of greater degrees of differentiation whereby each level should completely describe the other. When Freud set forth the concepts of Id, Ego and Super-Ego and then described their respective contents and functions, his method corresponds to a top-down approach.

2.7   Two Different Sciences – two Different Languages

Considering that the call for interdisciplinary co-operation as an indicator of scientific quality has become very much a matter of principle as well as a matter of course, academic circles are debating whether interdisciplinarity should qualify as a scientific paradigm. The meaning of the term paradigm in this context should not be understood in the sense of Thomas Kuhn (1962) as a set of generally shared guiding principles, questions and methods structuring the knowledge of one or more members of a scientific community, but instead it refers first and foremost to the (re-)organization of scientific work and thus to questions of communication, co-operation and institutional order (Wallner 1991).

It is important to draw a line between interdisciplinary and multidisciplinary work. Interdisciplinary work means among other things that methods are communicated among the disciplines and that problem solving strategies are not only developed through the exchange of results. A further line needs to be drawn with regard to intra-disciplinary co-operation. According to the definition that all the arts basically belong to a single discipline the co-operation of for example literary studies and the science of history would be intra-disciplinary work.

With regard to psychoanalysis, interdisciplinarity has a long tradition. In 1912 Freud initiated the founding of the journal "Imago. Zeitschrift für Anwendung der Psychoanalyse auf die Geisteswissenschaften" ("Journal for application of Psychoanalysis to the Humanities") which was devoted to the application of psychoanalysis to the arts and run by Otto Rank and Hans Sachs. In 1913 the journal "Scientia" published Freud's essay "The Claims of Psycho-Analysis to Scientific Interest" in which Freud discusses in great detail the psychological, linguistic, philosophical, biological, biogenetical, cultural historical, art historical, sociological and pedagogical interest in psychoanalysis. A few years later, in his "Epilogue" to "The Question of Lay Analysis" (Freud 1926e), Freud examines the requirements of a balanced psychoanalytical training and lists the various disciplines that should be part of the analyst's mindset: *"A scheme of training for analysts has still to be created. It must include elements from the mental sciences, from psychology, the history of civilization and sociology, as well as from anatomy, biology and the study of evolution"* (Freud 1926e, p. 252). That Freud himself integrated his past as a neuroscientist in his psychological work becomes particularly evident in his early treatise "Project for a Scientific Psychology" (Freud 1950c).

To give an example from our work in the ARS project, the different meanings of the word "energy" in the two sciences have to be mentioned. In technical sciences like electrical engineering, energy is the capability of a physical system to perform work. Within a closed system the total energy remains constant. Typical examples for this kind of energy are heat and electricity. Within the domain of psychoanalysis, energy defines the relevance of a representation or an urge. Hence, a representation which urges me strongly to perform a certain action is called highly cathected (with psychical energy).

3   Model

Since an important goal of the cooperation between psychoanalysts and engineers is to build a technically implementable unitary model of the mental apparatus, the following chapter introduces theories and concepts of defining a technical model. After a short introduction into the state of the art in perception and decision making of autonomous agents, the architecture defined within the research project ARS at the Institute of Computer Technology of the Vienna University of Technology[12] is described. The current research status is shown in the final sub-chapter.

3.1   Modeling a Decision Unit for Autonomous Agents

Theories provided by psychoanalysis are focused on human beings. What building automation needs are control systems – or in more detail – a computerised decision unit to control the system. Thus, the main task is to model this unit based upon neuropsychoanalytical insights. The purpose of a building automation system is to control processes in the building like temperature or access control constantly by evaluating data collected from hundreds of thousands of sensors and reacting properly to any given situation. To be able to extract meaningful information from this flood of data, new approaches to decision making are needed.

The tasks of a decision unit are processing sensory inputs, reasoning about the current situation and how to reach certain goals, maintaining a knowledge base and a history, and finally producing a meaningful response. As an example, in traditional Artificial Intelligence, the inputs are transformed into symbols[13]. The reasoning system evaluates those symbols and possibly generates new symbols based on the inputs. Using a database of symbols generated in the past, a knowledge base which contains facts about possible interconnections between symbols and a set of goals, can calculate the next steps. Depending on where the decision unit is situated, the execution of the next step could be a message sent to another system, a motor control command, etc. A decision unit does not necessarily need to be able to handle all previously mentioned tasks, nor are the tasks limited to this list. Nevertheless, such a unit should be at least able to handle input data and generate some output.

For each task different possible solutions exist. They range from simple if/then rule-based systems and statistical and mathematical approaches to bionic ap-

---

[12] The development of the architecture model was carried out in cooperation with the universities of Cape Town and New Brunswick.

[13] In the context of AI a symbol is best defined by „an arbitrary or conventional sign used in writing or printing relating to a particular field to represent operations, quantities, elements, relations, or qualities". Not to be confused with „an object or act representing something in the unconscious mind that has been repressed <phallic symbols>" (Definitions taken from Merriam-Webster Dictionary http://www.merriam-webster.com/dictionary/symbol).

proaches. As example, two approaches for the reasoning system are described in the following:

Belief, Desire, Intention (BDI) Architecture: This one is based on the philosophical work on the theory of human practical reasoning by Michael Bratman (1999) and developed by Georgeff and Ingrand (Georgeff & Ingrand 1989). It is a reasoning architecture for intelligent agents which are situated, goal directed, reactive, and social. Their world knowledge is modeled as beliefs. Desires represent the goals of the agent. To fulfill the desires, a list of possible action plans – the intentions – is available. An interpreter processes the input and selects output based upon the data provided by beliefs, desires, and intentions. This output is usually some kind of action with the purpose to change the environment. The agent tries to reach the current goals by selecting the most appropriate intention.

Subsumption Architecture: In contrast to the traditional symbolic Artificial Intelligence, Rodney Brooks focuses with this architecture (Brooks 1986) on behavior based robotics. The basic concept of this architecture is that intelligent behavior is an emergent property – without explicit representations (symbols), without explicit reasoning. It is a layered architecture. The topmost layer is the most complex one. Each lower layer is independent of the one above. For example, a subsumption architecture for path planning could be modeled with motion control as the lowest layer, obstacle avoidance as the second layer, and – as the third layer – optimal path planning. Motion control does not need obstacle avoidance in principle, but obstacle avoidance cannot operate without the lower layer motion control.

Both – BDI and subsumption architecture – focus on more or less the same target platform (robots can be seen as a special form of an agent). The platform is placed into an environment and able to sense and manipulate this environment. The environment does not necessarily need to be the real world or a simulated Cartesian world – it can also be a more abstract topology of computer nodes interconnected using a network. In building automation an agent could be seen for example as a kitchen. Not the space defined by the walls, rather the walls itself or inside the walls like a huge fungus. This fungus spreads across the walls with the wiring of the sensors and actuators. We (the system developers) can model an architecture for robots or agents situated in a Cartesian or even topological world easier than for a fungus. Robots and agents can be seen as somewhat like mammals. This makes it easier to understand their operations – who can project himself into a kitchen?

To circumvent this projection problem during the early stages of modeling a psychoanalytical-based decision unit, we decided to use agents situated in a simulated Cartesian world as an intermediate target platform. When we have reached a certain understanding of the system we can then start transferring it to the building automation domain and reason about which drives or feelings a kitchen has.

The following paragraphs summarize what constitutes an agent: An agent is an entity which operates in a predefined world to fulfill tasks within this world. Thus, it needs the ability to act and to perceive the environment. (Franklin & Graesser

1997) define an agent as follows: *"An autonomous agent is a system situated within and a part of an environment that senses that environment and acts on it, over time, in pursuit of its own agenda and so as to affect what it senses in the future"*. To be autonomous an agent needs to be driven by its own agenda, has to possess its own resources, which enable it to fulfill its goals independently. The perception of the environment is only partial (Ferber 1999, p. 9).

The requirement of autonomous action is essential but not sufficient for the definition of agents. For example, a simple thermostat controlling the room temperature by turning on the heating system if the temperature drops below a threshold would count as an agent, too. To avoid this inflationary usage of the term agent, an additional notion was added: flexibility (see (Wooldridge 2002)). A flexible autonomous agent has to have the additional abilities of reactivity, proactivity, and social abilities. Reactivity is the possibility of a system to respond to dynamic environment changes whenever necessary/at the right time. A purely reactive system lacks the ability to take chances and to affect the world in order to achieve a goal more easily. If a system not only reacts to outer events, but also can initiate actions by itself, it is called proactive. The last requirement – social abilities – has been left out by Artificial Intelligence until recently. The ability to cooperate and to communicate with other agents is necessary to fulfill more complex tasks. If two autonomous robots have to move a heavy block together for example, they have to have a common language and possibilities to interact with each other. Otherwise, if some unforeseen problem occurs, they cannot find a common strategy to overcome it.

Further, an agent needs to be embodied – hence it has a body and the ability to experience the world directly (Brooks 1991b). According to Rolf Pfeifer[14] not only the brain but also the body is needed for any form of intelligence. This body does not have to be a physical one; it can also be a virtual one. Furthermore, the border between body and environment can be abstracted to an interface. It needs sensors to perceive its environment (e.g. vision) and actuators to manipulate the environment (e.g. a robot hand) in order to satisfy, for example, bodily needs (e.g. hunger).

Brooks postulated two main ideas in (Brooks 1986) – first, situatedness and embodiment; second, intelligence and emergence. The first one defines that intelligence is situated within the world and not within a theorem prover or an expert system. This thinking is continued in the second idea – intelligence can be found in the interaction of the agent with its environment and is not an isolated property of the agent.

In (Pfeifer and Scheier 1999, p. 303), several design principles for autonomous robots are given which can also be applied to embodied autonomous software agents. Among them are the complete agent principle, the principle of ecological balance, the principle of sensory-motor coordination, and the value principle.

---

[14] http://cordis.europa.eu/ictresults/index.cfm/section/news/tpl/article/BrowsingType/Features/ID/1485

The first one defines that the agent has to fulfill as many properties of agency as possible (e.g. autonomy, self-sufficiency, embodiment, social, situatedness). The principle of ecological balance defines that an agent can only be as "intelligent" as the complexity provided by the body. Thus, the internal decision unit's complexity has to match the complexity of the sensors and actuators provided. The principle of sensory-motor coordination defines that the agents have to be able to actively structure their own sensory input by manipulating the world (e.g. turning an object while observing it). The last principle to mention – the value principle – deals with the ability of the agent to judge what is "good" for it.

A multi agent system consists of several interacting agents. Each agent has to have the ability to cooperate with other agents, to coordinate different tasks among agents and to negotiate for goods and tasks with other agents. In a system with agents lacking these three abilities, each agent is acting individually on behalf of a user. Even though two agents might work for the same user on the same goal, they would not be able to cooperate autonomously. Thus, the advantage of a multi agent system compared to a system consisting of independently operating agents is that the agents are solving different tasks without being told how to solve them or how to distribute subtasks among them.

To be able to cooperate, agents have to have the ability to reach agreements on certain topics. For example, which agent has to work on which task. These agreements can be simple auctions or more complex negotiations. After an agreement has been reached, the agents have to work together. Two different approaches can be identified: benevolent agents and self-interested agents. The first approach is applicable if the whole system is designed by one team. Each agent helps the others whenever it is asked for. In contrast, self-interested agents act on their own interests. This means that they might also act at expenses of others. Hence, conflict resolution is vital for a proper working system of such agents.

An important issue in designing a decision making unit for agents is the distinction between function, emergence, and behavior. Further important topics are information, categories of information processing functions and the interfaces used to exchange information.

Agents contain one or more modules for data processing and decision making. A module is a collection of functions that have a strongly correlated purpose. For example, a module could deal with the transformation of raw sensor data from different modalities into processable symbols. In contrast to a module, a function is a small collection of instructions and should be explainable with a single verb. The behavior of an agent emerges from the various functions. An ant colony, for example, is able to build an ant road based on very view simple rules like follow an odor or effuse odor. No ant has an idea of the concept of an ant road or a function which would be called "build ant road".

In sciences of the human mind, it is often difficult to distinguish between function, behavior, information, and categories of functions. In psychoanalysis, the term "(the) unconscious" is often used for different topics. Sometimes, it refers to the information stored in an area which is not accessible consciously. Or to some functions which belong to the category of primary processes. It also possibly refers to the result if some information has been processed by such functions. Another possibility is that it may refer to an unconscious action of the body (or behavior). Hence, it is important to clearly define what a function is, what information is, which attributes does the function/information have, and what behavior may be expected to possibly emerge from the functions.

Information has to be exchanged between modules. It is important to define an interface for each module which specifies what kind of information and in which form this information has to be transferred. It is not sufficient to state, that affects are sent from the Id to the Ego – the precise layout of this data packet is also required. Such a layout can be roughly compared to a well formed business letter with address, header, content, and signature. Otherwise, no interpretation of the exchanged information would be possible.

As stated above, many different solutions for various subsystems exist (e.g. Subsumption Architecture for decision making). Kismet (Breazeal 2002) is a robot designed to investigate human-robot-interaction. The robot was constructed using design hints from animals, humans, and infants (Breazeal 2002, pp. 42). The Artificial Intelligence used to control the robot is based upon concepts from behaviorism, psychology, and traditional Artificial Intelligence. While the achievements of Kismet are indisputable, the question has to be raised of whether the immense challenges postulated in this book can be solved using such an approach or not. Challenges like implementing a "theory of mind" or "empathy" (Breazeal 2002, pp. 235) cannot be solved by external observation as performed by behavior research. To solve them, consolidated findings from sciences like psychology and psychoanalysis are needed. However, combining theories from different sciences is difficult. Their compatibility is not always a given fact (see Part III Section 2.10.4). Thus, although many sub problems have been already solved using approaches like statistics or behaviorism, there might be intra-solution incoherencies. To bypass these kinds of problems, we aim to create a holistic architecture based on the psychoanalytic model of the mind.

To conclude this section, a problem of the final target application – building automation – should be addressed. The world/agent interaction is limited by the actuator types available in e.g. a kitchen. They usually consist of heating, air-conditioning, and lighting. Additionally, power control over devices in the room may be possible. These actuators are not sufficient for a "self experience" – the room cannot "feel" its body. This problem will be investigated in the future. Sometimes – with reference to the fungus example above – this may not be a problem at all.

In the next section (Section 3.2) the concept perception is explained for automation. These research results can also be applied to the agent perception. Two possible architectures for the decision making unit agent are introduced in Section 3.3 and Section 3.4. Later on, a multi agent system implementing the theoretical models presented in this chapter is presented in Section 4.2. All four sections are results of the research work of the project group Artificial Recognition System.

3.2   Perception in Automation

One very interesting difference in the definition of the fields of interest in automation and psychoanalysis is that of perception. Psychoanalysis expects perception to happen and to produce memory content, and psychoanalysis does not pay further attention to the functional organization of that memory. There are several points in literature where it is mentioned that humans need memory, however, psychoanalytic theory does not provide – or require – a holistic concept about what the perception-memory-motility control complex does.

While psychoanalysis simply acknowledges the existence of these functions, these topics are of great interest in automation. Perception is the key for each automation system or device to collect the information that allows it to control its actions. A robot without sensory perception, for example, would fulfill a certain task without knowing whether the actions it performs are appropriate. It would be like driving a car blind. For very simple applications (like: move 5 cm straight ahead; while knowing there cannot be an obstacle) such a robot may suffice, but it does not for more complex tasks.

In order to achieve the dedicated goal of this book – designing a technically implementable unitary model of the mental apparatus – the theory about thinking and decision making has to be enriched with a consistent model of perceiving and moving, hence motility control, as depicted in Fig. 3.2.1. The information about the world and the body is perceived and enters the inner world through the perception module. With the help of memory the mental apparatus evaluates these perceived inputs about the outside world, decides about the most appropriate actions, and plans and controls them. To carry out actions, motor control instructs the various actuators how to influence the outside.

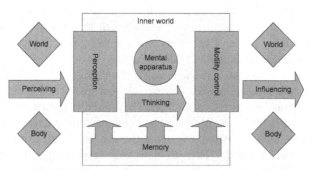

Fig. 3.2.1 Mental apparatus including its surrounding
perception – memory – motility control complex

The term perception has been used in computer and automation systems from the 1950's onwards, since Artificial Intelligence was founded. It means acquiring, interpreting, selecting, and organizing sensory information. The topic itself was not new to automation, but has gained a new quality since information processing could be separated from energy flow and performed in completely new ways.

The foreseen timeframe for revolutionary, human-like applications in Artificial Intelligence has been estimated to be roughly ten years – for the last 60 years. Engineers, psychologists, and neuropsychiatrists already cooperated during early phases of research in order to exploit synergies and achieve mutually fruitful results. Unfortunately, it soon turned out that loose coupling between scientists from different fields was not enough to find comprehensive solutions. In 1961, E. E. David wrote in his introductory words for an issue of Transactions on Information Theory: *"Not to say that cross-fertilization of engineering and the life-sciences should be scorned. But there must be more to these attempts than merely concocting a name, generating well-intentioned enthusiasm, speculating with the aid of brain-computer analogies, and holding symposia packed with 'preliminary' results from inconclusive experiments. A bona fide 'interdiscipline' draws its vitality from people of demonstrated achievement in the contributing disciplines, not from those who merely apply terminology of one field to another"* (David 1961). The authors here fully agree. It is of great importance to form teams with members from both sciences working together on a daily basis rather than meeting regularly presenting own theories about the other's profession. An indication that David's statement still remains true in many areas is that there are still no examples of human-like intelligence in automation.

The development in machine perception has taken two ways: one related with industrial process control, where machines are designed and built in order to increase productivity, reduce costs, as well as enhance quality and flexibility in the production process. These machines mostly need to perceive a well known environment and therefore consist of a selected number of dedicated sensors. The sum of sensor views composes the machines' view of the world. Numerous publications have been written explaining mathematical ways to cope with sensor data to fulfill the needs of automation in these respects.

The second development path was and is concerned with perception of humans and human activities on the one hand and with implementing perception systems imitating human perception for broader application areas on the other. The research fields involved are, among others, cognitive sciences, artificial intelligence, image processing, audio data processing, natural language processing, user interfaces, human-machine-interfaces.

Although the goal of perception systems was always to create human-like perception including all human sensory modalities, from the early beginnings of computer science until now, the user interface – the front end of the computer, through which computers can "perceive" humans – does not normally offer very human-like communication channels. User interfaces up to now, including today's so-called tangible interfaces, are still unintuitive (batch interfaces in the 1960s, command line interfaces in the 1970s, graphical user interfaces from the 1980s until today[15]).

The research field concerned with perceiving information about human users is called *context aware systems*. Context aware systems are used to build devices in the fields of intelligent environments or ubiquitous computing. The common view in these communities is that computers will not only become cheaper, smaller, and more powerful, they will more or less disappear and hide by becoming integrated into normal, everyday objects (Mattern 2004, Hainich 2006). Technology will become invisible and embedded into our surroundings. Smart objects will communicate, cooperate, and virtually amalgamate without explicit user interaction or commands to form consortia for offering or even fulfilling tasks for a user. They will be capable of not just sensing values, but also of deriving context information about the reasons, intentions, desires, and beliefs of the user. This information may be shared over networks – like the world wide web – and used to compare and classify activities, find connections to other people and/or devices, look up semantic databases, and much more. The uninterrupted information flow will make the world a "global village" and allow the user to access his explicitly or implicitly posed queries anywhere, anytime.

One of the key issues in contemporary research towards this vision is scenario recognition. Scenario recognition tries to find sequences of particular behavior in time and groups it in a way humans would. This can range from very simple examples like "a person walking along a corridor" up to "there is a football match taking place" in a stadium.

The research field of machine perception is concerned with building machines that can sense and interpret their environment. The problems that scientists are currently confronted with to achieve this goal show that this research area is still in its infancy. In contrast, humans generally perceive their environment without problems. These facts inspired R. Velik (2008) to develop a bionic model for human-like machine perception, which is based on neuroscientific and neuropsychological research findings about the structural organization and function of the perceptual system of the human brain.

To perceive objects, events, scenarios, and situations in an environment, a range of various types of sensors is necessary, with which either the environment itself or an entity navigating through this environment has to be equipped. The challenge that has to be faced is to merge and interpret the information coming from these diverse sources. For this purpose, an information processing principle called *neuro-symbolic information processing* is introduced using *neuro-symbols* as basic information processing units. The inspiration for the utilization of neuro-symbols came from the fact that humans think in terms of symbols, while the physiological foundation is the information processed by neurons. Examples for symbols are objects, characters, figures, sounds, or colors used to represent abstract ideas and concepts. Neurons can be regarded as basic information processing units

---

[15] http://en.wikipedia.org/wiki/User_interface

on a physiological basis and symbols as information processing units on a more abstract level. The important question is how these two levels of abstraction are connected. Actually, neurons were found in the human brain which respond exclusively to certain characteristic perceptions. These can be regarded as symbols. For example, neurons were discovered in the secondary visual cortex of the brain that responds exclusively to the perception of faces. This fact inspired the usage of neuro-symbols. Neuro-symbols represent perceptual images like for example a face, a person, or a voice and show a number of analogies to neurons. A neuro-symbol has an activation grade and is activated if the perceptual image that it represents is perceived in the environment. Neuro-symbols have a certain number of inputs and one output. Via the inputs, information about the activation grade of other connected neuro-symbols or triggered sensory receptors is received. All incoming activations are summed up. If this sum exceeds a certain threshold, the corresponding neuro-symbol is activated. The information about its activation grade is transmitted via the output to other neuro-symbols it is connected to. In order to perform complex perceptive tasks, a certain number of neuro-symbols have to be connected to a so-called neuro-symbolic network. An important question is what the structure should look like in which neuro-symbols are organized, since a random connection of neuro-symbols will obviously not lead to the desired results. To answer this question, the structural organization of the perceptual system of the brain is taken as archetype.

According to Luria (1973), perception starts with information coming from sensory receptors. Afterwards, this information is processed in three levels, which are referred to as primary cortex, secondary cortex, and tertiary cortex. Each sensory modality of human perception has its own primary and secondary cortex. This means that in the first two levels, information of different sensory modalities is processed separately and in parallel. In the tertiary cortex, information coming from all sensory modalities is merged and results in a unified, multimodal perception. Examples for perceptual images of the primary cortex of the visual system are simple features like edges, lines, or movements. Examples for the result of information processing in the primary cortex of the auditory system are sounds of a certain frequency. A perceptual image of the secondary cortex of the visual system could be a face, a person, or an object. Examples for perceptual images in the acoustic system on this level would be a voice or a melody. An example for a task performed in the tertiary cortex would be to merge the perceptual auditory image of a voice and the visual image of a face to the perception that a person is currently talking. Additionally, it has to be noted that the somatosensory system of the brain (commonly known as tactile system) comprises in fact a whole group of sensory systems, including the cutaneous sensations, proprioception, and kinesthesis.

In accordance to this modular hierarchical structure of the human brain, neuro-symbols are arranged in neuro-symbolic networks (see Fig. 3.2.2a). In a first processing step, simple so-called feature symbols are extracted from sensory raw data. Information processing in this level correlates with information processing performed in the primary cortex of the brain. In the next two steps, feature symbols are combined to sub-unimodal and unimodal symbols. These two levels correspond to the function of the secondary cortex of the brain. In connection with the somatosensory system of the brain it was mentioned that a sensory modality can consist of further sub-modalities. Therefore, there can exist a sub-unimodal level between the feature level and the unimodal level. On the multimodal level, information from all unimodal symbols is merged to multimodal symbols.

Within a neuro-symbolic layer, information is processed in parallel, which allows high performance. Connections and correlations between neuro-symbols can be acquired from examples in different learning phases. Besides sensor data processing, stored *memory* and *knowledge as well as a mechanism called focus of attention* influence perception by increasing or decreasing the activation grade of neuro symbols in order to resolve ambiguous sensory information and to devote processing power to relevant features (see Fig. 3.2.2b).

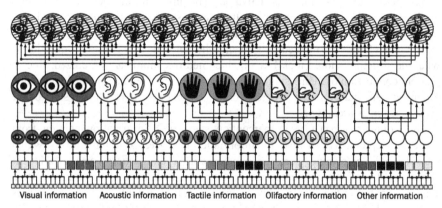

Visual information   Acoustic information   Tactile information   Olfactory information   Other information

(a)   Arrangement of neuro-symbols to neuro-symbolic networks

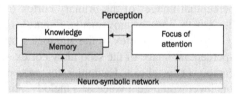

(b)   Interaction between neuro-symbolic networks, memory,
knowledge, and focus of attention

Fig. 3.2.2 Neuroscientifically and neuropsychologically inspired model
for human-like machine perception

3.3   Towards the new ARS-PA Model

Following a bionic approach, concepts of neuropsychoanalysis were used to create a model for a perception and decision unit for an autonomous, embodied agent. Concepts from Neurology were used to define a model for sensor fusion. In this case, we followed a bottom-up design approach to build a model, as described in the previous chapter. Based upon the work of G. Zucker (2006), W. Burgstaller (2007) used the model of A. Lurija and described a model for multi sensory data fusion and R. Velik (2008) enhanced the model by using neuro-symbolic networks for data processing.

Based upon these neurologically inspired theories, the system was able to perceive information about the world from symbolic representations that are ground to the sensor input. Following the research approaches of neuropsychoanalysis as described by M. Solms and O. Turnbull (2002, p. 30), there are two sources of possible information that can be perceived by humans: The *inner milieu* and the *environment,* both together named as the outer world. Fig. 3.3.1 shows the sensory part of the system that stops at the border between the *environment, the body* and the *mind.*

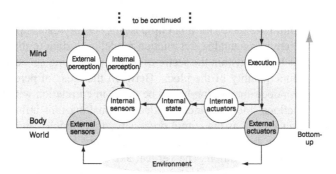

Fig. 3.3.1 Data processing towards the border between body and mind

The *external world* can be seen as the surroundings a person is placed in. Humans use their sensory systems to perceive information of the outer world. This information is perceived by different modalities: vision, auditory, somatosensory, gustatory, and olfactory. Apart from the complex task to create technical sensors that can smell it is no problem to apply this concept to building automation. The problem grows with the second part of possible perceptions, the *bodily senses.* The purpose of perceiving the bodily values is a continuous monitoring of the individual's body to determine whether we feel good or not. The values that are monitored are values that are in direct connection to the homeostasis of the human body, such as body temperature, heart beat frequency, hormonal levels, etc. As the necessity of keeping all these values within an optimum range to guarantee biological homeostasis is one, if not the basic key that drives us to react, to seek, to respond to the environment we are placed in, it was decided to imply this concept within our system. However, in this context the problem arises how to define a body within a building: Which values can be defined that represent the basis for a buildings homeostasis. After a closer examination, it was decided to first only take an intermediate step, where the research approaches of neuropsychoanalysis are

48

easier to apply. The system shall first be implemented and work within an embodied, autonomous agent. This approach has the advantage that the body can be easily defined, first of all because it is possible to describe the difference between an agent's body and the surroundings the agent is placed in.

In contrast to the kind of approach to sensor data processing that uses neurological concepts in a *bottom-up* design approach, it has been decided to design the decision unit for the autonomous agent in a top-down manner, using research approaches of neuropsychoanalysis and theories that describe functional units within a human decision making process. It is called top-down because a model of neuropsychoanalytical research approaches is taken as a top-level idea and then specified into detail until an implementation within the target platform – in this case the autonomous agent – is possible. The model of a decision unit is divided into two dimensions in the following section in order to visualize the concept followed during the design process (see Fig. 3.3.2). The horizontal axis represents the area where the functional blocks, which are the content of the diagram, are located. If a block is located within the sector for input processing, sensory data is processed to form a bodily and environmental representation model according to the perceived data. If the block is located within the sector of output processing, an action can be produced by the functionality of the block. Between the areas of perception and action, more or less decision making has to be done, in correlation with the complexity of the functional blocks. Since the decision making unit influences both the input processing as well as the output processing, this module can be found in both of the areas. The border between input and output has again been drawn purely out of technical motivation. The vertical axe shows the complexity of the module. The higher the functional block is located within the figure the higher the complexity of the decision path that has to be taken.

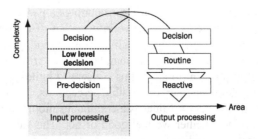

Fig. 3.3.2 Two folded structure of the decision unit

The following paragraphs will describe each of the blocks that are shown in Fig. 3.3.2, starting at the lowest level of perception and ending at the lowest level of actions.

The entrance point to the *mind* of the model is located, as mentioned previously, at the border between the body and the mind. There, the perceived information of both, body and external world is passed on to the *mind* in a pre-processed, symbolic, and symbol grounded form. The functional block that first processes the data is in this model the *pre-decision* unit. The basic purpose of this module is a first and fast evaluation of the incoming, perceived information. Basically, this unit consists of four aspects which will be discussed in detail further on: *image recognition*, the influence of *drives* on the behavior of the decision making process, the *basic emotional value system*, and the evocation of *reactive actions* (which is a part of the functional block 'Reactive' at the end of the cycle).

The idea behind the *image recognition* process described in further detail by Rösener (2007, p. 19) and Lang et al. (2007) is to enable the system to detect and to name an ongoing situation or simply parts of a situation the system is located in. A. Damasio (2005) describes a concept of data processing within the human mind, where he stresses the importance of creating images for further internal processing such as reasoning and decision-making. Within the proposed model a distinction is made between two different types of images. The *perceptual image* is – technically spoken – a container that holds all symbolic information that has been produced by the perceptive system at one single moment in time. It contains concrete changes of the environment and the organism itself (Solms & Turnbull 2002, p. 90), perceived by all sensor modalities. Supposing that the system runs within an autonomous agent, first all sensors are queried. After that the data is processed for symbolization and the result is stored within the perceptual image. Unlike perceptual images, *mental images* are not produced and discarded in every perceptive cycle. They are static, stored templates of already experienced situations or again, at least parts of these situations and they represent a part of the knowledgebase of the agent. Having an existing set of mental images, the ongoing situation can be categorized by comparing the perceptual image to each existing mental image. Some of the mental images may totally match the perceptual image, some of them may have more or less an affinity and some of them will totally differ. The result of this comparison can be seen as a list of mental images, sorted by the percentage of coverage.

Another important part of recognition that can be defined within a mental image are the bodily needs, represented as drives. In this model drives are defined as the basic motivations that can be experienced as emotions as explained by M. Solms and Turnbull (2002, p. 28). S. Freud (1915c) described the drive as a stimulus within the organism that is represented in the mind, describing a bodily demand. Since drives are the expression of demands of the body, they are perceived and considered within the perceptual image. When simulating an autonomous agent, drive-specific systems as defined by J. Panksepp (1998), who distinguishes between hunger, thirst, thermal balance, and sexual arousal, are utilized. A recognized mental image may influence the behavior of the system by producing an action and it may influence the values of the basic emotional value system within the module of pre-decision. In the system, this action and the influence on emotions have to be predefined.

Once again back to neurology, the concepts of J. Panksepp of a basic emotional value system have been taken as a reference model. According to Panksepp (1998) basic emotional systems have a commonly describable structure and are described on the basis of the following four emotions: fear, rage, panic and seeking. The seeking system, instead of being triggered generally by external stimuli as the others, contains the information of the drives, and forces the individual to search for satisfaction of their needs.

Each basic emotion is assigned a certain value in the pre-decision unit that can be influenced by identified mental images in a predefined way. Possible reactive actions are also attached to the corresponding mental images. After the described procedure, in the pre-decision process the perceived symbols are validated and converted into identified mental images; basic emotional systems are affected and first possible action tendencies are produced. The processing cycle is continued at the next higher cognitive level, the lowest part of the decision unit.

In the following, the decision unit within the input processing section is further divided into two parts, the lower of which contains *episode recognition* that frequently communicates with *episodic memory*, while the higher part manages complex emotions, desires and Super-Ego rules and provides functionality as described in the next chapter. *Episode recognition* can be easily compared to the recognition of mental images as described above, apart from one important fact: episodes are sequences of images in time. First, a specified mental image has to be perceived, and then the episode recognition process will be waiting for the next specified mental image to be recognized to step into the next state of the episode. The whole episode is recognized until the predefined sequence has been successfully passed. To define such episode patterns, the mental episode has been defined following the concept of mental images (Rösener et al. 2007). Within one mental episode, different paths can be defined that may yet lead to the same outcome: the complete recognition of the episode. When an episode is recognized, the system is aware of the situation it is placed in and is also able to include past procedures. Just as recognized mental images can influence *basic emotions* and produce *action tendencies;* recognized mental episodes can influence *complex emotions*, produce *action tendencies* and evoke *desires* within a system. This will be discussed later in this chapter.

In the context of episode recognition the functional block of *episodic memory* shall be discussed now. As postulated by Rösener (2007), researchers in artificial intelligence and other natural sciences agree that the human mind provides different types of memories. Although no consistent categorization is available, E. Tulving (1998) and A. Baddeley (1997) present a framework that describes the functions of episodic memory quite well. These theories have been used within our autonomous agent in the module for episodic memory and shall enable the agent to store once experienced situations and their effects and recall similar situations from the past in order to make decisions based on the knowledge of the past. The episodic memory acts like a videocassette recorder and contains all agent-specific, autobiographical, experienced episodes, including the emotional state of the agent. It also provides the agent with the possibility of prediction. It becomes possible to predict future trends of situations and to anticipate the impact of actions that have to be taken. The theories of E. Tulving are purely based upon neuroscientific theories without even touching the ideas of psychoanalysis. Although the episodic memory, in accordance with this concept, worked well when implemented, it must be further discussed how it can fit into the concept of a psychoanalytically inspired model. The problem is that a technical description of how a human's brain stores and retrieves information does not play a major role in the theories of psychoanalysis. Nevertheless, Sigmund Freud tried to explain the creation of memory traces and their interconnections, which will be discussed in the next chapter, where the new psychoanalytically inspired model will be described.

Until now, the functional blocks of the model that has been introduced are purely passive, except for the reactive action that is generated directly from recognized mental images. However, before describing the most deliberative areas of the model, one passive but very important ingredient is still missing: *complex emotions*. In contrast to the basic emotional system, which is the basis of human emotional evaluation, more complex situations evoke a broad spectrum of emotional flavors as described by Burgstaller et al. (2007). They represent a cognitive evaluation system highly influenced by social interaction with other individuals and social rules. Hope, joy, disappointment, gratitude, reproach, pride, shame, etc. are typical human complex emotions and are listed and described e.g. by A. Damasio (2003, p. 67), J. Panksepp (1998), M. Solms and Turnbull (2002), R. Plutchik (2001), and G. McDougall (Meyer 1999).

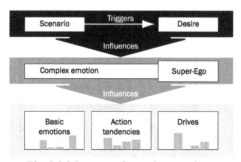

Fig. 3.3.3 Impacts of complex emotions

Fig. 3.3.3 gives a quick overview how complex emotions are changed and what they directly influence within the model. During runtime, the system detects and recognizes several scenarios that are happening at the moment. According to the experiences the system had in the past, these scenarios as well as the different states of each scenario are evaluated and flavored more or less by different complex emotions. Since a scenario can contain both, perception from the body as well as from the external world, a scenario includes basic emotional values and drives too. Specific recognized scenarios can trigger desires that also influence complex emotions, in a very similar way to scenarios. The complex emotions are therefore directly, but not exclusively influenced by basic emotions. Social rules are stored within a memory structure that we called the *Super-Ego* in our model. According to psychoanalysis and especially Sigmund Freud's second topographical model of the Id, Ego, and Super-Ego, the Super-Ego encloses values and prohibitions and is actively conscious, pre-conscious and unconscious.

The rising or decreasing value of one type of complex emotion is again repercussive on the value of the basic emotions and drives. This means that a complex emotion can amplify or inhibit basic emotions and drives. A complex emotion like disappointment can influence the basic emotion rage and may emphasize or inhibit drive values like hunger. It may also influence the behavior of the system and therefore the action tendencies the system has.

As described above, scenarios are implemented to conduct a purely passive recognition of observations of the body and the external world. Sequences in time can be detected and recognized. The required observed mental image brings the scenario recognition from one step to the next. In order to provide not only passive observation processes, but also active action procedures, a *desire* has been implemented. Compared to a scenario, a desire extends the functionality of the scenario with the possibility to additionally use *finished actions* as a transition. With this structure, it is possible within the system to define long-term goals and most importantly action plans that can lead to such a goal.

The underlying model for these desires is the Freudian wish, (Freud 1900a) and (Laplanche & Pontalis 1972). A desire contains information that is needed to initiate all demands that lead to an already known situation in the past where a need has been satisfied and therefore pleasure has been experienced. The main elements of the desire are depicted in Fig. 3.3.4.

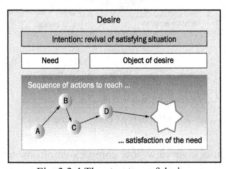

Fig. 3.3.4 The structure of desires

The desire contains the information what kind of demand can be satisfied. The object of desire describes the element that triggers the desire. It can be an object or a person in the external world as well as a remembered thought or emotion that appears in time within the psychic apparatus itself. The central basic part within the model of the structure of desires is the action plan (A-B-C-D-*) describing what kind of actions have to be taken to regain the situation that previously satisfied one or several needs. In fact, a desire is based upon an already experienced and processed cycle of perceptions and actions, which however does not necessarily mean that this pattern always leads to success. Mistakes have to be memorized and the action plans have to be updated. Additional ways have to be conceived to avoid possible obstacles. For evaluation, each developed plan contains the probability of success, estimated by the system. It depends on the current situation the system is currently placed in and the history that brought the system into the current situation. The reason why we cannot say that we are able to model the human mind with a state machine, as is often used in automation technologies, can be clearly given: every system that is highly influenced by its past – just as the human mind – will never be able to reach exactly the same state again that occurred in the past.

In the beginning, our project consisted of two project groups, the *perception-group*, working on the task of neurologically inspired symbolization and de-symbolization, and the *psychoanalytical group*, working on the task of creating a psychoanalytically inspired decision unit that can be used for autonomous agents. It was essential to define the proper technical interfaces between these two groups to guarantee an efficient, complete system. This interface was defined by using perceived symbols and action symbols as input and output parameters. Communication between the two blocks was only done using these types of symbols. In the design phase it was important to only use research approaches from the neuro-psychoanalytic society to develop the presented model. This concept made the decision unit unique as well as functional and is one main reason why parts of the concept have already been implemented in real world applications (observing scenarios in office buildings and nursing homes). Although the misunderstandings between the neuro-psychoanalytic and the technical ways of thinking have been dramatically reduced in the course of this co-operation, still different concepts of neurology, psychoanalysis, even psychology and behaviorism as well as artificial intelligence have occasionally been intentionally mixed up in some parts of the model. However, especially in these parts, it seemed that the more different models that were used, the harder it was to keep and establish a consistency without getting caught in the trap of "shopping for solution"-research. Since the system developed was able to perceive data and process it to arrive at an output action that fitted in well with most situations, the concept forms the basis for the project team's further research. Though how close is this solution to the basic theories of psychoanalysis and how far is it from human decision-making? In the next chapter, a new model will be introduced that focuses on one single theory to show a slightly different approach to modeling a decision unit: the second topographical model defined by Sigmund Freud.

3.4   The New Model and its Description: Top-Down-Design

Although the model as described in Section 3.3 is based on the novel research approach neuropsychoanalysis, it has a design flaw – it is a mixture of different theories which are sometimes incompatible. Neuropsychoanalysis aims at combining models from psychoanalysis with findings from neurology. As far as we know no unitary psychoanalytical model is currently available. The different models are highly detailed in some areas, but are incomplete or imprecise on various other topics. Some of them overlap or are even mutually exclusive. To fill the gaps in the theories or to clarify problems, findings from other sciences were taken into account – from neurology, over AI, to psychotherapy and even behaviorism. As discussed earlier in Section 2.2, even though the solutions may not be in conflict with psychoanalysis in principle, intra-solution incompatibilities might appear.

To overcome these problems, the ARS project team started to work on a new model in parallel. Based on the knowledge and experience gained while developing the first model, the decision was made to use Sigmund Freud's second topographical model (Freud 1923b) for the new ARS model. Thus, the new model is based upon Freud's original work and publications closely related to it. In contrast to the previous model, parts with gaps or imprecision are marked for further investigation once the model based on Freud's description is complete.

Another problem of the previous model is the unclear differentiation between the flow of information and flow of data. In psychoanalysis no clear distinction between these two has been defined. Language difficulties also contributed to this problem (see Section 2.7). Once these problems were identified a different approach for modeling had to be selected. The most suitable approach appeared to be a top-down design approach.

The top-down design is a modeling approach that starts the modeling process at the topmost level with a simple concept of the subject. This concept is divided into more detailed parts in the next level down. The process of dividing a part into more detailed subparts in the next level is repeated until the desired degree of detail has been reached. The sum of each level represents the idea as a whole but at different granularities. This design technique fits well into Sigmund Freud's description of the second topographical model – he too divided the mind into different parts.

This section is divided into two parts: a short introduction into Freud's second topographical model and secondly an overview of the current state of the ongoing research regarding the technical transformation of Freud's second topographical model.

3.4.1 Sigmund Freud's Second Topographical Model

As the new ARS model is based on Freud's second topographical model it shall briefly be described in more detail.

While in the first topographical model Freud made a distinction between the agencies unconscious, preconscious, and conscious, in 1923 – against the background of his theory of the unconscious conflict – he conceived the second topography which is also called the structural model.

Id, Ego and Super-Ego are now the three agencies of the psychic apparatus.

Id: The Id involves everything that psychically represents the drives. The characteristics of the Id are identical with those that Freud attributed to the system unconscious in the first topography. Unconscious processes follow the primary process that represents the modality for handling unorganized and contradictory processes. They are uncompromising with regard to the reality principle which would involve accounting for the demands and requirements of the external reality. In addition, the Id has no knowledge of abstraction, no idea of space, time and causality. Its reality is simply the psychic reality the desires and impulses of which follow the pleasure principle only: they demand immediate satisfaction without the least consideration for the givens of the external reality. The primary process is closely related to "body language" or "organ language", i.e. *"[...] it is situated between the cognitive processes of the consciousness and the physiological processes taking place in the central nervous system. At least by way of description according to Freud, the primary process can be understood as the connecting link between psyche and soma"* (Müller-Pozzi 2002).

The contents of the unconscious and Id exist in the form of so-called "thing presentations". They are the product of something concrete, of a sensual experience which is stored as a memory-trace of immediate experience independent from linguistic symbolization.

Ego: The concept of the Ego comprises all organizing, integrative and synthetic functions of the mind. In "An Outline of Psychoanalysis" Freud provides a summary of the Ego:

"In consequence of the pre-established connection between sense perception and muscular action, the Ego has voluntary movement at its command. It has the task of self-preservation. As regards external events, it performs that task by becoming aware of stimuli, by storing up experiences about them (in the memory), by avoiding excessively strong stimuli (through flight), by dealing with moderate stimuli (through adaptation) and finally by learning to bring about expedient changes in the external world to its own advantage (through activity). As regards internal events, in relation to the Id, it performs that task by gaining control over the demands of the instincts, by deciding whether they are to be allowed satisfaction, by postponing that satisfaction to times and circumstances favorable in the external world or by suppressing their excitations entirely. It is guided in its activity by consideration of the tensions produced by stimuli, whether these tensions are present in it or introduced into it. The raising of these tensions is in general felt as un-pleasure and their lowering as pleasure. It is probable, however, that what is felt as pleasure or un-pleasure is not the absolute height of this tension but something in the rhythm of the changes in them. The Ego strives after pleasure and seeks to avoid un-pleasure. An increase in un-pleasure that is expected and foreseen is met by a signal of anxiety; the occasion of such an increase, whether it threatens from without or within, is known as a danger. From time to time the Ego gives up its connection with the external world and withdraws into the state of sleep, in which it makes far-reaching changes in its organization"
(Freud 1940a).

In the Ego a thing presentation is linked to the word presentation according to the remembrance, memory, and ideational system. Conscious and preconscious presentations consist of thing presentations and their relevant word presentations. The difference between the two systems therefore lies in the potential of language, i.e. a system of symbols which is the pre-condition for the capacity of information to reach consciousness.

Super-Ego: The Freudian concept of the Super-Ego consists of the prohibitive/imperative and ideal functions. Freud first introduced it in "The Ego and the Id" (Freud 1923b) explaining that the critical function called Super-Ego represents an agency which has become separate from the Ego and seems to dominate it. As such it is linked to the Ego and preconscious parts of the person on the one hand and to the Id and the unconscious parts on the other: *"Notwithstanding different development processes and opposing interests the Super-Ego shares with the Id that with regard to their demands they are both inconsiderate of any relation to reality which must be established by the Ego"* (Köhler 1995, transl. by the author).

3.4.2 Functional Description of the Modules for the Three Topmost Levels

The new ARS model is based on the principles of Sigmund Freud's second topographical model mentioned in Section 3.4.1. Engineers and psychoanalysts try to find a way to translate Freud's definitions and explanations into technical language. The top-down approach, described in the introduction of Section 3.4, is used to analyze the second topographical model and to design a decision unit for autonomous agents. The first layer is formed by the brain. Fig. 3.4.1 shows the functional division of the *Brain*[16] module into the *Sensor interface*, the *Actuator interface* and the *Psychic apparatus* one layer beneath it. The *Sensor interface* represents the module which gathers information from the environment by the system body's sensor system. The incoming data is passed on to the module *Psychic apparatus* that generates resulting control decisions. The *Actuator interface* forms the connection between the psychic apparatus and the mechanical body elements. The *Psychic apparatus* forms the central part of the model. Therefore a closer look will be taken at it in this chapter.

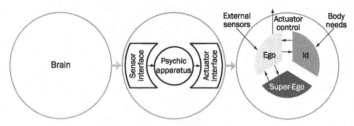

Fig. 3.4.1 New ARS model

---

[16] All terms that are written in *italic* style in Section 3.4.2 refer to a module of the ARS model.

Fig. 3.4.1 shows the sub modules of the *Psychic apparatus* that are named *Ego*, *Id*, and *Super-Ego* and agree with the second topographical model. The arrows show the information flow within the *Psychic apparatus* and to its environment. The *Ego* is able to receive environmental information as well as to send information to it. The *Id* receives bodily information. The *Super-Ego* does not dispose of an interface to the outer world but of an interface to the *Ego* that it receives environmental information from. All communication within the *Psychic apparatus* is centered in the *Ego*.

While the main focus of this section lies within the top-down description of the psychic apparatus and its sub-modules, another important issue – the interfaces between the modules – shall be addressed here briefly. The part of Fig. 3.4.1 to the very right shows the data flow between the modules *Ego*, *Id*, and *Super-Ego*. Each module has to provide an interface for the arrows which point to it. For example, the *Ego* needs an interface for the data from the external sensors, from the *Super-Ego*, and from the *Id*. The data transmitted by the sensors are perceptual images; data transferred from the *Super-Ego* are social rules, restrictions, and values stored as memory traces; *Id* transfers affects to the *Ego*. They have to be properly defined within the interface on each level of the model.

In the following, the modules *Ego*, *Id*, and *Super-Ego* and their functions are described in order to give the reader an overview of the new ARS model.

Id

According to Sigmund Freud's theory, the Id forms the source of the mental apparatus. It represents the source of psychic energy and upcoming drives. After birth a person's psychic apparatus consists of the Id only before the Ego and the Super-Ego evolve out of it.

Even if the concept of psychic energy is based on Hermann von Helmholtz's second principal of thermo dynamics and the theorem he established on the conservation of energy, it must not be mistaken for the psychic energy according to the physical definition (Laplanche & Pontalis 1973). The physical term energy is defined as a scalar quantity which generally defines the ability to work, while the psychoanalytic term corresponds to a quantity which defines the relevance of representations. Freud adopts Helmholtz's theory of conservation of energy, which states that the total amount of energy within an isolated system remains constant and cannot be created or destroyed. He established the theory that every mental process needs psychic energy for its functionality. Every person possesses a limited amount of psychic energy that is cathected to mental functions. One and the same part of psychic energy cannot be shared between two mental functions at the same time. If the mind is seen as an isolated system this definition is similar to the conservation of energy theorem.

As stated above the Id handles drives and its demands. A drive transfers organic processes to psychic processes by the use of representatives that are named affects and presentations. A drive evolves out of a source and gets discharged by reaching an aim by the use of the so called drive object. The source is the organ which signals the bodily need. The object is used to get rid of the excitement tension, the aim is satisfaction. A drive demand evolves in the mind if psychic energy is cathected to it. This process is initialized when a bodily stimulus arises, which leads to a psychic tension. By definition drives need to get discharged which is achieved by reducing the bodily need. (Freud 1915c)

The Id itself admits unorganized and contradictory processes. These processes are called primary processes and require to be executed regardless of other demands. Therefore numerous conflicting, unstructured processes come up within the human mind. Those which reach the Ego to get incorporated into the decision process get organized and are thus ruled by the secondary process. If a process which reduces a drive's tension cannot be executed, the mind tries to redirect this discharge through other processes, for which the Ego's functions are a prerequisite. In contrast to technical systems, conflicting processes can coexist here without cancelling out each other.

The top-down design requires well defined self-contained functional modules. As it is shown in Fig. 3.4.2, the *Id* is grouped into three modules named: *Physiological to psychic transformation*, *Selection of memory traces*, and *Quantification of drives*.

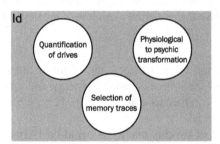

Fig. 3.4.2 The functional model of the Id

Physiological to psychic transformation: The module *Physiological to psychic transformation* deals with the conversion of physiological stimuli to psychic representatives – affects and presentations. As stated above this conversion is represented by the drive, which serves as input for the module *Quantification of drives* and represents an interconnection between physiological and psychic processes.

Quantification of drives: Drive representatives are cathected, i.e. allocated psychic energy and are thus evaluated and subsequently an attempt is made to pass them on to the Ego. Arranging and structuring these processes will be done by the Ego and is not included in the task of this module.

Selection of memory traces: The module *Selection of memory traces* deals with the generation of memory traces and their cathexis (Laplanche & Pontalis 1973). Before the term memory trace can be explained, the psychoanalytic term presentation has to be discussed.

The term presentation is defined as perceived object information (Freud 1915d). Sigmund Freud distinguished two types of presentations, thing presentations and word presentations. Thing presentations correspond to the sensorial characteristics of an object and mainly belong to the Id. Word presentations signify the description of an object, event, etc. via a set of symbols. Word presentations are processed in the Ego and form the basis of thinking. They make it possible to reflect on past, present, and future events.

The memory traces which are treated by the Id are formed out of thing presentations. Salient characteristics are extracted from the information received and stored in the mind as memory traces. They serve as shapes for subsequently perceived situations. If an object is identified, the corresponding memory trace is activated by cathecting psychic energy. This action is managed by the module *Selection of memory traces*. Depending on stimuli from the inner and outer world, the amount of energy, which is cathected to the memory traces, differs. Solms and Turnbull (2002, p. 28) describe the inner world as being defined by the mental apparatus while the outer world is represented by a person's environment and his or her physical body. Hence the brain interposes the outer world from the inner world. If the perceived input matches an existing memory trace this trace is activated. In general more than one memory trace will partly conform to the perceived input. Hence the one that has the highest percentage of accordance will get activated. If no match exists a new memory-trace is generated, if however two or more memory traces are activated at the same time they will be associated with each other. In Freud's theory this process is called facilitation (Laplanche & Pontalis 1973).

Super-Ego

The Super-Ego develops in early childhood and roots in the Ego and the Id. As a child, a person is confronted by numerous restrictions and demands by the parents. Therefore processes evolve that repress specific drive demands in order to satisfy internalized demands. In a parallel process a mental structure is formed on the basis of parental restrictions, demands, and awards. The Super-Ego manages all restrictions and demands and provides feedback on actions executed by the subject. This feedback can be given in a positive as well as in a negative way. Therefore the Super-Ego can also be interpreted as a kind of reward system. Id and reality demands are synthesized by the Super-Ego and the Ego, which results in control, satisfaction, and inhibition of impulsive actions. The Super-Ego adapts the individual's behavior and thinking to the demands of the reality, to social rules and expectances.

As shown in Fig. 3.4.3 the *Super-Ego* is divided into three modules – the *Management of demands*, the *Management of restrictions*, and the *Management of the Ego ideal*.

Management of demands: This module organizes demands based on social norms. This structure is formed by the individual's social background and permits ideal behavior and thinking.

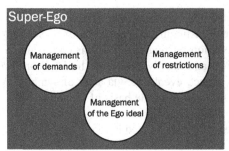

Fig. 3.4.3 The functional model of the Super-Ego

Management of restrictions: It organizes restrictions on behavior and thinking defined by the social background. This module becomes supplemented over the years, but it is mainly formed during early childhood.

Management of the Ego ideal: The module *Management of the Ego ideal* handles the ideal image of oneself. During childhood, the Ego ideal evolves out of the identification with the parents. This idealization is mapped to a structure and forms the base for the Ego ideal as a part of the Super-Ego.

Ego

The *Ego* forwards information about the perceived environment to *Id* and *Super-Ego*. According to the internally stored information, the perceived image evokes a preferred action in Id and Super-Ego. This action equals the idealistic conception of behavior and thinking and is communicated to the *Ego*. If the *Id* module's instinctual impulses and the *Super-Ego* module's rules disagree, conflicts arise which are ideally balanced by the *Ego*. Which content achieves acceptance depends among other things on the relevance of information, the environmental conditions, and the instinctual impulses and is elected by the *Decision management* module, which is placed in the *Ego*. The Ego takes over a mediation role between inner and outer world on the one hand and Id and Super-Ego on the other.

After birth the psychic apparatus is controlled by pleasure and unpleasure, it only consists of the Id. The Id tries to discharge its drive demands independently from the environmental condition. By and by, in the psychic apparatus the Ego and its functions develop out of the Id. A wide range of functions evolve, such as reality testing, focus of attention, management of affects, or executive motor functions. As described above, when the Super-Ego evolves, the Ego takes on the role of a mediator between the Id's drive demands, the environmental demands, and the Super-Ego's rules, demands, prohibitions, and positive as well as negative feedback. In addition to this internal conflict it has to factor the perceived environmental data into the decision making process. Therefore the Ego is responsible for balance in the system.

As shown in Fig. 3.4.4 the *Ego* module has been split up into two modules, which are entitled as *Psychic synthesis* and *Executive motor functions*.

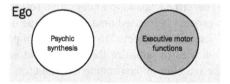

Fig. 3.4.4 The functional model of the Ego

(A) Psychic synthesis

The *Psychic synthesis* module receives information from the *Id* and the *Super-Ego*. In a given situation the *Id* informs the *Ego* about current drive demands; the *Super-Ego* provides relevant behavioral rules. Within the *Psychic synthesis* module, all functions deal with the mediation between *Id* and *Super-Ego* on the basis of perceived outer world conditions.

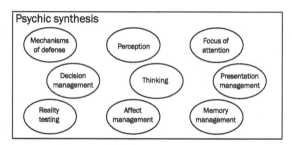

Fig. 3.4.5 The functional model of the *Psychic synthesis*

Since the *Psychic synthesis* module is responsible for finding a compromise between the instinctual impulses of the Id and the demands of the Super-Ego, a number of modules have to be defined to handle this process. As shown in Fig. 3.4.5 nine modules are defined which are entitled *Perception, Focus of attention, Affect management, Reality testing, Presentation management, Memory management, Decision management, Mechanisms of defense*, and *Thinking*.

Perception: The *Perception* module serves as a pool for outer and inner world information. In psychoanalytic theory, three modalities of perception are defined.

- Information is processed without activating a memory trace. This situation occurs if the perceived information is completely new to the subject. Therefore a fitting memory trace does not exist and has to be generated.
- In the perception process, memory traces are spontaneously activated.
- The last one is called attentive perception. This corresponds to observant thinking and is handled in the *Focus of attention* module.

Focus of attention: This module defines search for a required object in the outer world which corresponds to an activated presentation due to a drive demand. The information which is received by the module *Perception* is evaluated here. The attention process is described as a kind of cognitive selection. If the module is activated, the selection process is focused on a specific type of incoming information. This information is selected subjected to the amount of psychic energy that is cathected to it. Attention tries to optimize the information selection process and therefore accelerates the reaction on environmental changes. Furthermore the selected information is sent to the *Presentation management* module where it activates already existing memory traces.

Affect management: As a consequence of perception of an environmental or bodily process or a psychological process, psychic energy is cathected to a drive. According to Sigmund Freud, the term affect defines a quantifiable psychic representative of a drive (Freud 1915c). A drive is represented by two components in the Ego – one is the affect and the other the representation. These include joy, excitement, interest, fear, anger, or distress. The module *Affect management* forms the affects as drive representatives and forwards them to the designated Ego modules.

Presentation management: As noted above Sigmund Freud differentiated between two types of presentations – thing presentations and word presentations. The module *Presentation management* organizes the facilitations, which are mainly formed in early childhood, between thing presentations and word presentations. If a thing presentation is activated associated presentations are activated too.

Thinking: The main task of thinking is the structuring and organizing of the contents of primary processes. This organizational process proceeds on the base of a culturally influenced logic and results in the conversion to the secondary process. During the process of thinking, the current situation will be evaluated and compared to formerly experienced situations. Therefore the module *Thinking* uses already existing memory traces, facilitations, affects, and thing- and word presentations. In addition the thinking process can start the evolving of new memory traces. Thinking represses the direct execution of actions. Occurring situations can be tried out and planned within the mind, which avoids potential dangerous situations and results in an economical use of bodily energy.

Reality testing: In psychoanalysis, reality testing is used to differentiate between inner presentations and outer reality. Therefore it enables a differentiation between imagination and reality. As a consequence of a failure in reality testing the subject can get harmed, e.g. by having an accident because of the misinterpretation of a situation, or not being able to gain satisfaction and pleasure. *Reality testing* activates a presentation via inner world as well as outer world information. In the beginning the mental apparatus is not able to distinguish between the inner and the outer world. It changes during childhood. After an internal presentation has been activated by a drive demand, attention is focused on the drive object. If a perceived object seems to be the one in demand, it is compared to the activated presentation. The activation by inner and outer world information differs in the quantity of psychic energy that is cathected to the memory traces.

Memory management: The module *Memory management* organizes the memory traces the *Ego* is dealing with. In contrast to the *Id* the generation and activation of memory traces representing word presentations are managed too. Perceived information activates memory traces in this module or is redirected to the *Id* module's *Selection of memory traces.*

Decision management: The *Decision management* module handles the conflicts which occur within the *Psychic apparatus.* The incoming information, which is received from the *Reality testing, Perception* and *Super-Ego* modules, is analyzed. The information received from the *Id* is included in the information stream the *Ego* receives from the *Reality testing* module. A process evaluates and actuates the proper actions for the current situation. The module sends the generated information to the corresponding modules and initializes the above mentioned mediation process. If instinctual impulses are not discharged by the decision, other Ego functions such as mechanisms of defense will attend to them.

Mechanisms of defense: This module represses upcoming conflicts if the current environmental condition and the *Super-Ego* do not allow a drive discharge. This is a normal process in every individual. A mechanism of defense can be described as an inner fight on forbidden drive demands. The *Ego* decides which thoughts the mind is aware of and develops a censorship that divides allowed from forbidden stimuli. A variation of mechanisms of defense will be executed depending on the current outer and inner world. The module receives an input from the module *Decision management* if drive demands are repressed or distorted or the instinctual aim is displaced, suppressed, or inhibited. Specific mechanisms of defense are developed and activated according to the psychic structure.

(B) Executive motor functions

The *Executive motor functions* module defines the control of perception and motor activity. As shown in Fig. 3.4.6 the module is subdivided into *Focus of attention management* and *Motility control.*

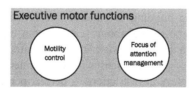

Fig. 3.4.6 The functional model of the *Executive motor functions*

Focus of attention management: This module is activated by *Psychic synthesis* processes which require specific information in order to be executed. The *Focus of attention management* of the *Executive motor function* module determines necessary environmental information by controlling the *Focus of attention* module of

Psychic synthesis. For instance, if a drive requires discharge *Focus of attention management* focuses the attention in order to locate the drive object. This process advances the reaction on condition changes and optimizes the execution of tasks.

Motility control: Information is received from the *Decision management* in the *Ego*. The incoming commands are used to control the body's motility and motor driven actions.

It is important to state that naturally, the work on defining and differentiating the various modules and their functions is not yet complete and therefore neither are the results presented above. Our current work focuses on further developing the different layers, defining the information flow between the modules, and modeling the interfaces between the *Psychic apparatus* and its environment. In addition a way must be found to connect the new ARS model with the sensor data processing theories explained in Section 3.2. This is an important step in order to be able to provide the decision unit with data from the outer world.

Currently the functionality of the new ARS model is evaluated using autonomous agents mentioned in Section 4.2. Section 4.3 discusses the setup for the evaluation of the ARS decision unit.

4   Implementation and Application

To verify the technical models described in Chapter 3, a test framework has to be defined. For this task the widely-used approach of autonomous embodied agents has been employed. The following chapter provides an overview on agent based social systems, the current simulation environment and the application of the model.

4.1   Differentiation between Modeling and Implementation

In Chapter 1, the general vision of this book and the project Artificial Recognition System was depicted, followed by a description of the sciences and theories this vision is based on in the next chapter. The transition from the theory to the model was dealt with in Chapter 3. This chapter discusses the issues of implementing these models in an agent based artificial life simulation. The task of this application is to verify the models using a concrete, task oriented implementation.

The vision is rooted in building automation. As mentioned in Section 3.1, a building can be divided into several rooms – each room can be seen as an agent. All agents are to share a common goal: increased safety and comfort for the people within the building. An agent senses its environment using its sensors (see also Section 3.1). Due to several restrictions, a perfect perception is not possible. No matter how many sensors are installed, some areas will not be covered. Each sensor type gathers a very specific type of information – a tactile floor sensor cannot give information regarding the color of the clothes of a person passing by. Object persistency for a once identified person reentering a room is an issue which can hardly be handled by one agent alone – especially if its sensors are not able to differentiate between individuals. Another reason is that sensors can fail. Thus, communication among the agents helps to complete their knowledge about the world and helps identify possible sensor failures. On the actuator side, possible actions for each room are very limited. A usual room is able to control heating, ventilation, light, and sometimes the doors are motor driven. In special purpose rooms like kitchens, potential dangerous devices like the stove can be turned off to attain a safe state.

The multi agent system based application described above bears a couple of problems regarding the vision of applying psychoanalytical concepts: "What emotions does a room have?" Not to mention the question of sexuality. As was shortly discussed in Section 3.1, an agent in a building automation system might be seen as a fungus spreading within the walls. This analogy does not help answer these questions. Instead, what is needed is a profound understanding of the models that were described in sections 3.3 and 3.4 for which verification and validation using an implementation of the model can be used. To implement the model, once again a profound knowledge of it is needed. This loop can only be escaped by introducing an intermediate step: an artificial life simulation where agents are modeled as strongly simplified humanlike creatures living in a simple environment. These artificial life forms have a body comparable to animals. Hence, using analogies from humans, drives and emotions can be easily identified. The social interaction among them is designed to be sufficiently complex to simulate certain situations needed for evaluation of e.g. the implemented emotions.

Using agents in an artificial life simulation offers an additional advantage: no ethical restrictions need to be considered. In contrast to humans, agents can be created, duplicated, altered, removed, and analyzed. The implementation of the model makes in-depth analysis of the processes possible, as any changes of the parameters in time can be recorded and interpreted. A question which is often raised – whether computer programs equipped with emotions, drives, and a sufficient degree of intelligence should be given human rights – can be answered in two parts: first, we do not build human minds (see Section 3.1); second, current research indicates that it will take a long time until we reach this level, if ever. Hence, this topic might be interesting, but from the current point of view only for philosophic discourses and authors of science fiction.

4.2   The Bubble-World

The Bubble Family Game (BFG) explained by Deutsch et al. (2008) is an artificial world simulator that tests the functionality of the ARS model. The configuration is quite familiar to the agent based social system that resulted from the Japanese psychologist Masano Toda's thought experiment – the Fungus-Eater (Toda 1982). It defines an agent based social system of small robots that have to collect uranium ore on a distant planet. They are rewarded for successfully collected ore. To survive the robots have to consume a certain kind of fungus. Equipped with a set of sensors, actuators and bodily needs, they act based on motivational subroutines linking cognition to action.

In the BFG, autonomous embodied agents – called Bubbles – roam the artificial world in their search for energy sources and social interaction.

The BFG consists of three functional blocks which are named *World simulator*, *Control unit architecture* and the *Agent body*:

The *World simulator* represents the artificial environment in which the software agents are placed. It provides the physics engine that implements a Newtonian model and therefore defines the influences of the environment on the objects. In addition the *World simulator* includes a knowledge database on environment characteristics like number, position and type of the deployed entities.

The *Control unit architecture* represents the software implementation of the current ARS model, mentioned in Section 3.3. This control module is implemented in every single agent and is responsible for its behavior. The decision making is not only influenced by the environmental situation but also by the internal evaluation system. This evaluation enables the agent to properly react to changing environmental or internal states. Therefore a bionic approach including a psychoanalytic research approach to emotions, drives and desires is used. Rage, fear, panic, and seeking are defined as basic emotions, just as in humans, while complex emotions are defined as joy, reproach, or hope (Panksepp 1998). They are implemented as rating systems and triggered by external and internal conditions. Hence the decision process is mainly influenced by the emotional system.

The third part of the BFG is represented by the *Agent body*. It provides the interface between the inner world represented by the *Control unit architecture* and the outer world represented by the *World simulator* and the *Agent body*. A body is needed for an interaction between the agent's control module and its environment.

A number of different entities are defined and placed within the World Simulator. Two types of entities are defined – passive and active ones. Active entities work on the basis of the ARS model and have the ability to interact with other entities placed in the environment. Their configurations differ slightly from agent to agent in terms of emotional parameters, memory entries or available action patterns. Various combinations of these parameters are evaluated and compared in their emerging behavior.

Agents are subjected to the following conditions:
(1) To stay alive the agent has to find and consume energy.
(2) The total amount of energy existing within the environment is limited.
(3) Every action the agent sets will reduce its energy.
(4) If the energy expends the agent will become deactivated.

The autonomous embodied agents are able to execute five basic movements which are *promenade, flee, attack, eat*, and *dance*. Although they operate autonomously they are not independent from each other. In contrast to Masano Toda's Fungus-Eater the BFG implements conditions which enable the need for social interaction among the agents. The Fungus-Eater had to balance its duty to collect uranium ore against its need to refill its energy sources. Within the BFG the agent's need for energy always remains active. The differences between the Fungus-Eater and the Bubbles are in their prime goal. Bubble agents form competing teams, the team with the last active agent win. Thus, the agent's duty consists of helping its team wins. It therefore has to balance taking care of its energy resources against helping other agents defend themselves from opponents. To enable a social system which differentiates between benevolent and egotistic behavior a social level for each agent is introduced. The chance for an agent to get assistance increases with the level of its social level.

Agents can communicate with their teammates. If another agent is located within a predefined radius it will receive information from the others. If a call for assistance gets rejected the team members that are placed within acoustic range will get a social reproach resulting in a decrease of social status.

Energy sources are passive entities: They cannot interact with the "Simulator World" but only produce energy. Two types of energy sources are defined: While one type of energy source can be consumed by single agents on their own, the other one requires two cooperating agents. This circumstance complicates the consumption of energy and increases the need for communication and social interaction among agents.

Obstacles like insurmountable objects and different types of landscapes influence the agents' route planning strategies. An agent has to sidestep 3-dimensional objects because a collision causes pain and a reduction of its energy level. Landscapes have different attributes. They consist of pathless ground or an expanse of water. Hence an agent has to spend more energy to get through this area or has to possess special skills like swimming. These configurations influence the agent in its action planning strategies.

Deutsch et al. (2007) defined a number of test cases that test social interaction and decision making processes. The BFG is designed to be adapted by expansion of the virtual environment and equipping the agent's body with a higher number and variety of sensors. Additionally, changes in the ARS model are passed on to the implementation of the *Behavioral Architecture*.

4.3   Applying the Model

Before going into detail how a feeling can be implemented in a machine, the question whether or not it is justifiable to talk of machines with feelings has to be discussed. Although many different definitions of terms like feeling/affect exist (from psychoanalysis, psychology, philosophy, etc.), they all share the fact that machines do not have – and may never have – feelings. If we apply the quote from Prof. Zemanek [17] "If a computer is intelligent – I am something different", to feelings, this helps understanding how it will be possible to implement them. The concept is taken e.g. from psychoanalysis and successively transformed to be applied in a technical system. This system has – despite the theoretical background – little in common with feelings as present in humans.

---

[17] Prof. Zemanek is professor emeritus at the Institute of Computer Technology, Vienna University of Technology. He is a computer pioneer from the early days.

McCarthy (1979) follows a philosophical approach in applying mental qualities to machines. Starting from Artificial Intelligence, he argues that higher level functions of a computer have to deal with incomplete world knowledge and non-deterministic behavior. Mental qualities help to model these functions. Wooldridge (2002) generalizes this topic to "the more we know about a system, the less we need to rely on animistic, intentional explanations of its behavior". This is comparable to the development of the human child– children use animistic explanations for objects like light switches until they understand the underlying process (compare with Piaget's (1937) child development model).

To conclude, the goal is not to enhance a computer by adding humanlike feelings. It is to synthesize the underlying concept into a mathematically describable construct and to use the mechanisms of feelings for a machine.

Bionic is a design approach where concepts and functions found in nature are implemented in technical systems. Hence, the goal described above is a bionic design approach.

One of the oldest attempts to copy nature is mankind's dream of being able to fly. In mythology, Daedalus and his son Icarus – both trapped inside a labyrinth – escaped by attaching feathers to their arms. The human body is not equipped with arms designed to fly like birds. Thus, despite approaches from Abbas Ibn Firnas and Leonardo da Vinci) it took several thousands of years until in the nineteenth century the concept of flight was successfully adapted from birds. This was made possible by moving the attempts from building moveable birdlike wings to fixed glider wings. With new techniques and better understanding of the anatomy of birds, moveable wings might be built in the future.

A major problem when dealing with bionics is that it often takes a long time until a thorough understanding of the processes observed in nature is achieved. This applies to mechanical as well as to mental processes.

Returning to agents and the psychoanalytical model – concepts such as feelings are implemented reflecting our understanding of how they work within the human mental apparatus. Each agent is an individual – thus even if all agents start with the same set of predefined knowledge and rules, they will evolve into different personalities due to their individual reduced perception of the world and the different experiences they have.

The term "psychic energy" often used to describe the force driving mental processes is not energy in the physical sense; moreover it is a term to describe information processing and might be better described as "importance". Nevertheless, the urgency of doing something is represented by a quantity.

A first approach to implementing such a quantity is to model it as a numerical value. For example, the property, urgency, can be attached to the drive, hunger. If the urgency is above a certain threshold some actions are initiated. If energy is consumed, the value is reduced.

The processes of the mental apparatus can be approximated by mathematical formulas. A perfect approximation would be indistinguishable from the true processes. Whether or not it is possible to reach such a degree and whether it is at all necessary to reach it is and will be subject to further research.

4.4   Possible Future Benefits for the Humanities

In physical science measurement is the estimation of the magnitude of a quantitative attribute like length in centimeters or the intensity of current in amperes. The process of measurement can also be used to verify a theory. To do this, one has to define which quantitative attributes (and the meaning of their qualitative values) should be used as input for the system. Using the formula representing the theory, the expected output is forecast. A measurement device can then be used to determine the difference between the expected and the real values (within the measurement error). If the difference is within an acceptable range, it can be said that the theory represents the physical process with reasonable accuracy.

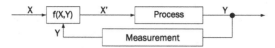

Fig. 4.4.1 System with feedback loop

On the other hand, it is a common task in engineering to control a process in order to achieve a desired output. The usual approach is to measure the actual output and to influence the process if the actual and the desired outputs do not match. Such a system is called a "feedback loop" (see Fig. 4.4.1). Since the output may exert influence on itself at any time via the feedback loop, understanding the system and the mathematical handling thereof are complicated.

To create a formula representing this process, the closed feedback loop has to be opened. The formula must not only represent the process but also the combination of the values X and Y. Thus, two input parameters have to be processed – the current new input X and the previous result Y. In Fig. 4.4.1 the previous result Y is determined using a measurement device. This is the ideal position to open the loop. The distinction between new and previous indicates that the process has to have a time index. Written as a mathematical formula, it is $Y_t = f(X_t, Y_{t-1})$. Here X_t is the current input, Y_t the resulting output, Y_{t-1} is the previous result now used as an input, and $f(X_t, Y_{t-1})$ is the formula or function representing the process and the feedback loop.

This little excursion into measurement and feedback loops shows that the problem when dealing with mutual influences – "everything is connected with everything and thereafter influences everything" – is well known in engineering and techniques have been found to overcome it. The technique is to deliberately cut open mutual influences and model them consecutively, one after another.

Psychoanalysis focuses on a subjective and introspective type of judgment. This is valid for working on the theoretical framework as well as working clinically with patients. A physically represented quantitative attribute is not available. In the representational theory of measurement, measurement is defined as "the correlation of numbers with entities that are not numbers". Thereafter, a mapping from the subjective point of view to numbers and units has to be found to be able to perform quantitative measurements within psychoanalysis. Currently, such a mapping is not available.

As stated above, before being able to measure, the quantities which have to be measured have to be defined. They do not necessarily have to be physical attributes; they can also be patterns of motion or behavior. Hence, if a motion pattern can be defined as the expected outcome for a given situation in combination with a certain theory, it can be used for measurement. The magnitude may be defined as how many elements of the motion pattern have been perceived by the measurement device.

In the beginning the metric of how well the given patterns were observed is hard to define and it is difficult to implement such a metric in a computer program – especially, if the computer has not been used before in this scientific field. Nevertheless, this process has taken place in many other scientific fields. For example, in architecture it was not obvious what computers might be good for in the beginning. It started with crude approximations of 3D models of buildings or computers were mere working tools for drawing floor plans. The creative process remained outside the computer, building cardboard models and similar techniques were the tools of choice. Nowadays, models are almost always computer generated due to the vast possibilities it offers – e.g. walking through a virtual building at different times of the day and night. In many fields computer driven development, testing, and simulation of new ideas evolved into essential tools. The new airbus A-380 has been designed solely by computer. Thus, although it might not be clear where computers will play a crucial part in psychoanalysis yet, we are optimistic that like in other sciences it will find its place.

The aforementioned artificial life simulation is a platform where parts of the models can be verified. Each agent is an individual with its own personality, experiences, and perception. The quantitative attributes which can be measured range from somatic needs such as energy to complex behavior patterns. Although the project focuses on the individual, statistical group behavorial analysis based upon theories from behaviorism and psychology can be also integrated for model validation. As stated above, psychoanalysis focuses on subjective and introspective validation. Thus, automatically measured results cannot be directly used. They have to be critically analyzed using psychoanalytic methods. No metric is currently available which transforms a subjective observation into automatically measureable quantities.

If the measured results contradict the theories of psychoanalysis, several conclusions are possible. First – and usually the most common one – the results were misinterpreted. Another very common mistake is to choose the wrong quantitative attributes. Further, an implementation error may be present – the implemented agent does not behave as defined by the model. Another possibility is that the synthesis of the psychoanalytical model into a technically feasible model needs to be reworked. Finally, the use of a computer based approach to validate psychoanalytical theories could show that some of them might have to be refined. If this point is ever reached, one long term vision of this book will have come true – to support psychoanalytical research using computer simulations of the psychoanalytical models.

References

Accardo, P. J. (1982). Freud on diplegia. Commentary and translation. *American Journal of Diseases of Children, 136,* 452-456.

Amacher, P. (1965). Freud's neurological education and its influence on psychoanalytic theory. *Psychol. Issues,* 4, 4.

Baddeley, A. (1997). *Human Memory: Theory and Practice.* London: Psychology Press.

Bratman, M. (1999). *Intention, Plans and Practical Reason.* Center for the Study of Language and Information, Stanford University.

Breazeal, C. (2002). *Designing Sociable Robots* Cambridge: MIT Press.

Broca, P. (1861). Remarques sur le siege de la faculte du langage articule, suivies d'une observation d'aphasie (perte de la parole) [Remarks on the Seat of the Faculty of Articulated Language, Following an Observation of Aphemia (Loss of Speech)]. *Bulletin de la Société d'anatomie,* 36, 330-357.

Broca, P. (1865). Sur le siege de la faculte du langage articule [On the Seat of the Faculty of Articulated Language]. *Bull. Anthropol., 6:* 377-393

Brooks, R. A. (1986). A robust layered control system for a mobile robot. *IEEE Journal of Robotics and Automation,* 2, (1), pp. 14-23.

Burgstaller, W. (2007). *Interpretation of Situations in Buildings.* Doctoral dissertation, Technical University of Vienna, Austria.

Burgstaller, W., Lang, R., Pörscht, P., Velik, R. (2007). Technical Model for Basic and Complex Emotions. *INDIN 2007 Conference Proceedings,* 1033 - 1038.

Charcot, J.-M. (1889). Clinical Lectures on Diseases of the Nervous System (Savill, T., Trans.). (Vol. 3), London: New Sydenham Society.

Damasio, A. (2003). *The Feeling of What Happens: Body amd Emotion in the Making of Consciousness.* New York: Harcourt Brace.

Damasio, A. (2005). *Descartes'error: emotion, reason, and the human brain* (Rev. ed.). London: Penguine.

David, E. (1961). Bionics or electrology? An introduction to the sensory information processing issue. *IEEE Transactions on Information Theory,* 8, (2), pp. 74 – 77.

Deutsch, T., Zeilinger, H., Lang, R. (2007). Simulation Results for the ARSP Model. *Proceedings of 2007 IEEE International Conference of Industrial Informatics,* pp. 1021-1026.

Deutsch, T., Zia, T., Lang, R., Zeilinger, H. (2008). A Simulation Platform for Cognitive Agents. *Proceedings of the 6th IEEE International Conference on Industrial Informatics,* July 13-16 2008, Daejon, Korea, to be published.

Dietrich, D., Sauter, T. (2000). Evolution Potentials for Fieldbus Systems. *Proceedings of the 4th IEEE International Workshop on Factory Communication Systems,* pp. 343-352

ENF (2007). First international Engineering and Neuropsychoanalysis Forum, Vienna, July 23, 2007, http://www.indin2007.org/enf.

Franklin, S. & Graesser, A. (1997). Is it an Agent, or Just a Program?: A Taxonomy for Autonomous Agents. In *Proc. ECAI '96: Proceedings of the Workshop on Intelligent Agents III, Agent Theories, Architectures, and Languages,* London, UK, pp. 21-35.

Freud, S. (1956a [1886]). Report On My Studies In Paris And Berlin. In: J. Strachey (Ed. & Trans.) *The Standard Edition of the Complete Psychological Works of Sigmund Freud* (Vol. 1, pp. 5-15). London: Hogarth Press.

Freud, S. (1891b). On Aphasia: A Critical Study. New York: International Universities Press, 1953

Freud, S. (1892-93a). A case of Successful Treatment by Hypnotism: With Some Remarks on the Origin of Hysterical symptoms through "Counter-Will". In: J. Strachey (Ed. & Trans.) *The Standard Edition of the Complete Psychological Works of Sigmund Freud* (Vol. 1, pp. 117-128). London: Hogarth Press.

72

Freud, S. (1895d [1893-1895]). Studies on Hysteria. In: J. Strachey (Ed. & Trans.) *The Standard Edition of the Complete Psychological Works of Sigmund Freud* (Vol. 2). London: Hogarth Press.

Freud, S. (1893b). *Zur Kenntniss der cerebralen Diplegien des Kindesalters (im Anchluss die Little'sche Krankheit).* Vienna: Moritz Perles.

Freud, S. (1893f). Charcot. In: J. Strachey (Ed. & Trans.) *The Standard Edition of the Complete Psychological Works of Sigmund Freud* (Vol. 3, pp. 11-23). London: Hogarth Press.

Freud, S. (1894a). The Neuro-Psychoses of Defence. In: J. Strachey (Ed. & Trans.) *The Standard Edition of the Complete Psychological Works of Sigmund Freud* (Vol. 3, pp. 41-61). London: Hogarth Press.

Freud, S. (1950c [1895]). Project for a Scientific Psychology. In: J. Strachey (Ed. & Trans.) *The Standard Edition of the Complete Psychological Works of Sigmund Freud* (Vol. 1, pp. 281-391). London: Hogarth Press.

Freud, S. (1896c). The Aetiology of Hysteria. In: J. Strachey (Ed. & Trans.) *The Standard Edition of the Complete Psychological Works of Sigmund Freud* (Vol. 3, pp. 191-221). London: Hogarth Press.

Freud, S. (1896b). Further Remarks on the Neuro-Psychoses of Defence. In: J. Strachey (Ed. & Trans.), *The Standard Edition of the Complete Psychological Works of Sigmund Freud* (Vol. 3, pp. 162-185). London: Hogarth Press.

Freud, S. (1897a). *Die infantile Cerebrallaehmung.* Vienna: Alfred Hoelder.Freud, S. (1897). Abstracts of the Scientific Writings of Dr. Sigm. Freud 1877-1897. In: J. Strachey (Ed. & Trans.) *The Standard Edition of the Complete Psychological Works of Sigmund Freud* (Vol.3, pp. 223-257). London: Hogarth Press.

Freud, S. (1900a). The Interpretation of Dreams. In: J. Strachey (Ed. & Trans.) *The Standard Edition of the Complete Psychological Works of Sigmund Freud* (Vol. 4 & 5). London: Hogarth Press.

Freud, S. (1905d). Three Essays on the Theory of Sexuality. In: J. Strachey (Ed. & Trans.) *The Standard Edition of the Complete Psychological Works of Sigmund Freud* (Vol. 7, pp.135-243). London, Hogarth Press.

Freud, S. (1911b). Formulations on the Two Principles of Mental Functioning. In: J. Strachey (Ed. & Trans.) *The Standard Edition of the Complete Psychological Works of Sigmund Freud* (Vol. 12, pp. 218-226). London: Hogarth Press.

Freud, S. (1915c). Instincts and their Vicissitudes. In: J. Strachey (Ed. & Trans.) *The Standard Edition of the Complete Psychological Works of Sigmund Freud* (Vol. 14, pp. 117-140). London: Hogarth Press.

Freud S. (1915d). Repression. In: J. Strachey (Ed. & Trans.) *The Standard Edition of the Complete Psychological Works of Sigmund Freud* (Vol. 14, pp.146-158). London: Hogarth Press.

Freud, S. (1915e). The Unconscious. In: J. Strachey (Ed. & Trans.) *The Standard Edition of the Complete Psychological Works of Sigmund Freud* (Vol. 14, pp.166-204). London: Hogarth Press.

Freud, S. (1923a). Two Encyclopaedia Articles. In: J. Strachey (Ed. & Trans.) *The Standard Edition of the Complete Psychological Works of Sigmund Freud* (Vol. 18, pp.235-259). London: Hogarth Press.

Freud, S. (1923b). The Ego and the Id. In: J. Strachey (Ed. & Trans.) *The Standard Edition of the Complete Psychological Works of Sigmund Freud* (Vol. 19, pp.12-59). London: Hogarth Press.

Freud, S. (1913j). The Claims of Psycho-Analysis to Scientific Interest. *The Standard Edition of the Complete Psychological Works of Sigmund Freud* (Vol. 13, pp.165-190). London: Hogarth Press.

Freud, S. (1926e). The Question of Lay Analysis. In: J. Strachey (Ed. & Trans.) *The Standard Edition of the Complete Psychological Works of Sigmund Freud* (Vol. 20, pp.177-258). London: Hogarth Press.

Freud, S. (1940a [1938]). An Outline of Psychoanalysis. In: J. Strachey (Ed. & Trans.) *The Standard Edition of the Complete Psychological Works of Sigmund Freud* (Vol. 23, pp. 144-207). London: Hogarth Press.

Freud, S. (1974). The Freud-Jung-Letters. In: McGuire, W. (Ed.). *The Correspondence between Sigmund Freud and C.G. Jung.* Princeton: Princeton Univ. Press.

Georgeff, M.P. & Ingrand, F. F. (1989). Decision-making in an embedded reasoning system, Proceedings of the Eleventh International Joint Conference on Artificial Intelligence, Detroit (Michigan). 972–978.

Gilbert, N. (2008). *Agent-Based Models* London: SAGE Publications.

Handlbauer, B. (1990). Die Adler-Freud-Kontroverse [The controversy between Adler and Freud]. Frankfurt: Fischer.

Harlow, J. (1868). Recovery from the passage of an iron bar through the head. Bulletin of the Massachusetts Medical Society, 2, 327-347.

Jackson, J. H. (1932 [1884]). Evolution and Dissolution of the Nervous System. In: J. Taylor (Ed.), *Selected Writings of John Hughlings Jackson*, (Vol. 2). London: Hodder and Stoughton.

Jellife, S.E. (1937). Sigmund Freud as a neurologist: some notes on his earlier neurobiological and clinical studies. *Journal of Nervous and Mental Disease.*, 85 (6), 696-711.

Jones, E. (1953). *Sigmund Freud: Life and Work*, Vol. 1. London: Hogarth Press.

Jung, C.G. (1976). Die Archetypen und das kollektive Unbewußte [Archetypes and Collective Unconscious]. In: Jung-Merker, L. & Rüf, E. (Eds.). *Gesammelte Werke* [Collected Works] (Vol. 9).Düsseldorf, Zürich: Walter.

Kandel, E.R. (1999). Biology and the Future of Psychoanalysis: A new intellectual Framework for Psychiatry. *American Journal of Psychiatry*, 156: 505 - 524

Kaplan-Solms, K. & Solms, M. (2000). *Clinical Studies in Neuro-Psychoanalysis. Introduction to a Depth Neuropsychology*. London: Karnac Books.

Klein, M. (1935). A Contribution to the Psychogenesis of Manic-Depressive States. *International Journal of Psycho-Analysis*,16, 145-174.

Khan, X. (2007). *Keeping up with Moore's Law: Recent Advances in Chip Design*. Keynote speech of the 5th International Workshop on Frontiers of Information Technology, FIT 2007, Islamabad, Pakistan, 17-18

Klein, M. (1946). Notes on Some Schizoid Mechanisms. *International Journal of Psycho-Analysis*,27, 99-110.

Kleist, K. (1934). *Gehirnpathologie*. Leipzig: Barth.

Köhler, T. (1995). *Freuds Psychoanalyse. Eine Einführung* [Freud's Psychoanalysis. An Introduction]. Stuttgart: Kohlhammer.

Kozulin, A. (1984). *Psychology in Utopia: Toward a Social History of Soviet Psychology*. Cambridge, : MIT.

Kuhn, T. (1962). *The Structure of Scientific Revolutions*. Chicago: University of Chigaco Press.

R. Lang, D. Bruckner, G. Pratl, R. Velik, T. Deutsch: Scenario Recognition in Modern Building Automation. In: Proceedings of the 7th IFAC International Conference on Fieldbuses & Networks in Industrial & Embedded Systems (FeT 2007), 2007, S. 305 - 312.

Laplanche, J. & Pontalis, J.B. (1972). *Das Vokabular der Psychoanalyse* [The Vocabulary of Psychoanalysis]- Frankfurt am Main: Suhrkamp.

Lichtheim, L. (1885). Ueber Aphasie. *Deutsches Archiv fuer klinische Medicin*, 36, 204-268.

Loy, D., Dietrich, D., Schweinzer, H.-J. (2001). OpenControl Networks - LonWorks/EIA 709 Technology; Kluwer Academic Publishers Sauter, T., Dietrich, D., Kastner, W. (eds.). (2001). *EIB Installation Bus System*. München: Publicis

Luria, A.R. (1970 [1947]). *Traumatic Aphasia: Its Syndromes, Psychology and Treatment*. The Hague: Mouton.

Luria, A.R. (1973). *Working Brain: An Introduction to Neuropsychology*. London: Penguin Books Ltd.

Luria, A.R. (1979). *The Making of Mind: A Personal Account of Soviet Psychology*. Cambridge: Harvard University Press.

Luria, A.R. (1980). *Higher Cortical Functions in Man*. New York: Basic Books.

Luria, A.R. (1987). Mind and brain: Luria's philosophy. In: Gregory, R.L. (Ed.), *The Oxford Companion to the Mind*. Oxford & New York: Oxford University Press.

McCarthy, J. (1979). Ascribing Mental Qualities to Machines. , In: Ringle, M. (Ed.) *Philosophical Perspectives in Artificial Intelligence*. Brighton: Harvester Press.

Meyer-Palmedo, I. & Fichtner, G. (1982). *Freud Bibliographie und Werkkonkordanz.* [Freud Bibliography]. Frankfurt: Fischer.

Meyer, W.-U., Schützwohl, A., Reisenzein, R. (1999). *Einführung in die Emotionspsychologie, Band II, Evolutionspschologische Emotionstheorien* [Introduction to Emotional Psychology, Vol.2, Emotion Theories from Evolutionary Psychology] (2nd ed.), Bern: Huber.

74

Müller-Pozzi, H. (2002). *Psychoanalytisches Denken. Eine Einführung* [Psychoanalytic Thinking. An Introduction. Bern: Huber.

Panksepp, J. (1998). *Affective Neuroscience, the Foundations of Human and Animal Emotions.* New York: Oxford University Press.

Plutchik, R. (2001). The nature of emotions. *American Scientist,* 89(4), 344-350.

Pratl G. (2006). Processing and Symbolization of Ambient Sensor Data, PhD Thesis, Institute of Computer Technology, Technical University of Vienna, 2006

Rapaport, D. (1960). *The Structure of Psychoanalytic Theory. A Systematizing Attempt.* Psychological Issues Vol. II No. 2. New York: Int. University Press.

Riese, W. (1959). *A History of Neurology.* New York: M.D. Press.

Rösener, C., Lang, R., Deutsch, T., Gruber, A., Palensky, B. (2007). Action planning model for autonomous mobile robots. *Proceedings of 2007 IEEE International Conference on Industrial Informatics INDIN07,* 1009 - 1014.

Rösener, C. (2007). *Adaptive Behavior Arbitration for Mobile Service Robots in Building Automation.* Doctoral dissertation, Technical University of Vienna, Austria.

Russ G. (2003). Situation-dependent Behavior in Building Automation, PhD Thesis, Institute of Computer Technology, Technical University of Vienna,, 2003

Schelling, T. (1971). Dynamic Models of Segregation. *Journal of Mathematical Sociology,* 1(2), 143-186.

Solms, M. (1995). Is the brain more real than the mind? *Psychoanal. Psychother., 9,* 107-120.

Solms, M. (2004). 'Freud returns.' *Scientific American,* 56-62.

Solms, M. & Saling, M. (1986). On psychoanalysis and neuroscience: Freud's attitude to the localizationist tradition *International Journal of Psychoanalysis.* 67:397-416

Solms, M. & Turnbull, O. (2002). *The Brain and the Inner World.* London: Karnac

Sulloway, F. J. (1979). *Freud, Biologist of the Mind. Beyond the Psychoanalytic Legend,* London: Burnett Books.

Tamarit C. (2003). Automation System Perception – First Step towards Perceptive Awareness, PhD Thesis, Institute of Computer Technology, Technical University of Vienna, 2003

Toda, M. (1982). *Man, Robot and Society.* The Hague: Nijhoff.

Triarhou, L. & Del Cerro, M. (1985). Freud's contribution to neuroanatomy. *Archives of Neurology, 4,* 282-287.

Tulving, E. (1998). *Elements of Episodic Memory.* Oxford: Clarendon Press.

Van der Veer, R. & Valsiner, J. (1991). *Understanding Vygotsky: A Quest for Synthesis.* Oxford: Blackwell.

Velik, R. (2008). *A Bionic Model for Human-like Machine Perception.* Doctoral dissertation, Technical University of Vienna, Austria.

Vogel, P. (1955). Sigmund Freuds Beitrag zur Gehirnpathologie [Sigmund Freud's Contribution to the Field of Brain Pathology]. *Psyche, 9,* 42-53.

Vogel, P. (1956). Der Wert der neurologischen Studien Sigmund Freud fuer die Gehirnpathologie and Psychiatrie [The Value of the Neurological Studies of Sigmund Freud for the Field of Brain Pathology and Psychiatry]. *Schweizer Archiv fuer die Neurologie and Psychiatrie, 78:* 274-287.

Waelder, R. (1983). *Die Grundlagen der Psychoanalyse* [The Fundamentals of Psychoanalysis]. Stuttgart: Klett-Cotta

Wallner, F. (1991). *Acht Vorlesungen über den Konstruktiven Realismus* [Eight lectures on Constructive Realism]. Wien: WUV.

Weaver, W. (1948). Science and Complexity. *American Scientist,* 36, 536. (http://www.ceptualinstitute.com/genre/weaver/weaver-1947b.htm)

Wiener, N. (1948). *Cybernetics.* New York: John Wiley

Wooldridge, M. (2002). *Introduction to MultiAgent Systems.* New York: John Wiley & Sons.

Part II Proceedings of Emulating the Mind (ENF 2007)

List of Authors in Alphabetic Order

Etienne	Barnard	Meraka Institute, Pretoria, South Africa
Elisabeth	Brainin	Vienna Psychoanalytic Society, Austria
Andrea	Clarici	University of Trieste, Italy
Dietmar	Dietrich	Vienna University of Technology, Austria
Georg	Fodor	Vienna Psychoanalytic Society, Austria; University of Cape Town, South Africa
Wolfgang	Kastner	Vienna University of Technology, Austria
David	Olds	Columbia University, New York, United States
Brigitte	Palensky (née Lorenz)	Vienna University of Technology, Austria
Peter	Palensky	University of Pretoria, South Africa; Berkeley National Laboratory, CA, United States
Jaak	Panksepp	Veterinary Medicine Washington State University, United States
Aaron	Sloman	The University Of Birmingham, United Kingdom
Mark	Solms	International Neuropsychoanalysis Centre, London, UK; University of Cape Town, South Africa
Mihaela	Ulieru	The University Of New Brunswick, Canada
Yoram	Yovell	University of Haifa, Israel
Gerhard	Zucker (né Pratl)	Vienna University of Technology, Austria

76

1   Session 1

1.1   A Brief Overview of Artificial Intelligence Focusing on Computational Models of Emotions

Brigitte Palensky (née Lorenz), Etienne Barnard

Computational models of emotions promise to extend the capabilities of artificial intelligence in a number of ways. We review several such models, and show how these models emphasize different aspects of the interactions between emotion and behavior. These recent developments are placed in the context of earlier approaches, ranging from those derived from symbolic logic, through statistical models, to the more recent interest in embodied agents. We present evidence that significant deficiencies of these approaches may be overcome through the use of suitable emotion-based models.

1.1.1   Introduction

The endeavor to develop automata with intelligent behavior has a long and conflicted history [1], [2]. The philosophical importance and practical benefits of such an automaton have served as inspiration for innumerable philosophers, scientists and engineers; but the difficulties experienced in its development have inspired a comparable number of charlatans to produce faked solutions, exaggerated claims, and much obfuscation. Thus, for every intricate wind-up toy or pneumatic duck, intended as a serious model of intelligence, there has been a dwarf hidden inside a chess machine or a parlor trick spuriously producing the appearance of intelligence.

Both of these tendencies gained considerable momentum with the appearance of the programmable computer in the middle of the twentieth century. Suddenly, engineers and scientists had at their disposal an automaton that, by its very nature, could respond to commands (software programs), and seemed but a short distance away from the foundations of intelligent behavior. It is therefore not surprising that pioneers such as Turing and Von Neumann saw the development of intelligent software as one of the main potential applications of the newly developed computer systems.

The current paper gives a whirlwind tour of the current status of AI. The past 50 years have seen far too many developments to be mentioned in a single review; our aim is therefore to give a flavor of the state of the field, with relevant references, so that the interested reader can pursue these topics in more depth.

For the purposes of exposition, we distinguish between 'symbolic AI' (Section 1.1.2), 'statistical AI' (Section 1.1.3), and 'behavior-based AI' (Section 1.1.4) even though this distinction is often blurred in practice. After describing the characteristics of these three approaches, and some typical algorithms they employ, we focus on the specific role and modeling of emotions in AI (sections 1.1.5 and 1.1.6). A number of successful applications of AI are discussed in Section 1.1.7.

Section 1.1.8 reviews how far the entire enterprise has progressed, and makes some predictions on the road ahead.

1.1.2   Symbolic Approaches to Artificial Intelligence

The symbolic approach to AI is motivated by the (introspectively suggested) hypothesis that humans process information by abstracting facts, perceptions, goals, and the like into 'symbols', and then manipulating those symbols. For example, in order to play chess, one would have symbols corresponding to the white and black pieces and the available squares on the board. To choose an appropriate move, these symbols are manipulated according to the rules of chess, and the various outcomes evaluated (subject to time constraints). The most favorable outcome is then selected as the player's next move.

This example suggests the two basic building blocks of the symbolic approach: *knowledge representation* (in order to capture information about the rules of the game, the current board position, clock time remaining, etc.) and *search* (to generate and compare the different alternatives that arise in playing the game). Both of these topics have produced an unexpected range of challenges as well as a number of breakthroughs, which have lead to several useful algorithms.

Search algorithms can usefully be distinguished in terms of the amount of domain-specific information that they use. *Uninformed search algorithms* search through a range of possibilities without information beyond the basic structure on how these possibilities are related to one another. Such search tends to be very expensive in both memory usage and computational cycles consumed: for non-trivial problems, complete algorithms – those guaranteed to find an optimal solution if one exists – usually depend exponentially on some fundamental parameter (such as the number of moves through which the chess player looks ahead in the example above).

To overcome this obstacle, *informed search algorithms* employ additional information about the particular problem being solved (for example, 'heuristic' suggestions on the comparative quality of different partial solutions). The improvement that results from such domain-specific knowledge obviously depends on the quality of the available knowledge. For interesting problems, one finds that optimal solutions often continue to have exponential computational costs, but good approximations can typically be found with a more tolerable burden. Algorithms developed for this purpose – most notably variants of the so-called A-star search, and various forms of hill-climbing algorithms, including the biologically-inspired genetic algorithms – have found use far outside the bounds of conventional AI.

Knowledge representation similarly sails a course between efficient, domain-specific representations and more costly, more general domain independent solutions. The latter category is perhaps best typified by the use of logic to represent knowledge: in a simple domain, statements of fact can be represented as the propositions of propositional calculus, whereas more complex relationships can be modeled using first-order (predicate) logic. Although a tremendous amount of work has been done on theoretical and practical aspects of knowledge representation using logic, the research community has by-and-large concluded that it is simply too complex a task to represent real-world domains within the constraints of these formalisms. Other domain-independent approaches to knowledge representation, such as semantic networks and frames, have therefore attracted many adherents, but these have also not solved the 'knowledge engineering' problem (that is, how to capture information about a new domain in the representation) to the satisfaction of most.

In practice, current systems for symbolic AI therefore tend to employ ad-hoc representations that are tailored to the characteristics of the domain being modeled. By combining such representations with sophisticated search strategies, systems have been built which have become an intrinsic element of the modern economy – from scheduling the activities of operators to planning the logistics of corporate fleets and armies, such systems are now rarely thought of as 'intelligent' (see Section 1.1.7 for a more detailed example). This practical progress has not been accompanied by comparable insights on the fundamental nature of intelligence, and the balance of scientific opinion would probably favor approaches arising from the statistical perspective to produce the next round of significant conceptual breakthroughs.

1.1.3   Statistical Approaches to Artificial Intelligence

Early symbolic approaches made rapid progress in a number of 'toy' domains (such as robotics in a strictly controlled environment, or 'mathematics' with a limited, well-circumscribed set of axioms). Time and again, however, the apparently mechanical task of extending these successes to real-world operations ended in abject failure. It became clear that the information required to solve realistic problems generally differs from that in the toy domains along a number of dimensions: exceptions abound, apparently unrelated phenomena can influence one another profoundly, and the sheer number of factors relevant to a particular problem can easily overwhelm even the most sophisticated symbolic approach.

How does biological intelligence deal with these difficulties? Although the full answer is still unknown, there are strong hints that a crucial aspect of the biological approach is the flexibility that comes from *learning* about a domain, even as problems are being solved. A learning organism does not need a complete description of the domain before attempting a solution; it can be robust to exceptions, and can learn about the interactions between different variables in the process of interacting with the world. For all these reasons, the development of learning algorithms has long been considered a key component in the quest to improve on the performance of AI systems. Since these algorithms attempt to extract information from the statistical properties of the environment, they are collectively known as the *statistical approach* to AI.

Many of the early statistical algorithms drew their inspiration from known facts about biological learning. Most importantly, the learning of biological organisms seems to express itself as changes in the connections strengths between simple computational units known as *neurons*. Much attention has therefore been devoted to *artificial neural networks*, which are abstractions based on some of the properties of their biological counterparts.

Numerous algorithms have been developed to 'learn' connections between these artificial neurons [3] from training examples that exemplify the desired behavior of the network. Some of these algorithms have very useful properties – they generalize well to unseen cases, and are robust to noise and errors in the training data. However, successful learning in these networks still requires careful pre-processing of the learning data. (In particular, all inputs must be presented as *features* in a format that is suitable for effective learning, and outputs must generally be limited to a small group of predetermined classes.)

These limitations have spurred the development of other approaches that are less directly tied to biological facts, building on theoretical insights into learning theory instead. Support vector machines [4], for example, are an algorithmic embodiment of the principle known as Occam's razor – that the best model for a particular data set is the simplest model which described it accurately. Bayesian networks [5] build on the insight that conditional independence is a core principle in simplifying the description of data: whereas the number of interactions between a set of variables grows exponentially with the number of variables in the set, this can be limited to linear growth is some variables can be shown to be independent of others (when conditioned on a third set of variables).

Algorithms such as neural networks, Bayesian networks, and Hidden Markov Models (a particular class of Bayesian networks that are well-suited for modeling temporal variability) have become the mainstay of AI systems that deal with 'sensory' input – vision, speech and other acoustic signals, and the like. These algorithms are also widely used in other applications where the ability to generalize from a training set to unseen examples is useful. Credit scoring companies, for example, use these algorithms to assess the applicants' creditworthiness based on measurements that have been found relevant to this task, and Internet search engines learn the characteristics of useful search 'hits' in similar fashion.

1.1.4   Embodied Approaches To Artificial Intelligence

In the early days of artificial intelligence, expectations that computers were only a small step away from producing intelligent behavior were widespread. Computers even became a metaphor for the brain. By the end of the 1980s, some researchers from artificial intelligence as well as brain and cognitive science realized that this view was maybe misguided or at least too limited. The brain was not designed to 'run programs' for specialized purposes, like performing logic or playing chess. Instead, as evolutionary theory tells us, the brain has evolved to control our behavior such that we can survive in highly dynamic environments. One of the key elements of the new perspective is that intelligence always has to manifest itself in *behavior*. Rodney Brooks, one of the founders of the field suggested to do away with thinking and reasoning and focus on the interaction of the organism or artifact with the real world [6]. He called the new view *behavior-based robotics*. This indicates a shift of the role model of intelligence, away from passive computer programs which completely abstract away any bodily issues to-wards active, physically embedded and embodied *agents*. Agents can be biologi-cal, or robots, or pieces of software, but, in any case, they have to be affected by the influences of their environment – whether real or virtual – and they have to in-teract with their environment. This implies that they must have sensors via which they can acquire relevant information about their environment, a motor system to act on their environment, and an internal structure with internal states. Agents are called *autonomous* when they are able to control their actions and internal states on their own without the direct intervention of humans or other agents. Addition-ally agents shall be able to perform not only a single task, but a variety of them – all they need to survive self-sufficiently in their environment. Agents that are si-tuated, embodied, autonomous, and self-sufficient are sometimes referred to as 'complete' agents [7].

Prototypical Embodied Agents: One advantage of the new approach is that the heavy use of the system-environment interaction minimizes the required amount of world modeling. In the mid-1980s, R. Brooks argued that the sense-model-plan-act paradigm of traditional AI was less appropriate for autonomous robots as the process of building world models and reasoning using explicit symbolic represen-tational knowledge often took too long to produce timely responses. As an alterna-tive, he developed a layered control system with a new functional decomposition into task-achieving modules called the *subsumption architecture*. Each of these modules can produce a specific behavior, for example wandering around, relative-ly independently of the others. Therefore, when extending the system, new mod-ules can be incrementally added on top of the others without having to alter the existing ones. However, the modules are connected to each other by connections that can suppress input to modules or inhibit output. From an engineering point of view, the subsumption architecture is attractive because it is robust and easily ex-tendable. From a cognitive science perspective, it contributes to the idea that intel-ligence can arise or *emerge* from a large number of simple, loosely coupled paral-lel processes.

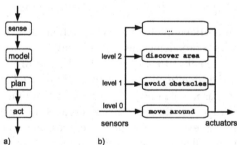

Fig. 1.1.1 The traditional decomposition of an AI control architecture versus the 'task-achieving' approach of the subsumption perspective (based on [8])

Another famous type of agents that is ideally suited to study the relationship between mechanisms and arising behavior are *Braitenberg vehicles*, named after their inventor, the neuroscientist Valentino Braitenberg [9]. Vehicles are very simple machines or robots situated in an environment containing heat or light sources (Fig. 1.1.2). The architecture of a vehicle mainly consists of almost direct connections between sensors and motors. By making small variations of how to connect sensors and motors – either laterally or counter-laterally – and how to relate sensor intensity to motor intensity – either positively or negatively – different kinds of behavior arise. Thus, vehicles show behavioral patterns which are not directly programmed. These behaviors can look quite complex.

The earliest role model for complete agents comes from the Japanese psychologist Toda [10]. Already in the 1950's, he was discontent that cognitive psychology was only focusing on complex planning and decision making strategies in order to analyze intelligent behavior.

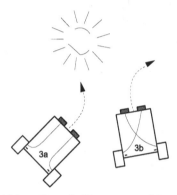

Fig. 1.1.2 Braitenberg vehicles of type 3. The sensors of these vehicles exert an inhibitory influence on the motors: The more the sensors are activated, the less power is delivered to the motors. Vehicle 3a turns towards the light source (and stops when it is close enough), whereas 3b turns away. Both of them stay longer in the surroundings of the light source (based on [9]).

As an alternative, he designed an autonomous agent, the *Fungus Eater* robot, together with an artificial, science-fictionally inspired environment. Sent on a distant planet, the robot has the task to collect uranium ore. Any activity it does requires energy. In order to survive, it has to eat wild fungi growing on the surface of the planet. Both resources, ore and fungi, are distributed randomly over the foreign planet. The agent is equipped with a relatively simple but *complete* set of rules – including everything it needs to survive, like sensory, locomotion, decision making, and collecting capabilities. Additionally, Toda provides the robots with 'urges'. Urges are small, pre-defined subroutines that are activated once a situation has been identified as being relevant to some vital concern. Each urge is linked to a specific action. Toda describes a whole set of those urges and calls them 'fear', 'anxiety', 'love', etc. He claims that urges would make the fungus eaters 'emotional' and that being emotional would make their behavior more intelligent. With his concept of the fungus eater, Toda was one of the first to argue that emotions can contribute to intelligent behavior.

1.1.5   Emotion-Based Approaches to Artificial Intelligence

Emotions have often been considered solely as disturbing influences to 'rational' thinking. When trying to create intelligent artifacts, AI researchers therefore neglected emotions completely and focused instead exclusively on 'higher-level' cognitive capabilities such as thinking, planning, and language. With the advent of embodied cognitive science, perceptual and motor abilities were given more attention within the artificial intelligence research community. The same is true for emotions which, in an informal sense, can be seen as mechanisms to 'enhance' acquired information with values and valences. The insight that emotions may be an important facet of intelligent behavior was accepted only recently. The neuroscientist Damasio argues convincingly that emotions are crucially intertwined with cognitive problem solving and decision making [11]. Besides, as for example elaborated by the psychologist Frijda [12] among many others, emotions are an essential part for the establishment of social behavior. However, the task of defining emotions in technical terms has often been accounted as not feasible. The main reason for this lies in the difficulty to reach a profound comprehension of emotional behavior. Nevertheless, an increasing number of researchers in the software agent and robotic community believe that computational models of emotions will be needed for the design of intelligent autonomous agents in general, and for the creation of a new generation of robots able to socially interact with each other or people in special. As M. Minsky puts it: *'The question is not whether intelligent machines can have emotions, but whether machines can be intelligent without any emotions'* [13].

During the last years, various kinds of systems trying to computationally model, implement, and investigate emotions have been developed. The systems differ significantly regarding their aims and assumptions. They refer in various degrees to existing theories of emotions. An overview of the many different emotion theories in psychology can be found in [14]. The concept of emotion is very broad and covers various aspects including physiological, motivational, and expressive ones, as well as the subjective experience of emotional states in the form of feelings, and the ability to cognitively reason about emotions. So far, most of the computational emotion systems focus only on a *subset* of the above aspects. The existing work can roughly be divided into 'communication-driven' approaches that focus on the surface manifestation of emotions and their influence on human-computer interaction, and 'process-driven' approaches that attempt to model and simulate the mechanisms of emotion as they unfold. This distinction, however, is not clear and mutually exclusive. Some of the systems address both perspectives. In the following, a small number of key projects mainly focusing on the *process perspective* of emotions will be discussed. Concerning the question of algorithmic representation, all of the previously presented approaches to AI are utilized. Some of the computational models of emotion only rely on one approach (be it either symbolic, statistical, or embodied). Some systems are based on hybrid approaches.

1.1.6   A Closer Look on Computational Models of Emotions

In this section, some prototypic attempts to use the concept of emotion in the context of artificial search, replace all intelligence are presented. They are divided into ethology-inspired, learning-related, appraisal-based and architectural approaches. This is a very rough division as there is no general dimension along which a clear distinction could be made.

Ethology-Inspired Models

Emotions clearly possess bodily aspects. From an evolutionary perspective, to protect the body's integrity and wellbeing most probably is the primary task for which emotional mechanisms have been evolved in the first place. The view of emotions as control mechanisms that have been efficiently governing behavior long before higher-level cognitive mechanisms showed up is often utilized when constructing selection architectures for autonomous adaptive agents. This approach leads to ethology-inspired architectures for which agents with a physically embedded body – whether real or simulated – and mechanisms to prioritize the usage of limited resources are important design aspects.

The Model of Cañamero: As an example, the architecture proposed by Cañamero will be presented [15]. It deals with robots that have the task to survive as long as possible in an environment containing various types of resources, obstacles, and predators. In order to maintain their well-being (their internal 'milieu'), the robots have to carry out different activities. The behavior selection architecture of the robots includes the following components:

- A *physiology* consisting of synthetic survival-related variables ("energy", "blood sugar", "vascular volume"), and "hormones";
- A set of *motivations* (hunger, aggression, curiosity, fatigue, self-protection, etc.);
- A set of *behaviors* (attack, withdraw, eat, drink, play, rest);
- A set of *basic emotions* (fear, anger, happiness, interest, boredom, and sadness).

Motivations (needs) are activated when the survival-related variables depart from their homeostatic regime. For example, when a robot is too warm, its motivation to decrease its temperature is invoked. Each motivation has an intensity and the one with the highest intensity controls the behavior of the robot. Motivation intensity, and therefore behaviors, are influenced by the emotions of the robots. Emotions can be triggered by the presence of external objects, or by the occurrence of internal changes or patterns. Such an internal pattern is for example when one or more of the motivations are continuously too high. Then the robot becomes angry. Activated emotions release hormones that have an effect on the robot's physiology, attention, and perception. Cañamero concedes that the described architecture incorporates only very simple and low-level mechanisms and that issues such as for example learning or the development of strategies are not addressed. However, what can be accommodated into the model are different reward and punishment mechanisms, and those are thought by Cañamero as well as many other researchers to be very important for learning. This will be discussed in the following section.

Emotion-Related Learning in Autonomous Agents

Emotions can be viewed as evaluations that signal what is 'good' and what is 'bad' for the organism of an autonomous agent. Thus, emotions can be utilized for learning. In the machine learning community, learning based on positive or negative feedback is called *reinforcement learning* and widely used, however without referring to the concept of emotions. Compared to the traditional reinforcement learning approach, the introduction of explicit emotional mechanisms is argued to provide additional flexibility that cannot be found in simple stimulus-response models of learning. This is supported by the work of Mowrer [16] who demonstrated that emotions enable a two-stage learning process. In a first step, the agent learns to respond to a special stimulus with a special emotional state ('classical conditioning'), for example to 'fear' a certain sound if it is always coupled with pain. In a second step, it learns to associate a *behavior* with its influence on the emotional state ('operant conditioning'), for example that going away from the sound reduces the fear. Simultaneously, the introduction of an emotional state such as 'fear' *motivates* the agent to actively seek for different means of behavior if the originally conditioned one is not useful any more. Another advantage of explicit emotional states is that a singular emotion can influence several processes at the same time. For example, frustration can direct the focus of attention, trigger reassessment of a situation, activate predictions about how to improve the situation, etc.

Cathexis: J. Velásquez from MIT developed a connectionist model of emotion synthesis and learning called Cathexis [17]. He physically implemented it in a robot dog called Yuppy (a YAMAHA puppy). Apart from six basic emotions (anger, fear, sadness, happiness, disgust, and surprise), the architecture of the robot contains a perceptual, a drive, a behavior, and a motor system. The model has four types of elicitors (releasers) of emotions:
- *Neural elicitors* (neurotransmitters and other neurochemical processes);
- *Sensorimotor elicitors* (facial expressions, postures, muscular tension, etc.);
- *Motivational elicitors* (drives, pain and pleasure, other emotions, etc.);
- *Cognitive elicitors* (appraisal-based reasoning, attribution memory, etc.).

Each emotion is calculated separately, but according to an update-rule that takes the same general form for each type of emotion (only the exact values of the parameters may vary from emotion to emotion): The new intensity of each emotion is a function of its elicitors, its decayed previous value, and influences from other emotions. The resulting intensity is compared to an emotion-specific activation threshold. Only when this is passed, the corresponding emotion influences the behavior system as well as other emotions. The behavioral repertoire includes action-oriented and communicative emotional behaviors ('smiling', 'tail-wagging', etc). Emotions can result from interactions with the drive system, the environment, or people. When rewarded or disciplined by persons, Yuppy can learn this in the form of 'secondary' emotions. For example, it can learn to 'fear' the sound of a flute when the occurrence of this sound is often enough paired with the occurrence of 'pain'. The learned associations are stored in the form of new cognitive releasers.

An achievement of the model of Velásquez is to be one of the first to incorporate at least an approximation of all the types of influences known to be involved in human emotion synthesis. However, apart from the ability to learn associations and the usage of some cognitive elicitors, the model remains still rather low-level. The focus is on basic control circuits implemented via connectionist networks. No explicit symbolic representations are used.

Appraisal-Based Models

While acknowledging the importance of physiology and behavior on emotional processes, appraisal-based models of emotions largely ignore these factors and focus instead solely on the *cognitive* structures and mechanisms that are involved in the generation of emotions. In these models, affective reactions are the *result* of cognitive appraisal processes that map the features of a situation onto a set of output emotions. To do so, high-level rule-based representations of goals, preferences, and situations are used.

	Positive Reactions	*Negative Reactions*	
Event	Because something good happened (joy, happy-for, gloating)	Because something bad happened (distress, sorry-for, envy)	Goal
	About the possibility of something good happening (hope)	About the possibility of something bad happening (fear)	
	Because a feared bad thing did not happen (relief)	Because a hoped-for good thing did not happen (disappointment, sadness)	
Action	About a self-initiated praiseworthy act (pride, gratification = pride + joy)	About a self-initiated blameworthy act (shame, remorse = shame + distress)	Standard
	About a other-initiated praiseworthy act (admiration, gratitude = admiration + joy)	About a other-initiated blameworthy act (reproach, anger = reproach + distress)	
Object	Because one finds something appealing (love, liking)	Because one finds someone/-thing unappealing (hate, dislike)	Taste

Table 1 The OCC Structure of valenced Reactions (based on [19, p. 194])

The OCC Model: One early appraisal-based model that has served as the basis for the implementation of several computational models of emotions is the OCC Cognitive Model of emotions by the emotion theorists Ortony, Clore, Collins [18]. Originally, the model was not intended for emotion synthesis but to enable AI systems to *reason about* emotions, a capability thought to be useful especially for applications such as natural language understanding and dialog systems. The model does not use basic emotions, but groups emotions according to a scheme of cognitive eliciting conditions. Depending on the stimuli that cause the emotion, three classes of emotions are distinguished: those induced by events, those induced by agents, and those induced by objects. Using this structure, 22 emotion types are specified (Table 1).

A complex set of rules couples the features of a situation with the agent's beliefs and goals. More specifically, events are coupled with goals, (moral) standards are coupled with agents, and objects are coupled with preferences. To do so, intervening structures and variables are used. To illustrate this by an example, if $D(a, e, t)$ is the desirability that agent a assigns to event e at time t then the potential for generating a state of joy P_j is given by a joy-specific function f_j that depends on the assigned desirability of the event and a combination of some global intensity variables (e.g., expectedness, reality, proximity) represented by $I_g(a, e, t)$:

$IF\ D\ (a,\ e,\ t)\ \_\ 0\ THEN$

$set\ P_j(a,\ e,\ t) = f_j(D(a,\ e,\ t),\ I_g(a,\ e,\ t))$

The rule above does not directly cause a state of joy. There is another rule that activates joy with a certain intensity I_j only if a joy-specific threshold $T_j(a,\ t)$ is exceeded.

Concerning details like what values to use for the thresholds, or how emotions interact, mix, and change their intensity, the model is not very specific. Basically, the OCC model is a knowledge-based system to generate different types of emotions. Hence, it is not the most flexible one. Moreover, it has a limited capability for emulating 'hot', i.e. emotionally affected, cognitions as it more or less solely focuses on how cognitions influence emotions and not vice versa. There is almost no feedback from the emotional system to the cognitive system.

Em: An emotion generating computational system based on a subset of the OCC appraisal theory of emotion is the system Em by S. Reilly and J. Bates [20]. Em is part of a larger project called 'Oz' whose goal is it to create *believable* agents that *appear* to be emotional. These agents are synthetic characters inspired by Disney figures that live in virtual worlds as for example the simulated house cat named Lyotard or the ball-like creatures called Woggles. The full architecture of these characters integrates not just emotions, but also rudimentary perception, goal-directed behavior, and language. The importance of goals influences the intensities of the generated emotions. There are thresholds for the emotions as well as functions that model emotion decay. Most importantly, Em's emotions are separated into positive and negative ones. By combining all the positive emotions, e.g., joy, hope, relief, etc., and all the negative emotions, e.g., distress, hate, shame, etc., respectively, a state called 'mood' can be determined that is either good or bad. The summing of the intensities within a group is done using a logarithmic formula. Emotions generated by Em can influence behavior and perception as well as some cognitive activities like the generation of new goals. For example, a character might be so angry that it generates a goal to get revenge. Most of Em's rules are hard-coded including social rules of which some researchers think that they should be more flexible. However, in the project where Em is part of the creators of characters do not want to give up deliberative control over their creatures.

The Affective Reasoner: Another implementation of (an extended version of) the OCC model is the Affective Reasoner by Elliott [21]. It focuses on the generation of emotions among characters with social relationships, and on the *deduction* of other character's affective states. Again, the conditions to synthesize each of the 26 emotion types of the model are implemented as rules. Additionally, characters or agents are given a personality in the form of a set of symbolic appraisal frames that contain the agent's goals, preferences, principles, as well as current moods. The personality of an agent exerts a two-fold influence. First, it addresses which emotional state is derived based on each agent's individual appraisal frame. Second, it influences how an agent will express its emotional state. For example, an agent with an outgoing personality might express its joy verbally; a more inward type might simply enjoy an internal feeling of happiness. Concerning the derivation of other agents' emotions, an agent maintains an internal representation of the presumed ways in which others appraise the world. Based on this, it can perform forward logic-based reasoning from presumed appraisals, and events, to guesses about the emotions of others, and backward, case-based, reasoning from facts about the situation and the other agents' expressions to their presumed emotions. Generally, an agent can have three kinds of social relationships with other agents:

- *Friendship:* An agent generates similarly valenced emotions in response to another agent's emotions.
- *Animosity:* An agent generates oppositely valenced emotions in response to another agent's emotions.
- *Empathy:* An agent temporarily substitutes another agent's presumed goals, standards, and preferences for its own.

An agent can use its presumed representation about how others appraise the world to produce empathy. A typical dialog between an agent and a user looks as following [22]: (The agent has previously learned that the word "Ruth" is a female friend of the user.)

Agent: "How is Ruth?"
User: "Ruth is jealous of her rival."
Agent: "How much is Ruth feeling this resentment?"
User: "Very much."
Agent: "Perhaps you are worried about Ruth?"

WILL, EM: There are other appraisal theories than the OCC theory that have been adopted as basis for computational systems of emotions, for example the model of Frijda [12]. This model is the basis for a number of systems, including WILL [23]. An important feature of WILL is that it ties appraisal variables to an explicit model of *plans* whereby plans capture the causal relationships between actions and their effects. In a similar vein, the Emotion and Adaptation (EMA) model of Gratch and Marsella [24] focuses on how subjective appraisals can alter plans, goals, beliefs, etc. Unless the majority of computational models of emotions that focus on using appraisals or emotions to guide action selection, EMA attempts to model the wider range of human coping strategies such as positive reinterpretation, denial, acceptance, shift blame, etc.

Architecture-Level Models – Exemplified by the Cognition and Affect Project (CogAff)

Apart from emotional computational systems that deal either with low-level mechanisms of emotion or with schemes of how to cognitively elicit emotions, there are also some ambitious models that try to combine all these aspects. As a prominent prototype of such an architecture-level model, the approach of A. Sloman and his colleagues to emotions and intelligence will be described in detail.

A. Sloman, professor at the School of Computer Science of the University of Birmingham, belongs to the most influential theoreticians concerning computer models of emotions. With his 1981 article *'Why robots will have emotions'*, he was one of the first to write to the Artificial Intelligence community about computers having emotions [25]. Regardless of this fact, and some other papers he wrote on emotions, he is not particularly interested in emotions, at least not in creating artifacts that just *simulate the effects* of emotional behavior. What he is interested in is the construction of a general intelligent system. Within this system, emotions would be just one part among others embedded in an overall architecture. Apart from the generation and management of emotions and motives, such a global architecture of the mind also has to include mechanisms for perception, learning, making plans, drawing inferences, deciding actions etc. All these mechanisms have to be implemented within a resource-bounded agent. For Sloman it is more important to design a *complete* system that may be shallow than isolated modules with great depth. Sloman refers to this view as *design-based* or *architecture-based* approach [26].

Design space and niche space. According to Sloman, there is not just one 'right' kind of architecture, but a wide variety of architectures. To clarify this view, he defines the terms *design space* – the space of possible architectures – and *niche space* – the space of sets of requirements for architectures. Design and niche space are not independent from each other but connected via a diverse set of relationships. The collection of requirements to be satisfied by a functional system (its 'niche') determines a range of architectures that can be possibly used for its implementation. A given architecture may fit a given niche more or less well. For example, whether a robot needs or should have emotions depends on the tasks and environment the robot is intended for. Thus, the requirements to be fulfilled by the robot determine the kinds of 'emotional' mechanisms necessary or useful for the robot, and the structure of the architecture determines whether these mechanisms can be met by the architecture. The same interdependence of design and niche space not only applies to emotions but also to other cognitive abilities. Consequently, different information processing architectures not only support different classes of emotions, but also different varieties of perception, different forms of mental reasoning, different forms of consciousness, etc.

During the last years, Sloman and his colleagues have proposed various drafts for a cognitive architecture. All their drafts are based on an architectural scheme that is very generic, called the CogAff scheme [27]. This scheme is able to integrate various kinds of emotional and non-emotional mechanisms (Fig. 1.1.3). So far, only some of the mechanisms that principally fit into the scheme have been explicitly elaborated, actually implemented in software and thoroughly evaluated. One fundamental conjecture for the generic framework is that it has to be multi-layered whereby each layer provides a different level of abstraction. Sloman thinks that a cognitive architecture for the human mind needs at least three layers:

- *Reactive layer;*
- *Deliberative layer;*
- *Self-monitoring layer (Meta-management layer).*

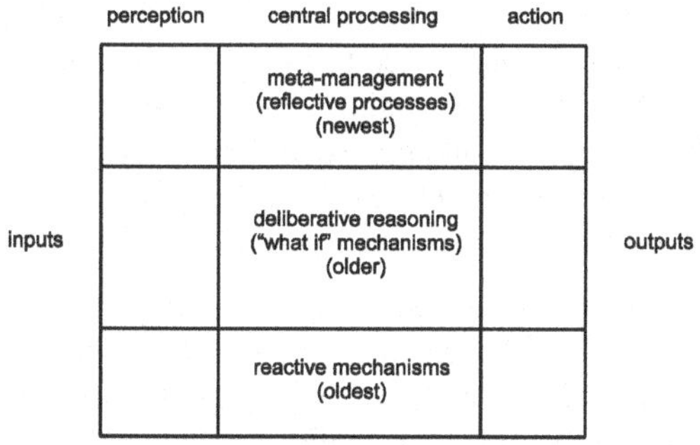

	perception	central processing	action	
		meta-management (reflective processes) (newest)		
inputs		deliberative reasoning ("what if" mechanisms) (older)		outputs
		reactive mechanisms (oldest)		

Fig. 1.1.3 The CogAff Scheme [26, p. 20] defines a crude, first-draft division of mechanisms by superimposing two three-fold distinctions. Many information flow-paths between boxes are possible.

These layers are ordered according their complexity, starting with the simplest one. This 'vertical' order is combined with a 'horizontal' perspective decomposing cognitive processing into the stages of perception, central processing, and action. The combination of both ordering schemes leads to a grid-like structure.

- *Reactive Layer:* The reactive layer is the oldest one in evolutionary age. Systems with just a reactive layer can only produce relatively simple and predictable behavior. The reactive layer is based on the detection of characteristic stimuli in the environment. In turn, rather automatic behavioral responses are elicited. Reflexes based on direct connections from sensors to motors can be counted as the most basic form of reactive mechanisms.

Although most of the reactive mechanisms are hard-wired, some simple form of conditional learning is also possible on this layer. However, reactive agents cannot make plans or modify their behavior to a larger extent. Thus, they are not very flexible. But they are very fast as the underlying mechanisms can be implemented in a highly parallel way using a mixture of analog and simple rule-based algorithms.

- *Deliberative Layer:* The deliberative layer introduces 'what-if' mechanisms. It contains formalisms for combining existing behaviors in new ways, making plans, describing alternative possibilities and evaluating them before execution. The deliberative layer is also capable of learning generalizations. The plans created on this layer need some form of long-term memory for the storage of the various behavioral sequences and all the consequences of the alternative behavioral patterns. The step-by-step, knowledge-based nature of the construction of plan makes the involved processes serial and slow. Therefore, the question of how to allocate limited resources becomes an important issue. However, serial processing also presents some advantages especially when it comes to the task of associating rewards with previous actions.

- *Meta-Management Layer:* The meta-management layer provides mechanisms for monitoring and evaluating internal states and processes. It can control, reject, modify, and generalize the processes of the lower layers in order to keep them from interfering with each other and to increase their efficiency. The categories and procedures on this layer are largely culturally determined. As the meta-management layer has no complete access to all the internal states and processes a perfect self-evaluation is impossible.

Concerning the question of representation, the suggested architecture is intended to be implemented by a hybrid symbolic/sub-symbolic modeling approach. The automatic, pre-attentive modules that correspond to the very old parts of the brain are thought to make use of sub-symbolic, distributed representations that enable highly parallel processing of incoming sensor data. However, higher modules are argued to require explicit symbol manipulating capabilities. All the three layers are thought to work in parallel. Between the processes on the different layers, lots of interactions take place. Thus, there are concurrently occurring processes. To manage them efficiently, a variety of *control states* is required. Control states can be based on different realizations, reaching from simple physical signals to complex processes acting on an informational level. Some of them are very short, and some of them endure over a longer period of time. Each control state has a structure (syntax), a content referring to a certain state of affairs (semantics), and a functional role (i.e. dispositional powers) to determine internal and external actions (pragmatics). Control states can be found on each layer of the architecture, and they can circulate within the hierarchy, gain or lose influence, split up etc. Usually different control states are mutually dependent. The connections between control states can be supportive or suppressive – even leading to deadlocks. For Sloman, one useful form of such control states are emotions.

Emotions and Affects: Sloman views emotions as a subset of the broader class of 'affective phenomena'. On several occasions, he has written on the confusion arising from the lack of generally agreed on definitions of concepts like 'emotions' or 'affects'. In his terminology, the word 'affect' is used as an umbrella term. It includes 'ordinary' emotions like fear or anger but also such concepts as desires, pleasures, pains, goals, values, attitudes, preferences, and moods. According to Sloman, 'affective states' roughly refer to whatever initiates, preserves, prevents, modulates or selects between actions, and, without affect, there is no reason to do anything. As this still makes up a very vague definition, Sloman introduces the notions of desire-like and belief-like states [26]. Both notions are functionally defined referring to the needs of an information-processing system or architecture. 'Desire-like' states are those which have the function of detecting needs of the system so that the state can act as an initiator of action designed to produce changes or prevent changes in a manner that serves the need. Examples would be pleasures, pains, preferences, attitudes, goals, intentions, moods, and emotions. 'Belief-like' states 'just' provide information to enable the desire-like states to fulfill their function. Examples include percepts, fact-sensor states, and memories. Emotions as a subset of the above defined desire-like states are of course also given an architecture-based definition. Sloman uses the idea of an alarm system as a basis for a general definition of 'emotions'. All emotional states depend on the ability of a system – after one of its parts has detected an 'abnormal' state – to interrupt, re-direct or modulate the 'normal' processing of another part of the system. While the system is involved in some sophisticated processing, a situation may arise that needs fast handling. Therefore, alarms are necessary. Different types of emotions can be distinguished by the sources of their alarm triggers, the components which they can modulate, the time-scales on which they are operating, etc. In principle, emotions can arise on each level of the architecture. A full three level architecture – an example of which, the H-CogAff architecture by Sloman et al. [27], can be seen in Fig. 1.1.4 – leads to at least three classes of emotions, namely:

- *Primary emotions* – initiated in the reactive layer;
- *Secondary emotions* – initiated in the deliberative layer;
- *Tertiary emotions* – initiated in the meta-management layer.

Primary emotions of the reactive layer are triggered by simple stimuli and based on innate, largely hard-wired mechanisms. As they arise without prior cognitive appraisal, they are very fast. Examples are being startled by a loud noise, or being disgusted by some vile food. Secondary emotions are triggered by events in the deliberative layer. They usually require some reasoning about goals, objects, and events. Unlike primary emotions which can only be triggered by actual occurrences, they can also be initiated by thinking what might have happened or what did not happen, etc. Examples are being relieved that something bad did not happen, or being worried about not to reach a goal. Tertiary emotions situated on the meta-management layer are the ones for which the notion of 'self' is relevant. They disrupt high-level *self*-monitoring and control mechanisms, and can thus perturb thought processes. Sloman suggests that emotions associated with this layer include shame, grief, jealousy, humiliation, and the like. In complex situations, all three kinds of emotions can coexist. In humans, primary emotions often immediately trigger some higher order kinds of emotions. Therefore, emotions cannot easily be attached to a certain level. The whole architecture does not contain a dedicated emotions module. Instead, emotions of various types are thought to emerge on all levels of the architecture from various types of interactions between many mechanisms that serve different purposes.

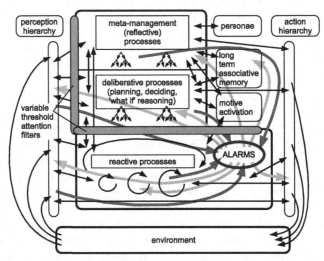

Fig. 1.1.4 The H-CogAff architecture [26, p. 23]. It is a special case of the CogAff schema in which all the 'boxes' have many mechanisms performing different sorts of tasks concurrently, with complex interactions between them.

Apart from emotions, Sloman introduces 'motivators' as another form of control states. Motivations can only be generated when there are goals. For Sloman, *goals* are representations that try to initiate behavior that adapts reality to its representation in contrast to *beliefs* that are representations that try to adapt to reality. Motivators have a structure that consists of several attributes (e.g., importance value, urgency, commitment status, beliefs about possible states, plans, etc.), and they are generated by a mechanism called motivator generator. Motivators compete for attention. There are filters that define a certain threshold to be passed to be able to recruit attentional capacities. These thresholds are variable. They can be modified by learning. As more than one motivator can pass the filter at one time, motivators need to be managed which includes processes like adoption assessment, scheduling, expansion, and meta-management.

Test implementations. During the last years, Sloman and his colleagues have developed a generic cognitive framework that is able to accommodate a wide variety of types of organisms and machines. However, at presence, there is no complete implementation of the hypothesized architecture, although some of the aspects of the architecture have been implemented in some sorts of test-beds. One of the oldest is the minder scenario [28]. In a nursery, a minder or nursemaid has to look after baby-robots keeping them out of trouble until they are old enough to leave. Possible sources of troubles are for example ditches the babies could fall in, or the danger that they are not recharged regularly.

Although it has not been thoroughly implemented and evaluated in computers, Sloman's approach is very important for affective computing and especially for the problem of emotion synthesis. As emotions are likely to evolve in resource-constrained environments, they are likely to prove useful for certain AI applications even as there is no physiological body. Sloman generally argues that there should not be put too much focus on physiological processes. To which degree the bodily aspects of emotions can be neglected may be debatable.

Discussion of Computational Models of Emotions

In recent years, an increased interest in modeling emotion within cognitive models of human intelligence and behavior-based agent architectures – whether software or robotic – can be witnessed. As attempts to computationally model intelligence move beyond simple, isolated, non-adaptive, and non-social problem solving algorithms, the challenge of how to focus and allocate mental resources has to be faced – especially when considering competing goals, parallel, asynchronous mental functions, and a highly demanding, constantly changing environment. Recent findings from affective neuroscience, contemporary psychology and evolutionary biology indicate that emotions as well as drives and other kinds of non-rational behaviors service such resource allocation needs for biological organisms, including humans. 100 years ago, Sigmund Freud has already stressed the fact that human decision making is not only based on deliberate, conscious judgments, but also on non-conscious mechanisms such as drives and 'affects' that these mechanisms inform the mind of bodily needs. The described progress in emotion research as well as in computer, automation and agent technology has led to the development of a number of emotion-based computational architectures and applications. Some prototypical approaches have been described in this paper. On the whole, they all deal with one or more key factors of emotional intelligence. However, often the work is carried out in a rather 'ad hoc' manner. Although most of the developed applications or systems somehow refer to an existing theory of emotion, in most cases only some elements of the theories are taken, whereas, on the other hand, additional components are introduced without neither explicitly stating so, nor arguing the choice. Of course, there is the problem that there is not *one* generally agreed on theory of emotion, but a great variety of them, most of them highlighting some aspects of emotions more than others. In general, emotion theories that are good for operationalization are preferred as a basis for computational models of emotion. Most of the existing models and architectures cover either the low-level bodily aspects of emotions or the high-level cognitive ones. There are some ambitious approaches that try to integrate both aspects, but they are either not general enough or too vague concerning the details of all the emotional and cognitive mechanisms that can be accommodated in principle within the architecture. What is missing is a *comprehensive model* that nevertheless makes clear statements concerning the included mechanisms and modules and *how to put them together*. Of course, there is always the big problem, that the developed models must not get 'too complex' so that an implementation in current software or in a robot is still feasible. Often, to get a running system, only some selected parts of the developed architectures are actually implemented and tested. Another huge shortcoming of the present state of research is the lack of systematic studies of architectural trade-offs and general standards for a sound validation practice. What is necessary to improve the scientific character of the field is an intensified cooperation between engineers, neuroscientists and scientists dealing with the human psyche.

Emotions and Consciousness

The subjective experience of emotions is not covered by current computational models of emotions. This is a significant issue, and introduces the problem of consciousness. Damasio speaks of 'feelings' when referring to the conscious part of emotions [11]. For him, the essence of a feeling is a combined perception of bodily states together with the thoughts to which they are juxtaposed. Thus, feelings are just as cognitive as any other perceptual image, but they are first and foremost about the body. In a similar vein goes Edelman's view of consciousness [29]. For him also the same two kinds of signals are critical: those from self, constituting value systems and regulatory elements of the brain and body, and those from non-self, signals from the world. Both kinds of signals are correlated, leading to a value-category memory. Previous value-laden experiences can then be linked by *"reentrant paths"* to the current perception of world signals. This linkage is the critical evolutionary development for consciousness.

Oatley and Johnson-Laird emphasize that consciousness occurs only at the top level of cognitions, and that it has an integrating function, receiving messages from below and assessing them within a model of goals, desires and knowledge [30]. This level is also concerned with the construction of a model of the *self*. Emotions are also important as they inform the self about changed evaluations. Emotions can originate from bodily needs, but also from higher-level 'needs'. Only if an emotion is associated with a cognitive interpretation of events, can it give a clue to its cause. As the various levels have limited access of one another, it is possible that an emotional control signal, but not its *semantic content*, reaches consciousness. In this case, one would feel an emotional state without having clarity why it occurred.

Although research into the phenomenon of consciousness is still far away from complete understanding (i.e. to the extent that it can be implemented in an artificial system) some of the necessary elements for consciousness to arise have already been discovered. Conscious states are high-level, integrated, temporally ordered, serial states that reflect subjective feelings. Their existence requires a *self-monitoring* system that develops its own ways of categorizing its internal states and relating them to external events. The resulting value-category memory needs to be constantly connected with the current perception. It has to link together a great variety of sensory modalities, and it needs to be explicitly and not only implicitly accessed. Some of these aspects are already addressed in existing architectural models of emotion and cognition. Still, there is a long way to go and there are numerous open questions concerning the details of the mechanisms that lead to the emergence of consciousness and how they could be represented in an artificial system.

1.1.7   State of the Art Applications of Artificial Intelligence

A surprising aspect of the history of artificial intelligence is the rapid uptake of its algorithms into mainstream computer science. Everyday computer tools such as compilers – which take software written in comparatively natural language and convert them to machine-executable instructions – and relational databases – which can retrieve data within a large memory store based on a limited description – have significant roots within AI, even though they have long since taken on an independent existence. In this section, we summarize two applications of AI that are currently closer to the heart of the field, though these are also in the process of shedding their connections to the major thrusts in general-purpose AI.

Resource Planning

The efficient assignment of resources in order to accomplish complicated tasks is well suited to the symbolic approach to AI. Such tasks are often accurately modeled by a well-defined set of variables; the interactions between those variables can also be modeled with good precision (although these interactions may be many in number, they tend to be fairly regular); finally, the sheer size of many resource-planning tasks typically overwhelms human capabilities, so that human planners have to resort to fairly crude heuristics in order to achieve feasible solutions.

A well-known example of successful resource planning using AI techniques is the Dynamic Analysis and Replanning Tool (DART), which was used by US military forces during the Persian Gulf crisis of 1991 [31]. DART performed logistics planning and scheduling of transportation, involving up to 50 000 'objects' (vehicles, people, or cargo); a subsequent analysis by the US Defense Advanced Research Project Agency suggested that this application alone more than repaid the investment that the US military had made in AI during the preceding 30 years. More constructively, AI techniques also played a useful role in scheduling observations for the Hubble telescope; according to Russell and Norvig [2], the time to schedule a week of observations using manual techniques had been three weeks; using appropriate search algorithms, this was reduced to around 10 minutes.

Speech Recognition

One of the most alluring goals of AI is the development of computers that respond to spoken input. In the popular mind, speech recognition is often viewed as the fundamental aim of AI, and motion pictures such as *Space Odyssey 2001* and *Star Wars* have reinforced the association between speech recognition and intelligent computers. It turns out that speech recognition is both a good test bed for the development of AI algorithms, and a practically useful tool.

Software that enables personal computers to perform useful control and dictation with speaker-dependent speech recognition is available today, as commercial software packages. However, these offerings suffer from a number of limitations. Most importantly, they still lack the 'common sense' that humans use so effortlessly to understand one another. Thus, speech-recognition enabled computers commonly make mistakes no human would, and so appear to be fragile. Highly motivated users (e.g. those with disabilities that prevent typing) are currently the main users of personal dictation systems, and typically require more training than users who simply use a keyboard and mouse.

Speech recognition has in fact been more useful in telephone-based applications, where keyboard and mouse are simply not available. Worldwide, telephone-based spoken dialogue systems answer millions of calls every day. In applications such as airline reservations and share trading, these systems have set a new standard for customer care. Currently, such systems only work for relatively sophisticated users, and within a narrow application domain. Within these boundaries, speech recognition generally functions very well, and in countries such as America or Japan people regularly use such systems without giving the issue of machine intelligence a second thought.

The heart of most modern speech-recognition systems is a trainable statistical technique – typically a Hidden Markov Model or neural network. These algorithms convert raw acoustic measurements to probabilities about underlying linguistic events, and a search algorithm is typically used to match these linguistic entities to dictionaries and grammars that describe the domain of interaction.

1.1.8 *Conclusion and Outlook*

This brief overview suggests two apparently contradictory conclusions. On the one hand, tremendous progress has been made during the past 50 years, and large tracts of modern economies depend on algorithms that have their origin in these developments. On the other hand, the fundamental processes that lie at the root of biological intelligence are as mysterious as ever: we now understand reasonably well why apparently promising concepts in the symbolic, statistical and behavior-based domains do not, in fact, produce intelligence that approaches general human capabilities, but these insights have not produced alternative approaches which have a high likelihood of leading us to human-like AI.

Research into emotion and its models has recently progressed in a number of significant ways, and casts a novel light on issues that had previously largely been unexplored in this field. It is not clear, however, how far this research will take us towards achieving human-level performance in computational systems. The progress that we can expect from AI in the coming decades therefore remains a matter of intense speculation. One school of thought holds that the persistent gap between biological and artificial intelligence results from a fundamental limitation in algorithmic computation (see, for example, the arguments of Searle [32] on this matter). The opposite opinion is also widely held: many believe that the various steps towards AI that have been taken will continue to multiply, and obtain capabilities similar to those of biological systems as the speed and memory capacity of computer systems continue to improve. It could very well be that neither of these opinions is correct – that a conventional algorithmic description is sufficient to account for the capabilities of biological intelligence, but that a fundamental breakthrough within that framework is required to make a quantum jump towards those capabilities. Choosing between these competing perspectives is likely to be one the most enlightening intellectual pursuits of the twentyfirst century.

References

[1] R. Kurzweil, The Age of Intelligent Machines. MIT Press, 1990.
[2] S. J. Russell and P. Norvig, Artificial Intelligence: A Modern Approach (2nd Edition). Prentice Hall, 2004.
[3] S. Haykin, *Neural Networks: A Comprehensive Foundation (2nd Edition)*. Prentice Hall, 1998.
[4] C. J. C. Burges, "A tutorial on support vector machines for pattern recognition," *Data Mining and Knowledge Discovery*, vol. 2, no. 2, pp. 121–167, 1998. [Online]. Available: citeseer.ist.psu.edu/burges98tutorial.html
[5] R. E. Neapolitan, Learning Bayesian Networks. Prentice Hall, 2003.
[6] R. A. Brooks, "Intelligence without reason," in Proceedings of the International Joint Conference on Artificial Intelligence, 1991, pp. 569– 595.
[7] R. Pfeifer and C. Schreier, Understanding Intelligence. MIT Press, 1999.
[8] R. Brooks, "A robust layered control system for a mobile robot," *IEEE Journal of Robotics and Automation*, vol. RA-2 (1), pp. 14–23, 1986.
[9] V. Braitenberg, *Vehicles: Experiments in Synthetic Psychology*. Cambridge MA: MIT Press, 1984.
[10] M. Toda, *Man, Robot and Society*. Martinus Nijhoff Publishing, 1982.
[11] A. Damasio, *Looking for Spinoza: Joy, Sorrow, and the Feeling Brain*. Harvest Books, 2003.
[12] N. Frijda, "The psychologist's point of view," in *Handbook of Emotions*, 2nd ed. The Guilford Press, 2004, ch. 5, pp. 59–74.
[13] M. Minsky, *The Society in Mind*. Simon and Schuster New York, 1986.
[14] K. T. Strongman, *The Psychology of Emotion*, 5th ed. John Wiley, 2003.
[15] L. Cañamero, "Modeling motivations and emotions as a basis for intelligent behavior," in *Proc. of the First International Conference on Autonomous Agents*, W. L. Johnson, Ed. New York: The ACM Press, 1997, pp. 148–155.
[16] O. H. Mowrer, *Learning Theory and Behavior*. New York: John Wiley, 1960.
[17] J. Vel'asquez, "Modeling emotion-based decision-making," in *Emotional and Intelligent: The Tangled Knot of Cognition*. D. Cañamero , 1998b, pp. 164–169.
[18] A. Ortony, G. L. Clore, and A. Collins, *The Cognitive Structure of Emotions*. Cambridge, MA: Cambridge University Press, 1988.

[19] A. Ortony, "On making believable emotional agents believable," in *Emotions in humans and artifacts*, R. Trappl and P. Petta, Eds. MIT Press, Cambridge, MA, 2003.

[20] W. S. Reilly, "Believable social and emotional agents," Ph.D. dissertation, School of Computer Science, Carnegie Mellon University, 1996.

[21] C. Elliott, "The affective Reasoner: A process model of emotions in a multi-agent system," Ph.D. dissertation, Northwestern University, Evanston, Illinois, 1992.

[22] C. Elliott, "Components of two-way emotion communication between humans and computers using a broad, rudimentary model of affect and personality," *Cognitive Studies: Bulletin of the Japanese Cognitive Science Society*, vol. 1(2), pp. 16–30, 1994.

[23] D. Moffat and N. Frijda, "Where there's a will there's an agent," *Intelligent Agents: ECAI-94 Workshop on Agent Theories, Architectures, and Languages*, pp. 245–260, 1995.

[24] J. Gratch and S. Marsella, "A domain-independent framework for modeling emotion," *Journal of Cognitive Systems Research*, vol. 5(4), pp. 269–306, 2004.

[25] A. Sloman and M. Croucher, "Why robots will have emotions," in *Proceedings of the 7th Int. Joint Conference on Artificial Intelligence*, 1981, pp. 197–202.

[26] A. Sloman, R. Chrisley, and M. Scheutz, "The architectural basis of affective states and processes," in *Who Needs Emotions? – The Brain Meets the Machine*, M. Arbib and J.-M. Fellous, Eds. Oxford University Press, Oxford, New York, 2004.

[27] A. Sloman, "Beyond shallow models of emotion?" *International Quarterly of Cognitive Science*, vol. 2(1), pp. 177–198, 2001.

[28] A. Sloman, "What sort of control system is able to have a personality?" in *Creating Personalities for Synthetic Actors*, R. Trappl and P. Petta, Eds. Lecture notes in AI, Springer Berlin Heidelberg New York, 1997, pp. 166–208.

[29] G. M. Edelman, "Naturalizing consciousness: A theoretical framework," *PNAS*, vol. 100, pp. 5520–5524, 2003.

[30] K. Oatley and P. N. Johnson-Laird, "Towards a cognitive theory of emotions," *Cognition and Emotion*, vol. 1(1), pp. 29–50, 1987.

[31] S. E. Cross and E. Walker, "Dart: applying knowledge based planning and scheduling to crisis action planning," in *Intelligent scheduling*, M. Zweben and M. S. Fox, Eds. Morgan Kaufman, 1994, pp. 711–729.

[32] J. Searle, "Is the brain's mind a computer program?" *Scientific American*, vol. 262, pp. 26–31, 1990.

1.2 Considering a Technical Realization of a Neuropsychoanalytical Model of the Mind - A Theoretical Framework

Dietmar Dietrich, Georg Fodor, Wolfgang Kastner and Mihaela Ulieru[18]

As foundation for a paradigm shift in artificial intelligence we propose a bionic model that encapsulates psychoanalytic principles of the human mind based on which we map Sigmund Freud's model of the "psychical apparatus" in combination with Luria's dynamic neuropsychology into a machine. Motivated by the first paper of this book which outlined the state-of-the-art in artificial intelligence we suggest future research directions and obstacles that need to be overcome when moving forward towards building conscious machines that will be even able to perceive and act on emotions and feelings. This paper outlines the motivation behind our joint effort where scientists of the fields of psychology, pedagogy and psychoanalysis on the one hand, and engineers on the other hand are involved. As first outcome of this joint work, a model for a technical realization of a neuro-psychoanalytical model of the mind is presented. Ongoing activities and research results based on this model are shown in the following parts of this book.

1.2.1 Motivation

Today's automation systems demand for a high number of data points (sensors and actuators) and controllers (intelligent units[19]) to meet all the requirements of the underlying process [1], [2]. For this reason, they can no longer be based on standalone (central) systems, but have to be handled by a multitude of sub-systems leading to a distributed approach. Specific considerations for control units of (high dynamic) sub-processes (such as safety issues for drive control systems dedicated for airplanes or cars) are not within the focus of this paper. This work has its origin in the "Smart Kitchen Project" started in 1998. The initial idea was to perceive scenarios typically found in a kitchen and adequately react to dangerous situations. Special emphasis was put on the use of readily available technology (i.e., fieldbus systems for data collection, and databases for storing scenarios). Several people (e.g., [3]–[7]) contributed to answer the following two questions:

(1) What technology can be used for a straightforward realization?

(2) Where is basic research necessary?

The "Smart Kitchen Project" was followed by a European Union funded project called SENSE ("Smart Embedded Network of Sensing Entities"), which started 2006. Beyond this successor the team has started several European independent projects, presented in the following papers. From our point of view, it is mandatory to get neuro-psychoanalysts involved in our efforts at the cutting edge.

---

[18] This work was supported by the HarrisonMcCain Foundation.

[19] With the expression "intelligent" the authors mean in this case the technical definition and not the psychological meaning.

Therefore the integration of such experts is a condition-sine-qua-non for all further work in this field [8].

In response to such needs we aim to develop a holistic model for automatic control of processes that tightly interact with human beings and their environment, such as robots that support, for example, persons suffering of dementia. Here, decisions concerning the overall context (e.g., "safe cooking") are necessary. Such decisions can, mainly for two reasons, not be obtained through the traditional way when defining control algorithms. Firstly, the involved parameters are numerous and can sometimes not even be described formally. The second reason is that the memory systems of traditional control systems are much too insufficient for the kind of control algorithms needed in this project.

Researchers from the communities of AI (Artificial Intelligence) and CI (Cognitive Intelligence) have taken a similar approach, as described in the previous paper [9] [20]. At the beginning they adapted the psychological principle of symbolization resulting in knowledge representation. Then, in a second phase, statistical methods and learning algorithms were applied [6]. During a third phase they concentrated on the term "embodying", realizing that the human mind depends on an individual body (independent whether it is virtual or not). At that time, one goal was to implement a representation of the world into robots where data from the outside are captured by sensors and robots operate relying on their internal knowledge base. The current fourth phase can be characterized by the search for definitions of emotions and corresponding feelings, and a possible way to implement them.

A holistic model integrating emotions and feelings (terms as e.g. used by Damasio in [10]) into the technical realizations of intelligence is still not existent. We aim to include psycho- analytical models, by this starting a fifth phase. Our work is inspired by Sigmund Freud who was the first to develop a model of a "psychical apparatus" and its behavior. We aim to obtain a technical realization of the psychoanalytical model of the mind, thus following a bionic approach.

1.2.2 Premises

Within the last decade efforts have been intensified to correlate psychoanalytic models with modern neuroscientific concepts [11] (this intricate step has been appreciated and accepted.) The International Society of Neuro-Psychoanalysis was founded as a scientific society dedicated exclusively to this mission [12].

A further linkage between completely different scientific fields, namely neuro-psychoanalysis and the engineering science, will possibly lead to "cultural" difficulties. Psychoanalysis is still facing strong (also politically formed) reservations in the scientific community. Moreover fundamental principles and approaches from AI and CI have to be re-considered for this project. Therefore, it seems necessary to define several premises for this endeavor.

---

[20] Following the demands of the Engineering and Neuropsychoanalysis Forum (ENF) three papers shall provide the basis for this endeavor.

• Premise 1: Eventually, all functions of the brain/psyche will be understood. In the long term a modeling of all functions will be possible [13].

• Premise 2: Science in general and specifically neuro- science have traditionally declared that subjectivity can- not be studied in a scientific way [14]. Indeed, attempting to study subjectivity one is confronted with the enormous challenge that subjective experience is directly accessible for the subject only and can never be directly observed and measured by an outside (objective) entity. Nevertheless we share the psychoanalysts' opinion that those subjective processes, which were left out by the neurosciences, have an immense and crucial meaning for the understanding of mental life. We acknowledge the extraordinary difficulty of the scientific approach to subjectivity by the psychoanalytical method of indirect observation and interpretation. The resulting model of a psychical apparatus is in our opinion still the best available model and shall serve as the base for further studies in this field. Thus, science has to take on the challenge.

• Premise 3: The final engineering model has to be cooperatively developed by engineers working together with experts in neurology, psychoanalysis, pedagogy and psychology. We have to consider that each community has its own culture, methods and ways of thinking, and make every effort to ease the inter-community communication difficulties. The scientific methods and concepts of the respective scientific worlds must be mutually acknowledged and respected. The task of the engineers has to be to study possibilities of simulating or even emulating the psychoanalytic models, and if successful, to find methods to put them into practice. The task of the experts in the neuro- psychoanalytic field must be to define, together with the engineers, a model, which satisfies the requirements of engineering (Fig. 1.2.1).

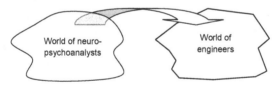

World of neuro-
psychoanalysts

World of
engineers

Fig. 1.2.1. Transfer of the models into the engineer's world

• Premise 4: If a complex organ, such as the psychical apparatus, is to be simulated or even emulated, a consistent model must be available. It is not acceptable to combine different descriptions of functions and types of behavior from various scientific approaches to the understanding of the mind (e.g., [15]) without evaluating the consistency of their combination.

• Premise 5: Sigmund Freud designed a functional model of the psychical apparatus with disregard to all anatomical and physiological correlates. He expected his contemporary colleagues to share this approach while working with or on the model. Similarly, the engineers working on this working with or on the model. Similarly, the engineers working on this project expect the psychoanalysts, in their work-groups, to form purely functional descriptions and to disregard the actual technological creation. It is quite evident that the engineers do not work with RNA, transmitters, and neurons but rather with silicon, transistors, controllers and computers. As much as the models rendered by the neurosciences are valuable inspirations for the actual technological realization, engineers would mostly refrain from any ambition to tackle the Sisyphus task of copying the biological systems.

1.2.3   Models

Algorithms used in AI and CI are based on symbolic logics and mathematical principles, respectively, which give them a key advantage: they are understandable and can – to some extent – be verified. Thus, contradictions can be excluded.

Experiments in psychoanalysis are not repeatable in a direct, trivial way. As already mentioned, subjectivity cannot be directly observed and reproduced in an objective way, although broad patterns can be extracted using statistics encapsulating the essence of several observations. Scientists have therefore decided to leave out subjectivity which they thought was not fit for scientific inquiry and opted for using statistical methods which allow distinct predictions. On this base specific behavior can be described in a clear way, but with the disadvantage of losing sight of the overall context. A unified and holistic model of the functionality of the mental processes, and how they work is still lacking. In the field of psychology and pedagogy many facets are explainable, but these facets cannot be put together to an overall model. For simulation or emulation an overall model is necessary.

The neurologist Sigmund Freud came to understand the high complexity and significance of the psychological processes that he found in his patients when focusing on their subjective experiences [16]. At the same time the neurological knowledge and the means of technical investigation of his time seemed to him to be much too insufficient to allow for a correlation of subjectively observed psychological processes with anatomical structures and physiological events in the brain. He therefore decided to observe the subjective psychological processes and interpret them in order to design a functional model of what he called the "psychical apparatus" completely disregarding the physical side of the equation. He named the so founded field of science "psychoanalysis". Departing from the existing schools of thought of his time, and specifically from the reductionist attempt of narrow localization, Freud defined instances of his apparatus like Ego and Id and their dynamic interplay. In his research work he analyzed the behavior of human beings and tried to explain their emotional and motivational aspects. Freud felt that the necessary correlation of his functional model of the psyche with neurological processes must be postponed until more knowledge and suitable technology was available.

Indeed, increasing efforts were launched to find such correlations. For example, A.R. Luria [17] developed his dynamic neuropsychology on the base of Freud's neuropsychological thoughts e.g., in his aphasiology. Luria's "objective" research is accepted and held in high regard in the neuroscience community.

Luria's work is of high importance for the integration of psychoanalytic and modern neuroscientific concepts specifically focused on by M. Solms [18], [19]. The results of these activities are of crucial importance for the authors when trying to bridge the gap between the research fields of the sensor and actuator areas and the "higher functions". If the idea is to develop a technical concept, this part is an important component for the whole chain of units.

If we are ready to accept these ideas, we have precise constraints which allow tackling the realization of the assumed model[21]. We aim to describe the brain with only one consistent model integrating three models.

Model 1: Neurological model (as a base for a model of communication, information flow and simple control functions of the human body)

The central nervous system can be roughly differentiated into two units, the brain and the peripheral nervous system. The brain, i.e., the "master station", is, from the point of view of computer technology, totally decentralized. Each neuron can be seen as an autonomous controller. The peripheral nervous system, connecting sensors, actuators and the brain, has communicating, but to a limited extent also computational functions.

To get an understanding of the operating principles of the brain, neurologists and biologists analyzed the topological structures and networking. Their results were previously the base for AI to design neural networks. However, this method offers only a limited chance for huge complex systems, a serious drawback if we realize that a human brain has billions of neurons. In the area of biology it is an efficient way to understand the information system of animals like bugs, flies or worms, which have only a small mass of neurons. Nevertheless, it is illusionary for the not so distant future to hope that scientists will be able to understand the connection between hardware and the higher functions of a brain like consciousness, if they follow this way of thinking – regardless of wild speculations.

The neurological knowledge can serve as model, to understand the lower level of the brain – if we assume a hierarchical brain system for the lower levels [20].

Model 2: Psychoanalytical model (as a base for a functional model of the psychical apparatus)

As mentioned above neurologists such as Freud and Luria recognized very early [17], [21] that the higher brain functions must be modeled as functional, dynamic systems. The psychoanalytic theory is based on the idea of a psychical apparatus being a functional system but as mentioned before contains no models of anatomical systems or physiological processes correlating with the mental processes.

A considerable gap between both models opens up which makes a further model necessary to bind model 1 and 2.

---

[21] It is necessary to point out again that this paper of the ENF should not explain the whole context and all the details, which were worked out up to day. The goal is only to present the idea of the new research step and the vision of it.

Model 3: Link between the neuron system and the psychoanalytical model (between model 1 and 2)

A decisive point in modeling the mental system is the correlation between the psychological processes and the physiological processes of the nervous system. Knowledge about this will considerably increase our understanding of the mind.

We will adhere to Freud's idea of this relationship which is fundamental to his entire psychoanalytical concept. The fundamental proposition, so elaborately discussed by Solms [22] is that mental processes are in themselves unconscious. Consciousness is a mere reflection or perception of mental activity. The psychical apparatus has two perceptual surfaces generating the totality of conscious experience: One surface is directed towards external objects and the processes they are involved in representing the existence of things including our physical body and proprioception. The other surface is directed towards the inside perceiving psychical states that represent processes occurring inside ourselves.

Consciousness serves the perception of both classes of sensory input. These two classes are registered on two different perceptual surfaces, one facing outwards, the other one inward. These surfaces are hierarchically equivalent. One does not produce the other. They have rather qualitatively different ways of registering reality, which, as I. Kant reminds us, is and will always remain unknowable in itself. Therefore what we perceive as a physical object on the outside – the brain – appears as a subjective psychical world viewed from the inside; one and the same thing – the psychical apparatus – perceived from two different perceptual surfaces (cf. [23], p. 141).

Based on this proposition the authors of [24] have tried to find a first bridging theory. Today this approach must be tackled in a more differentiated and modified manner.

In the field of computer technology a connection between hardware and software is relatively easy to define unambiguously, if one only considers the aspect of the system description. It means that this part of the system can be defined as a link (or interface) if it is possible to describe this part as hardware as well as software (both are abstract formalisms). The micro program control unit of the microprocessor can be regarded as such a part. This unit – because it is pure hardware – can be described by a hardware description language. On the other hand – because of efficiency reasons – the program of this unit is nowadays usually described in a micro program language, which is specific for the respective microprocessor. So, in this way the micro program unit represents the link between hardware and software.

The micro program unit is the complex control unit for the microprocessor, and represents the base on which the higher level software is mounted. These function levels (above the micro program) are the drivers, which are part of the operating system, the operating system itself and the application units[22] of the computer.

---

[22] Application units are systems like word processing applications or programs which control machines.

Beside the hierarchical configuration of the functional units in a computer system, the different languages can also be ordered in a hierarchical way. The machine respectively the assembler language is situated above the micro program. All of them are hardware specific. Above them the high languages are defined, and in the next level the functional languages [25], [26].

According to [11], [17], [27], one can regard the lower functions of the brain as a system of (abstract) hierarchical levels. The cortical regions can be differentiated into three areas (cf. Fig. 1.2.2): Luria defined the primary region as the *projection field*, which receives data from sensors, and sends commands to actuators; for higher order processing, the association *field* and as a next higher level the *comprehensive regions*. This classification may help to describe different levels of abstraction and integration from neurological units to higher functions in a way analogous to how computer systems are designed[23].

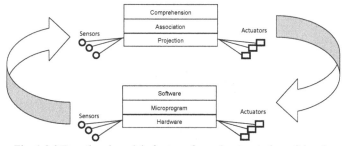

Fig. 1.2.2 Functional models for transfer and computation of data in both worlds with different levels of abstraction

Hardware and software are functional units combined by a hardware/software interface (micro program). In order to allow matching the computer model and the model of a human mind, both models have to be defined in a transparent and strongly modularized way. However, we have to keep in mind, that we will never be able to compare each level 1:1. The unified view on both models must be comprehensible for all parties involved, the engineers and the neuropsychoanalysts.

Our proposal is to use symbolization as the interface between model 1 and model 2 which will be explained in detail in the next section.

1.2.4 Symbolization of the Outer World

Invoking ideas of [27]–[29] a model which was already partly presented in [1], [24] is useful. Human beings' actions are based on experiences and their own behavior learned in past scenarios. The infant acquires knowledge of the outer world

---

[23] We have to consider that computer systems are described in such an accurate way because the designer possesses tools for all different abstraction and language levels. The computer expert usually synthesizes systems and does not analyze them. The neurologist and psychoanalyst try to understand nature which means they have to go the opposite way which is incomparably more difficult.

by learning processes. This means that the embryo and infant cannot perceive raw data from the outside world. The flood of data coming from all its sensors is initially extensively diffused. The infant has to learn to transform perceptions of outside objects and the processes they are involved in into symbols of rising levels of abstraction (Fig. 1.2.3). The representation of the Self and the outside world is increasingly composed by the process of symbolization (in the projection field) producing integrated images (in the association field). Thus, two function units can be differentiated: one unit where all symbolized objects are memorized and the representation unit which will be described in more detail in the next chapter.

To understand Fig. 1.2.3 one has to consider that the output of the eye[24] (after the neuron layers in the eye) are not pixels like a camera, but only characteristic values like areas, edges or arcs, which the brain combines to form images. Images of the inner world may thus be regarded as a matrix (collection) of symbols [29], which are again assembled in the representation unit.

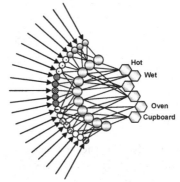

Fig. 1.2.3 The process of symbolization: condensation of
sensory input to higher order symbols

According to [28] it was an immense achievement to find out that the embryo and the infant first have to learn, from the diverse, incoming dataflow, to compose all the images. In this case we do not only speak about optical and acoustic stimuli, but also about images of the olfactory organ, sense of touch and images of the motion of the own muscles. The embryo and infant are not able to differentiate between the physical inner and outer world. In the beginning they only understand one holistic image of the inner and outer world as one whole object. The perception of objects and their dynamic behavior are a computational "work" of associated images by an incoming data flow and the data from memory (Fig. 1.2.4). The human being "sees" a virtual image which is the result of a complex computational neural process of matching incoming data against stored information/knowledge.

---

[24] The following statements are valid for all senses.

We refer to the representation process, including the association of images as "the image handling" (in the association field). It can be regarded as the base for all higher functions of the mental process which are described by the psychoanalytical model. How this image handling could work on the base of neural networks has to be investigated. However, this is not within the scope of this article.

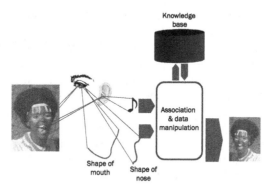

Fig. 1.2.4 Assembling of images in the inner world [30]

It is important to differentiate between "perception" and "recognition". Perception uses symbolization only. In contrast to human beings, bugs and worms only have few neurons. Therefore, it seems to be reasonable to assume that these creatures rely on (simple) symbolization, which is also a principle of the lower level functions of human beings. These lower level functions can be described by means of mathematical algorithms such as fuzzy logic algorithms or statistical methods at the sensors level, and with symbolization above it [6], [7]. For the association of the pictures AI offers different procedures, which can be partly solved using hardware and/or software. To find optimal solutions for this will require much effort and will still remain a difficult task. In our model "recognition", in contrast to perception, involves feelings.

1.2.5 Extended Model: Association and Projection Fields – Emotions and Feelings

The models of the psychoanalysts and their modularization of the psychical apparatus may be sufficient for their work, and it was also the base for our first research steps. However, we had to realize soon that the functional units, as they will be described in the following papers of this forum, were not differentiated and distinct enough for a technical mapping because the single units were too complex for a clear technical definition. Psychoanalysts have solved the problem for themselves by using different models, which they defined from different point of views. As the engineers are only able to work with one *unified model*, their task is

108

now to look for further concepts. These concepts should further refine modulariza-tion of the psychoanalytic entities without contradicting them [31][25].

Solms differentiates between simple or primary consciousness (PC) and reflex-ive or extended consciousness (EC) (cf. [11], p. 95). This hierarchical concept is brilliant and extends Luria's structure. The primary consciousness fulfils the re-quirements of Luria's projection and partly association field. The extended con-sciousness corresponds to Luria's comprehensive field. All information, which is captured by sensors, will be supplied to the primary consciousness. Perception is in this meaning the presentation of sensory data condensed by a symbolization process like the symbolic information *wet* or *hot* (cf. Fig 1.2.3) or like a data field of characteristic values of an optical or an acoustical image of the outer world like the object *wardrobe*. Images of the inner and outer world are composed by sym-bols. The different images of the optical, acoustical or olfactory channels are in such a way pure logically and mathematically computed data, composed by sym-bols of higher abstraction levels (Fig. 1.2.5). Scenarios are short sequences of im-ages and will also be memorized in the brain like the images. A much higher level of intelligence, the comprehensive field [5], is necessary for acts which are represented in sequences of scenarios, which composing must be a much more complex function.

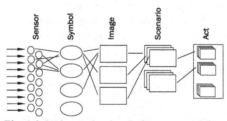

Fig. 1.2.5 Abstraction levels for computed data

Furthermore, a representation field (representation layer), which is the percep-tion unit of intelligent creatures (as explained in the previous section), is situated in the association field. This unit gets data from the outer world and elicit the as-sociation of images and scenarios at the same time which were developed in the past. Both these input resources, the channel with data coming from outside and the data associated, are combined and develop these images and scenarios which we believe to perceive. The process can be seen as a pure mathematical procedure. The result is assembled unconsciously. What we perceive is therefore not the opti-cal picture, passing the lens of the eye, or the sounds, passing the eardrum, but an image of the world which the intelligent subject has developed by means of a complex procedure with different kinds of data (Fig. 1.2.4). The science of today

---

[25] We believe that mixing several models will not lead to a feasible solution. Although interesting. for further technical realizations, such "mixed" modeling would bring to great confusions from the psychoanalytical perspective.

is not able to say how much percent of the input is coming from outside and how much from the database. We assume that the biggest part of these images comes from the "knowledge base" of our brain, because the throughput of sensor data is poor in opposite to all the images and scenarios which we are able to see in a fast sequence.

As mentioned before, insects only have a few neurons in contrast to human beings, and similarly perhaps a projection or association field but never a comprehensive field. They react in a purely "mechanical" way, similar to our concept of a robot. This means in the language of electrical engineering that only control loops (mathematical models) and/or if-then-rules (logical reasoning) are the base of their intelligence. Damasio wrote in [32], that a typical representative of such an operating mode was Mr. Spock a well-known science fiction character from the Star Trek series. It is interesting that he was presented to be superior to human beings. "He was (nearly) not influenced by feelings." In contradiction to this Damasio wrote, that nature – in a Darwinistic view – tells us exactly the opposite. Feelings are the base for the comprehensive field and therefore the base for a high intelligence. The behavior of Mr. Spock is only understandable in a mathematic/logical reasoning. In comparison to the human beings' Spock's performance is much poorer. A human being is capable of much more which turns out to be a selection advantage.

Nature developed the comprehensive field. For the evolution of human beings it was a very important step to set the comprehensive field above the projection and association field which allows the formation of the Self [11].

When constructing a hierarchical resolution, the reflex arc could be placed as a simple control loop into the lowest "intelligence level", as Norbert Wiener has already described it [33]. The projection and association field could be placed into the next higher level, the comprehensive field into the highest intelligence level. It is obvious that especially the upper two levels have to be differentiated and further modularized into sub-levels and sub modules. The comprehensive field is even more complex than the projection field [34]. The core functions of the projection and association fields, namely representation, memorizing of images and scenarios, and symbolization occur in the comprehensive field too. However, the decisive aspect is that the comprehensive field needs to be constructed with at least two representative layers, which means that in our approach the mental system will include three representation layers at least: the first being the emotional representative layer in the association field, next the representation of the outer world, and finally a representation of the Self. Thus, consciousness also means that the subject sets itself in relationship with the outer world. The human being is seeing itself as a person vis-à-vis of the outer world. He or she is something distinct from it. Because the representation of the Self can only be developed by images, it is easy to assume that the Self is again nothing else but a vast collection of images and scenarios [35], and the representation of the Self at any given moment is only a short snapshot of a huge number of various images and scenarios permanently changing [26]. These images and scenarios can be associated "all the time" from our knowledge base, triggered by symbols, coming from the inside world but also formed by data coming from outside. This is what makes the human being so difficult to describe. He cannot be represented by only one image or a single algorithm. A lot of contradictory images can be memorized. The image of the Self is specifically formed by images and scenarios, which were laid down as memories in the beginning of a human being's life and are never forgotten [11].

The representation of the Self has two disadvantages, one being that the images are from the past all the time (in contrast to the outer world) and the other, that it is very difficult to superimpose them with newer ones. There is always a difference in how one sees oneself compared to reality. This poses a special problem for engineers during the phase of implementation.

To be able to take the first step into the direction of a psychoanalytic inspired bionic system we will start from the question: How can feelings be defined in contrast to emotions and how do emotions work in the comprehensive field [36]? As explained above, the research team in Vienna [5] defined emotions as symbolized data (in the projection and partly in the association field). They are value-free and inform the "process", representing the creature, about physical states and behavior. The snail senses whether it is wet outside, the fly senses whether the air flow is rough.

Feelings, as opposed to emotions, are valuations. In the comprehensive field, symbols connect feelings with valuated images and scenarios. Symbols, images and scenarios are memorized in a "weighted" way. If a symbol is formed, it associates not only a reaction image in the projection and association field but also images from the inner world, which are images from the past. They are evaluated and then linked with the Self (a term as used in psychoanalytic theory) creating a new feeling, which is a complex cumulative value. Depending on the current inner state an input from outside initiates a particular feeling which depends on the knowledge of the past. This is why repeatability in experience with the human psyche is so hard to obtain.

With these notions it is also possible to differentiate between perception and recognition. The Vienna team applies the definition that perception (achieved by sensing) is situated on the level of projection and association field and recognition on the level of comprehensive field and this has something to do with feelings.

1.2.6 Open Questions and Proposal for Technical Realization

Three representation layers have been identified by the Vienna team (see Section 1.2.5): one for the projection and association field and two for the comprehensive field. It is a hypothetical model and needs to be proven. The idea was: Following our definition that the projection and association fields do not include

---

[26] Here it becomes understandable that because of the huge number of memorized various outer worlds and the Self images and scenarios, experiments with human beings are not repeatable like physical experiments.

feelings, the representation layer for both fields can be seen as a simple architecture [3], [5]. Incoming symbols are classified according to elementary sensations, e.g., color, brightness, loudness level or heat [37]. After a first "computation" – for fast reactions – they can trigger an action depending on the scenario which is recognized (which also means that the outside scenario must be similar to the memorized scenario). For these steps, not so complex algorithms can be applied [3].

The upper representation layers are much more complicated. Feelings are involved. The scientific literature of neuro-psychoanalysts does not provide an answer clear enough for the kind of model engineers need for technical realization. We know that the tasks of these representation fields are to take care for what we can "see" or "smell". We engineers understand that these representation fields play a decisive role for the human being's consciousness. One representation layer is responsible to "see" the Self, the other for the physical body in the outside world (Fig. 1.2.6). In this sense the Self is a virtual person, an understanding of oneself, strongly influenced by the own homoeostasis. However, there are a lot of upcoming questions, which cannot be answered at present. How are the images and scenarios for the representation layer of the Self composed? What stands behind the symbol "Σ" in Fig. 1.2.6? Must be differentiated between emotions and feelings in the model? What is a representation field? Solms writes in [11]: "If the brain is dreaming, the data channel coming from the sensors are turned back to the knowledge base, and the brain initiate itself to deliver images and scenarios to the representation field". Who determines the first images and scenarios? What affects the course of the dream? The answers to these questions psychoanalysts can provide are not sufficient for technical realization.

We have to find solutions for all these questions to be able to synthesize the model with the different facets.

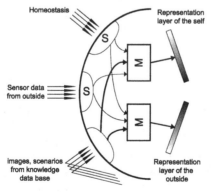

Fig. 1.2.6 The two different representation layers of the comprehensive field (S: Symbolization)

Excluded Issues

More difficult topics like learning, forgetting, sex-specific differences and sexuality are not addressed for the moment. Today, the team concentrates on the simpler aspects of the psychoanalytic models, which seem to be technically solvable.

Constraints of the Model

Following a bionic way of thinking, we as engineers try to emulate the architectures found in nature. However, it is not within our goal to copy them. As such it is definitely not the goal of our research work to copy the human being. We consider this issue to be a matter of philosophical and ethical concern.

1.2.7 Conclusion

The present project has the goal to adapt, simulate and to emulate parts of the psychoanalytic model for bionics applications. If we try to formulate and prove all upcoming questions carefully, then a possibility to map the principles into the world of engineers is realistic.

Currently, control systems lack the possibility to perceive complex scenarios. Damasio's idea that the development of a comprehensive field was a necessary step is consistent with Darwin's principles [10]. So far, engineers have used bionic approaches. Also, AI was successful in taking over neurological principles. Starting from the current state of the art we aim to go one step further and analyze higher functions of the brain together with neuro-psychoanalysts. In anticipate that this will help to describe complex scenarios for all kinds of automation systems in a better way.

Besides automation systems, we also expect the applicability of our model for various other application fields. Take the long standing research field of speech recognition as an example. Often, research institutes have announced a breakthrough, but each time the real success was more than modest. Elaborate algorithms based on semantic rules proved insufficient to tackle this complex problem. Thus, the success story of speech recognition systems is still limited, although they are applied in restricted domains (e.g., medicine). A universally applicable machine which understands sentences from independent speakers does not exist to date. We would like to support a radical shift in approaching complex problems such as speech recognition by bringing to the attention of the engineering community works overlooked from the field of psycho-analysis, such as Sigmund Freud's "Zur Auffassung der Aphasien – eine kritische Studie" [21] in which he criticized 100 years ago the neurological models of that time. Without question, neurology has made big progress up to date. However, we believe that if engineers read this fundamental paper of Sigmund Freud they would have considered the speech recognition problem in a different light which involves feelings and perception as delineated in this paper.

We claim this would have been a better starting point not only in the development of speech recognition systems.

References

[1] D. Dietrich and T. Sauter, "Evolution potentials for fieldbus systems," in *Proc. IEEE International Workshop on Factory Communication Systems (WFCS'00)*, Porto, Portugal, Sept. 2000, pp. 343–350.

[2] G. Pratl, W. Penzhorn, D. Dietrich, and W. Burstaller, "Perceptive aware- ness in building automation," in *Proc. IEEE International Conference on Computational Cybernetics (ICCC)'05*, Port Luis, Mauritius, Apr. 2005, pp. 259–264.

[3] G. Russ, "Situation-dependent behavior in building automation," Ph.D. dissertation, Univ. of Technology Vienna, Vienna, 2003.

[4] C. Tamarit-Fuertes, "Automation system perception," Ph.D. dissertation, Univ. of Technology Vienna, Vienna, 2003.

[5] G. Pratl, "Processing and symbolization of ambient sensor data," Ph.D. dissertation, Univ. of Technology Vienna, Vienna, 2006.

[6] D. Bruckner, "Probabilistic models in building automation – recognizing scnearios with statistical methods," Ph.D. dissertation, Univ. of Technology Vienna, Vienna, 2007.

[7] W. Burgstaller, "Interpretation of situations in buildings," Ph.D. dissertation, Univ. of Technology Vienna, Vienna, expected May, 2007.

[8] The Artificial Intelligence Recognition System (ARS) Project. [Online]. (2007) Available: http://ars.ict.tuwien.ac.at/

[9] B. Lorenz and E. Barnard, "Artificial intelligence – paradigms and applications," in *Proc. 1st International Engineering & Neuro- Psychoanalysis Forum (ENF'07)*, Vienna, Austria, July 2007, p. in print.

[10] Damasio, *The Feeling of What Happens, Body and Emotion in the Making of Consciousness*. New York, NY: Harcourt Trade Publishers, 1999.

[11] M. Solms and O. Turnbull, *The Brain and the Inner World: An Introduction to the Neuroscience of Subjective Experience*. New York, NY: Karnac Books Ltd., 2003.

[12] The International Neuro-Psychoanalysis Centre and Society. [Online]. (2007) Available: http://www.neuro-psa.org.uk/

[13] W. Freeman, *Society of Brains – A Study in the Neuroscience of Love and Hate*. Hillsdale, NJ: Lawrence Erlbaum Associates Publishers, 1995.

[14] W. Balzer, *Die Wissenschaft und ihre Methoden. Grundbegriffe der Wissenschaftstheorie*. Munich, Germany: Alber Verlag, 1997.

[15] C. L. Breazeal, *Designing Sociable Robots*. Cambridge, MA: The MIT Press, 2002.

[16] M. Solms, "Freud, luria, and the clinical method," in *Psychoanal. and History*, 2000.

[17] R. Luria, *Working Brain: An Introduction to Neuropsychology*. London, UK: Penguin Books Ltd., 1973.

[18] M. Solms and K. Kaplan-Solms, *Clinical Studies in Neuropsychoanalysis*. London, UK: Karnac Books Ltd., 2000.

[19] M. Solms, *"The Neuropsychology of Dreams – A Clinico-Anatomical Study (Institute for Research in Behavioral Neuroscience)"*, Hillsdale, NJ: Lawrence Erlbaum Associates Publishers, 1998.

[20] G. Pratl and P. Palensky, "The project ars – the next step towards an intelligent environment," in *Proc. IEE International Conference on Intelliegent Environments*, Essex, UK, June 2005, pp. 55–62.

[21] P. Vogel, *"Sigmund Freud zur Auffassung der Aphasien – eine kritische Studie (2. Auflage)"*, Frankfurt am Main, Germany: Fischer Taschenbuchverlag, 2001.

[22] M. Solms, "What is consciousness," *Journal of the American Psychoanalytic Association*, vol. 45/3, pp. 681–703, 1997.

[23] S. Freud, *"An Outline of Psychoanalysis"*, London, UK: Hogarth Press, 1940.

[24] D. Dietrich, W. Kastner, and H. Schweinzer, "Wahrnehmungsbewusstsein in der Automation - ein neuer bionischer Denkansatz," *at*, vol. 52, pp. 107–116, Mar. 2004.

[25] J. P. Hayes, *"Computer Architecture and Organization (2nd ed.)"*, New York, NY: McGraw-Hill Publishing Company, 1988.

[26] Olsen, O. Faergemand, B. Moller-Pedersen, R. Reed, and J. Smith, *"Systems Engineering Using SDL92"*, Amsterdam, Netherlands: Elsevier, 1994.

[27] O. Sacks, *"The Man Who Mistook His Wife for a Hat"*, New York, NY: Simon & Schuster, 1987.

[28] M. Dornes, *"Der kompetente Säugling - Die präverbale Entwicklung des Menschen"*. Frankfurt am Main, Germany: Fischer Taschenbuch Verlag, 2001.

[29] H. Förster, *"Wissen und Gewissen"* Frankfurt am Main, Germany: Suhrkamp Taschenbuch Wissenschaft, 1993.

[30] E. Brainin, D. Dietrich, W. Kastner, P. Palensky, and C. Rösener, "Neuro- bionic architecture of automation – obstacles and challenges," in *Proc. IEEE Africon*, Gaborone, Botswana, Sept. 2004, pp. 1219–1222.

[31] C. Rösener, B. Lorenz, K. Vock, and G. Fodor, "Emotional behavior arbitration for automation and robotic systems," *in Proc. IEEE Inter- national Conference on Industrial Informatics (INDIN'06), Singapore*, Singapore, Aug. 2006, pp. 423–428.

[32] Damasio, *"Looking for Spinoza; Joy, Sorrow and the Feeling Brain"*. New York, NY: Harcourt Trade Publishers, 2003.

[33] N. Wiener, *"Cybernetics, or Control and Communication in the Animal and Machine"*. Cambridge, MA: The Massachusetts Institute of Technology, 1948.

[34] B. Palensky-Lorenz, "A neuro-psychoanalytically inspired cognitive architecture for autonomous systems," Ph.D. dissertation, Univ. of Technology Vienna, Vienna, expected April, 2007.

[35] T. Deutsch, R. Lang, G. Pratl, E. Brainin, and S. Teicher, "Applying psychoanalytical and neuroscientific models to automation," in *Proc. International Conference on Intelliegent Environments*, Athens, Greece, July 2006, pp. 111–118

[36] W. Burgstaller, R. Lang, and P. Pörscht, "Technical model for basic and complex emotions," in *Proc. IEEE International Conference on Industrial Informatics (INDIN'07)*, Vienna, Austria, July 2007, p. submitted.

[37] G. Pratl, *"A bionic approach to artificial perception and representation in factory automation,"* in *Proc. IEEE International Conference on Emerging Technologies and Factory Automation (ETFA'05)*, Catania, Italy, Sept. 2005, pp. 251–254.

1.3 What is the "Mind"? A Neuro-Psychoanalytical Approach

Mark Solms

The brain is the object of neurological science. The object of psychological science is the mind. Few people would disagree that the mind and the brain are ontologically indistinguishable. This begs the question: what is the 'mind' and how does it differ from the brain? In my view, the mind is distinguishable from the brain only in terms of observational perspective: the mind is the brain perceived subjectively. *Psychoanalysis is a branch of psychology that has taken this perspective seriously.*

Psychoanalytic study of subjective experience has resulted in a model of the mind which can be reduced to five components. (1) The driving principle of life is survival in the service of reproduction. (2) The function of the mind is to register survival/reproductive needs and satisfy them in the world. (3) Since the same could be said of the brain, the mind comes into its own by registering such satisfactions through feelings. *Feelings – pleasures and unpleasures – register the brain's biological successes and failures. This is the basis of consciousness. (4) Feelings generate the values from which* intentionality *is derived. Intentions boil down to wishes to repeat previous pleasurable experiences. This requires memory. (5) Experience, registered in memory, demands increasingly complex decisions about how pleasures can be obtained in reality. This in turn demands response flexibility, which is achieved through thinking. Thinking is experimental action. It depends fundamentally upon response inhibition. This is the basis of 'agency'. Agency is the freedom* not *to act.*

Attempts to manufacture artificial minds must replicate these functional principles.

1.3.1 Introduction

(1)   Since our engineering 'colleagues' ultimate aim seems to be the construction of an artificial mind – and since they wish to use our (neuro-psychoanalytic) knowledge in this regard – it is an ideal opportunity to address the question I have framed in my title: *what* is a 'mind'? In the process of addressing this question I will of necessity also consider two related questions: *where* do minds occur in nature? (Localization); and *why* do they exist? (Function)

(2)   It is one thing to address such questions, and another to do so neuro-*psychoanalytically*. What is special about the neuro-psychoanalytical approach to this question? The engineering colleagues to whom I am responding (Dietrich et al [1]) enthusiastically adopt the psychoanalytic approach, but they do not tell us why they have turned specifically to *this* approach to the mind.

(3)   Freud [3] says that the psychoanalytic approach consists in three things: it is a *therapy* for treating mental disorders, a *theory* about the mind and its workings, and a *method* for investigating those workings.

(4)   Most people (Dietrich et al included) seem to think of psychoanalysis mainly as a *theory*. There are in fact, especially nowadays, many different theories that are called 'psychoanalytic'. To call a theory 'psychoanalytic' therefore begs the question. Dietrich et al (who have no reason to bother themselves with such matters) implicitly answer the question by limiting themselves to *Freud's* psychoanalytic theory. But Freud frequently reminded us that scientific theories are of necessity provisional things – they are always subject to revision. The problem therefore remains: what about Freud's theory cannot be revised if it is to still qualify as 'psychoanalytic'? What is *enduringly* psychoanalytic about a theory?

(5)   Many people know that psychoanalysis is a special form of *therapy* (the 'talking cure'). But this aspect of psychoanalysis is no less subject to revision than its theory. Indeed, the therapeutic technique of psychoanalysis is largely derived from, and only makes sense in relation to, its theory.

(6)   This aspect of psychoanalysis is perhaps least relevant in the present context; our engineering colleagues want to construct an efficient mind, not to fix a broken one. I will therefore not discuss the mechanisms of psychopathology and techniques of psychotherapy here. I will only say in passing that, to the extent that the engineers succeed in accurately emulating the human mind, to that extent they will find that their model is prone to certain types of malfunctioning. One is almost tempted to use this as a criterion of their success!

(7)   Psychoanalysis is least well known (and perhaps least well respected) as a *method* of scientific enquiry. However, in my opinion, this is the most fundamental aspect of psychoanalysis – its most enduring and most unique feature. The psychoanalytic method consists essentially in a certain mode of listening to the (ideally unedited) introspective observations of a person concerning his/her current mental experiences.

(8)   The raw data produced by such introspective reports provides information about the mind that cannot be obtained in any other way. Such data (derived from *subjective* observation) represents nothing less than half of what can be known about the mind (the other half being derived from *objective* observation – observation of the brain and behavior).

(9)   The last-mentioned conjunction makes it possible to have such a thing as *neuro*-psychoanalysis.

1.3.2   Subjectivity

(10)   The mind is unique in this respect: it is the only part of nature that can be known directly – that is, *subjectively*. Apparently the mind alone knows what it feels like to be itself. Moreover, in the case of the human mind, it can provide a verbal report about this subjective state of being. This is the rationale for the psychoanalytic method.

(11) It is surely not a matter of indifference that the mind has this unique property. The use and investigation of this property is likely to reveal something of fundamental importance about the mind – something which exposes how and why it differs from all other biological things (indeed from all other things in general). It will therefore be of considerable importance, in relation to the question we have set ourselves ('what is the mind?'), to review what we in psychoanalysis have discovered about this unique property: subjectivity.

1.3.3 The Unconscious

(12) Before doing so, however, I must point out that the psychoanalytic method consists in more than merely gathering introspective reports. That would be phenomenology. What sets psychoanalysis apart from phenomenology is not just the nuanced complexities of psychoanalytic listening (evenly suspended attention, free association, empathy, transference, counter-transference, interpretation, and the like) but rather a presupposition that underpins all these technicalities.

(13) It is important to recognise that this fundamental feature of psychoanalysis does indeed boil down to a *presupposition* – and one which blurs the distinction between psychoanalytic method and psychoanalytic theory. Although this presupposition can be justified both empirically and logically [2] it is nevertheless also true that it literally *creates* the domain of psychoanalytic enquiry.

(14) In psychoanalysis we presuppose that the events that occur in the gaps and discontinuities in introspective experience (e.g., the events that cause unbidden thoughts or memories to occur, as if from nowhere) deserve no less than the continuous, conscious remainder to be described as 'mental'. This is the starting point of all psychoanalytic enquiries: the inference that a large portion of what we call the 'mind' is *unconscious*. This conceptual innovation creates a universe of natural phenomena (a complete chain of causes and effects) that can be studied scientifically like any other aspect of nature. Without this presupposition, subjective experience can not be studied scientifically. It (or at least a good portion of it) would always have to be translated into some other ontological plane, outside of the mental sphere – probably into something neurological. With that, the notion of a 'mental' science would effectively disappear.

(15) The possibility of a purely mental science is thus created by the psychoanalytic method, the technical peculiarities of which are aimed at inferring the missing (unconscious) mental events that best explain the observed discontinuities in subjective experience. The missing material is thus translated into the language of experience ('interpretation').

(16) What is essentially psychoanalytical therefore is the attempt to systematically study, understand and influence the part of nature that we call 'mind' *in purely mental ways*.

(17)   There is a complication relating to this fundamental feature of psychoanalysis that I hesitate to introduce; but it cannot be avoided. It is this: we do not believe that the mind is mental *in itself* (in Kant's sense). We believe that the subjective universe created by the above-described fundamental proposition is an *observational perspective* on the mind, not the mind itself. Subjectivity (the introspective perspective) is mere perception. The mind itself is the thing *represented by* introspective perception.

(18)   Saying that the unique feature of the mind is subjectivity (point 10 above), and then saying that subjectivity is mere perception of the mind (point 17 above), leads to confusion. But this apparent confusion is only semantic and will be rectified (point 21 below).

(19)   The philosophical complication just introduced is not a unique feature of subjectivity (introspective perception); it applies equally to objective perception. Perceptions, whether inwardly or outwardly derived, are always representations *of* something. What distinguishes inward (subjective) from outward (objective) perception is certain *empirical* features, first discovered by Freud, the most important of which is this:

(20)   Inward perception encounters *resistances*. These resistances are intimately bound up with the emotional feelings encountered by inward perception. (Emotional feelings are a unique feature of subjectivity; see point 40 below.) In general, subjects want to avoid things that generate unpleasure (this is Freud's 'pleasure principle'). When mental things that generate unpleasure cannot be avoided, the subject's primary tendency is to misrepresent them (this is the 'primary process'). Facing unpleasant internal facts requires special mental effort (this is the 'reality principle', governed by the 'secondary process', to be described below).

1.3.4   Different Meanings of 'Mind'

(21)   Returning now to where we left off (point 11 above): what is the 'mind' or the 'mental'? To answer this question, we make a semantic distinction. In neuro-psychoanalysis we distinguish between three grades:

(22)   mental *experience* (which is subjectively perceived – and provides the singular data of psychoanalytic enquiry);

(23)   the *organ* of the mind (the brain - which is objectively perceived, and provides the data for the various neurosciences)[27];

(24)   the mental *apparatus* (which is an abstraction – a model – a virtual thing, and is inferred from the two abovementioned empirical sources)

---

[27] Behaviour (the output of the brain) is likewise objectively perceived, and belongs in the same category.

(25)   The possibility of using the second set of data to correct viewpoint-dependent errors derived from the first set (and vice-versa) is the fundamental rationale for the existence of neuro-psychoanalysis. This is because the aspect of nature – the 'thing' in Kant's sense – that *both* psychoanalysis and neuroscience try to understand is the mental apparatus.

(26)   I am not saying anything too controversial, I hope, if I assert that mental experience (point 22) is what most people mean by 'mind'. This concomitant of the brain (point 23) is what sets it apart from all other organs of the body (and all other natural things).

1.3.5   Basic Features of the Mind

(27)   What then does mental experience consist in? It seems to involve three basic (but overlapping) features:

(28)   It is *conscious*. When the mind is not conscious it literally disappears as a *mind*. (This might seem to contradict point 14 above; but unconscious subjective events do not exist as *such* – as actual, empirically observed phenomena. They exist only as inferences – as hypothetical mental events. They therefore have the same ontological status as the 'mental apparatus' (see point 24).

(29)   It is *intentional*. Mental experience is always *about* something, always directed *toward* something. (Freud speaks here of 'wish'.)

(30)   It has *agency* (or free will). The experiencing mind makes decisions by and for itself: 'I shall do this'. (Even though the actual degrees of freedom are limited by the 'reality principle', the experiencing mind always retains potential for radical freedom – as is demonstrated by certain forms of insanity which disavow the reality principle.)

(31)   We shall see below that these three basic features of mental experience are intrinsically linked.

(32)   The brain possesses none of the above-mentioned special features. It is simply a bodily organ, no different in principle from the heart or the liver. Put differently, when the brain is considered from the subjective perspective (when it is considered from the perspective of consciousness, intentionality, and agency) it is no longer called the 'brain'; it is called the 'mind'.

(33)   These conclusions do not bode well for Dietrich et al's project. How does one begin to construct a machine with these features? Subjectivity is the very antithesis of machinery.

1.3.6   A Model of the Mental Apparatus

(34)   I can only suggest that we approach the problem from an indirect perspective. If the 'mental apparatus' (point 24) represents the set of laws we have derived about the mind and its workings from the two sets of empirical data

represented by points 22 and 23 above, and if it alone unites them (as they must be united, since they are perceptions of the same underlying thing), then the 'mental apparatus' should not be despised as something beyond reality. The mental apparatus is the mind *itself* – in Kant's absolute sense. Mental experience is mere appearance – mere perception – whereas the things we infer from the study of experience (and of the brain) are the ultimate stuff of the mind. Such ultimate things are always abstracted, unobservable in themselves (like 'gravity' and 'electricity' and 'quarks'). The mind in itself, like all ultimate things, can only be modelled; it cannot be observed directly.

(35)   What then are the *functional* principles governing the mental apparatus, insofar as we have been able to abstract them from psychoanalytic (subjective) observation? This is a functional question, but the anatomical point of view should not be ignored (cf. 'Localization'; point 1) and still less contradicted (point 25).

(36)   The localization of the mental apparatus suggests that its main function is to *mediate between the endogenous needs of the organism (the internal milieu) and the external objects of those needs (the environment)*. In psychoanalysis we call the pressure exerted on the mind by endogenous needs 'drives'. It is the most basic 'fact of life' that organisms cannot meet their needs (cannot satisfy their drives) autochtonously. If an organism is to stay alive and reproduce, which are the two essential things that an organism *must* do, then it needs to find, interact with, consume, copulate with, etc., things other than itself (with 'objects' from the viewpoint of the subject). This is why life is difficult, why it requires work.

To this end:
(37)   The mental apparatus receives information from two great sources: from the outside world and from the internal milieu. This information-receiving function is called 'perception'. The first function of the mind (flowing from point 36) is to bring these two things together. This confluence is being in the world, which may be divided into elementary units: "I am experiencing *this*". In such units, "I am" is the product of internal perception and "... experiencing this" is the product of external perception. [28]

(38)   The mind registers traces of previous perceptual confluences of this kind (previous units of experience) and, by sequencing them chronologically, it also registers cause-and-effect sequences of events. This is the function of 'memory', which greatly extends the function of perception - and renders cognition possible.

(39)   Memory and the cognition derived from it extend perceptual being in the world. This extended being is the 'ego'. However the immature ego has little control over the information that flows through it. The control that is normally equated with 'ego' (or agency) depends upon another function not yet described

---

[28] The "I am ..." statement referred to here relies upon a capacity for reflexive thinking, which depends on other functions not yet described (see footnote 3 and point 47 below).

(see point 48 below). The immature ego is at the mercy of its drives and the environment, and only gradually acquires its agency.

(40)   Perceptions, memories and cognitions acquire *meaning* not only by being ordered chronologically but also by being rated on a scale of biological *values*. This is the primary function of 'emotion'. Biologically successful confluences of drives and objects feel good (pleasure) and unsuccessful ones feel bad (unpleasure). This is their meaning. [29] In this way, quantitative events acquire quality. 'Biologically successful' here means 'advancing the cause of staying alive to reproduce' (cf. point 36) which is felt as *satisfying*. [30]

(41)   The values just described, much elaborated through development (memory and consequently cognition), will later form the bedrock of ethical sentiment.

(42)   This capacity to *feel* on the pleasure-unpleasure series is the biological purpose of consciousness. 'Consciousness' is the most essential feature of any feeling, and is inherently evaluative. Consciousness extends outwards from its subjective sources (evaluation of drive states) onto the external world (perceptions of objects): "I feel like *this* about *that*." Here we recognize the 'aboutness' – or *intentionality* – referred to above (cf. point 29). This links two of the basic features of mental experience: consciousness and intentionality. All that remains is agency.

(43)   Attaching meaning to units of experience (point 40) generates a simple motivational principle. Motivation consists in *desire to repeat previous experiences of satisfaction* (and *to avoid previous experiences of frustration*). This is the function of 'wishing', derived from the 'pleasure principle', both already mentioned. The wishes of the immature ego possess compulsive force.

(44)   Motivation filters the present through the past in order for the subject to reach *decisions* about what is to be done. The decisions of the immature ego are more-or-less obligatory. They are essentially dictated by its wishes (see point 39).

(45)   Decisions trigger *actions*, via the (motor) output channel of the mind.

(46)   Experience (memory) shows that actions can be satisfying in the short-term but still biologically unsuccessful in the long term. This requires the development of a capacity to tolerate unpleasure in the short term – the capacity to take a long-term view. This capacity for delay we equate with 'maturity'. It depends on *inhibition* of action.

---

[29] 'Pleasurable' versus 'unpleasurable' does not exhaust the range of meanings. The various 'basic emotions', which recognise certain primal situations of universal significance, arose during mammalian evolution from elaborations of the function described here (see [4]). Language is a further elaboration of this primitive meaning-making function.

[30] A pleasure-generating mechanism, once it exists, can of course be exploited independently of its original biological purpose. Herein lies the source of many psychological ills and social evils; but these does not concern us here.

(47)   Inhibition of action permits 'thinking'. Thinking is inhibited (imaginary) action, which permits evaluation of potential (imaginary) outcomes, notwithstanding current (actual) feelings. This function we call 'secondary process'.

(48)   Secondary process inhibition replaces compulsive, instinctual or stereotyped actions with considered ones. Considered action is the essence of agency. It implies *ownership* of action. Agency depends on inhibition. Free will, ironically, turns on the capacity to *not* act.

1.3.7   Conclusion

(49)   This, in barest outline, is the functional plan of the mental apparatus – according to the findings of psychoanalytic research. Neuro-psychoanalytic research, which recently began to sketch in broad brush-strokes the anatomical and physiological correlates of this living apparatus [5], and to correct the viewpoint-dependent errors of psychoanalysis [6], should be able to greatly assist Dietrich et al in their ambitious quest to design an artificial equivalent.

References

[1]   Dietrich, D., Fodor, G., Kastner, W. & Ulieru, M. (2007). "Considering a technical realization of a neuro-psychoanalytical model of the mind", These Proceedings
[2]   Freud, S. (1915) "The unconscious", SE, 14
[3]   Freud, S. (1922) "Two encyclopaedia articles", SE, 18
[4]   Panksepp, J. (1998) "Affective neuroscience", Oxford University Press
[5]   Solms, M. (1996) "Towards an anatomy of the unconscious", J. Clin. Psychoanal. 5: 331-367
[6]   Solms, M. (2006) "Sigmund Freud Heute: Eine neurowissenschaftliche Perspektive auf die Psychoanalyse", Psyche, 60: 829-59.

1.4 Discussion Chaired by Authors

The following is a summary of the plenary discussion after the presentations of the papers at the ENF 2007. In this context, "Q" is used for any type of question or comment, usually from the auditorium, while "A" stands for "answer", usually from the chairs.

Q1: There is one visible contradiction in these three presentations: psychoanalysis is very much about consciousness and unconsciousness, while artificial intelligence research cannot apply this concept.

A1: Consciousness can only be known from an introspective point of view. What can be done, however, is to construct models which point to what consciousness does and in the end design an apparatus that is able to do those things. Whether it feels conscious or not cannot be found out.

Q2: The Turing Machine had a model of a man with three abilities: to move his head, to read, and to write. Which ability would you like to add to these abilities in order to get a mind?

A2.1: The ability to learn.

A2.2: Learning is not a fundamental property of the mind. What is fundamental to the mind is consciousness, intentionality and agency.

Q3: In the presentations, there were hints of hierarchical thinking in the five principles and of modular theories, which is problematic. What should be discussed is the structure of a functional system, how functional systems are organized.

A3: Modularity and hierarchy move in the realms of ideology, which is problematic in itself: We do not like to think in terms of hierarchy, we do not like to think in terms of modularity. But in engineering, it is important to make something work: So, first, a plan is needed; if a hierarchical plan makes it work – great. If a modular system makes it work, it will be just as well. The ideology behind these concepts has nothing to do with this.

Q4: Dr. Solms, are you not concerned that your introspective view of the mind is inadequate, because evolutionarily there might be more primary processes which we do not have introspective access to, and the three processes which you named are features that have evolved evolutionarily recently, on top of these primary mechanisms?

A4: These are certainly more recent functional elaborations of a basic primary process drive mechanism, and it is necessary to understand this mechanism to understand how the thing works. From the point of view of introspective perception it is definitely the furthest removed from the observational surface and therefore the aspect which will be the most uncertainty. Therefore it is necessary to link psychoanalytic observations with the observations of related disciplines.

2 Session 2

2.1 Machines in the Ghost

Aaron Sloman[31]

Ideas developed by the author over the last 35 years, about relations between the study of natural minds and the design of artificial minds, and the requirements for both sorts of minds, are summarised. The most important point is that natural minds are information-processing virtual *machines produced by evolution. Much detailed investigation of the many kinds of things minds can do is required in order to determine what sort of information-processing machine a human mind is. It is not yet clear whether producing artificial minds with similar powers will require new kinds of computing machinery or merely much faster and bigger computers than we have now. Some things once thought hard to implement in artificial minds, such as affective states and processes, including emotions, can be construed as aspects of the* control *mechanisms of minds. This view of mind is largely compatible in principle with psychoanalytic theory, though some details are very different. Psychoanalytic therapy is analogous to run-time debugging of a virtual machine. In order to do psychotherapy well, we need to understand the architecture of the machine well enough to know what sorts of bugs can develop and which ones can be removed, or have their impact reduced, and how. Without such deep understanding treatment will be a hit-and-miss affair.*

2.1.1   The Design-Based Approach to Studying Minds

My primary goal is to understand natural minds of all kinds, not to make smart machines. I am more a philosopher than an engineer – though I have first- hand experience of software engineering. Deep scientific and philosophical understanding of natural minds requires us to describe minds with sufficient precision to enable our theories to be the basis for designs for working artificial minds like ours. Such designs can be compared with psychoanalytic and other theories about how minds work. If a theory about how natural minds work is to be taken seriously, it should be capable of providing the basis for the design of artificial working minds.

Part of the problem is describing what needs to be explained, and deciding which concepts to use in formulating explanatory theories. A very serious impediment to progress is the common assumption that we already know what needs to be explained and modelled, leaving only the problem of finding good theories and designs for working systems. Alas, what needs to be explained is itself a topic still requiring much research, as acknowledged by the Research Roadmap project in the euCognition network (www.eucognition.org).

A less obvious, but even more serious impediment to progress is the common assumption that our ordinary language is sufficient for describing everything that

[31] This work was partly supported by the EU CoSy project.

needs to be explained, leading to over-reliance on common-sense concepts. When scientists and engineers discuss what needs to be explained, or modelled, and when they report experimental observations, or propose explanatory theories, they often use concepts (such as 'conscious', 'unconscious', 'experience', 'emotion', 'learn', 'motive', 'memory') that evolved not for the purposes of science, but for use in informal discourse among people engaged in every day social interaction, like this:

- What does the infant/child/adult/chimp/crow (etc) perceive/understand/ learn/intend (etc)?
- What is he/she/it conscious of?
- What does he/she/it experience/enjoy/desire?
- What is he/she/it attending to?
- What sort of emotion is he/she/it having now?

These everyday usages may be fine for everyday chats, gossip, consulting rooms, or even law courts and medical reports, but it does not follow that the concepts used (e.g. 'conscious', 'experience', 'desire', 'attend', 'emotion') are any more adequate as concepts to be used in scientific theories than the everyday concepts of 'mud', 'cloudy', 'hot', 'vegetable', 'reddish' or 'smelly' are useful in theories about physics or chemistry.

From Vernacular to Deep Concepts

The history of the physical sciences shows that as we understand more about the architecture of matter and the variety of states, processes, and causal interactions that lie hidden from ordinary observation, the more we have to construct new systems of concepts and theories that make use of them (and thereby help to define the concepts), in order to obtain deep explanations of what we can observe and new more reliable and more precise predictions, regarding a wider range of phenomena. In particular, in physics, chemistry and biology, new theories have taught us that however useful our ordinary concepts are for ordinary everyday social interactions, they are often *grossly* inadequate: (a) for distinguishing all the phenomena that need to be distinguished, (b) for formulating precise and reliable predictions, and (c) for describing conditions (often unobservable conditions) that are the basis for reliable predictions.

So it may be unwise to go on using the old concepts of folk psychology when describing laboratory experiments or field observations, when formulating descriptions of what needs to be explained, and when formulating supposedly explanatory theories. But choosing alternatives to ordinary language needs great care. Recently attempts have been made to give these concepts scientific status by using modern technology to identify precise brain mechanisms, brain states, brain processes that correlate with the states and processes described in ordinary language. Compare trying to identify the precise location in a particular country where some national characteristic or process such as religious bigotry, scientific ignorance, or economic inflation is located.

The 'Design-Based' Approach

Is there any alternative to going on using pre-scientific contexts in our descriptions and theories? Yes, but it is not a *simple* alternative. We need to adopt what Dennett [1] calls 'the design stance', which involves constructing theories about how minds and brains actually work, which goes far beyond what we either experience of their working in ourselves or observe in others. (My own early attempts at doing this 30 years ago are presented in [2].) Moreover, our theories must account for the existence of many kinds of minds, with different designs.

In particular we cannot just start defining precise new explanatory concepts in terms of precise measurements and observations. Thinking that definitions come before theories is a common mistake. In the more advanced sciences, concepts are used (e.g. 'electron', 'charge', 'atomic weight', 'valency', 'oxidation', 'gene', etc.) that cannot be defined except by their role in the theories that employ them. The *structural relations* within the theory partially define the concepts. That is because a theory is a formal system, and, as is familiar from mathematics, and made more precise in the work of Tarski, the structure of a formal system determines which things are possible *models* of that system. Usually there are many possible models, but the set of possible models can be reduced by enriching the theory, thereby adding more constraints to be satisfied by any model. This may still leave different models, of which only one, or a subset is intended. Such residual ambiguities are reduced (but never completely removed) by links between the theory and methods of observation and experiment, and practical applications associated with the theory. However those links do not *define* the concepts, because old methods of observation and old experimental procedures can be replaced by new ones, while the old concepts endure. This loose, theory-mediated, connection between theoretical concepts and observable phenomena is referred to as 'symbol attachment' in [3] and 'symbol tethering' in [4]. Once this possibility is understood, the need for so-called 'symbol grounding' (deriving all concepts from experience) presented in [5], evaporates. (The impossibility of 'concept empiricism' was demonstrated by Kant [6] over 200 years ago.)

The ontology used by an individual or a community (i.e. the set of concepts used to describe things in the world) can be extended in two ways, either by definitional abbreviation or substantively. A definitional abbreviation merely introduces a new symbol as a short-hand for what could be expressed previously, whereas *substantive* ontology extensions introduce new concepts that cannot be defined in terms of pre-existing concepts. Such concepts are implicitly defined mainly by their role in explanatory theories, as explained above. So *substantive* ontology extension always requires theory construction, as has happened many times in the history of science and culture.

2.1.2 *Virtual and Physical Machines*

We can do for the study of mind and brain what was previously done for the study of physical matter and biological processes of evolution and development, if we understand that minds and brains are not just matter-manipulating, or energy-manipulating machines, but information-processing machines. Minds are information processing *virtual machines* while brains are *physical* machines, in which those virtual machines are *implemented* or some would say *realised*.

In computing systems we also have virtual machines, such as running operating systems, firewalls, email systems, spelling correctors, conflict resolution mechanisms, file optimisation mechanisms, all of which run on, i.e. are implemented in, the underlying physical hardware. There are no simple mappings between the components of the virtual and the physical machines. There are often several layers of implementation between high level virtual machines and physical machines. This provides enormous flexibility for re-use of hardware, both in different systems using the same hardware, and in one system performing different functions at different times. It seems that evolution discovered the power of virtual machines before we did and produced brains implementing hierarchies of mental (and indirectly social) virtual machines.

The various components of a complex virtual machine (such as an operating system distributed over a network of processors) need not run in synchrony. As a result, timing relations between processes can change from time to time. For this and other reasons, Turing machines do not provide good models for minds, as explained in [7]. Moreover if sophisticated memory management systems are used the mappings between components of virtual machines and physical parts of the system can also keep changing.

If such complex changing relationships are useful in systems we have designed and implemented we need to keep an open mind as to whether evolution, which had several billion years head start on us, also discovered the power and usefulness of such flexibility. If so, some apparently important searches for mind-brain correlations may turn out to be a waste of time.

Can Virtual Machines do Things?

It is sometimes thought that the virtual machines in a computer do nothing: they are just figments of the imagination of software engineers and computer scientists. On this view, only the physicists and electronic engineers really know what exists and interacts in the machine. That derives from a widely held theory of causality, which assumes either that only physical events can really be causes that produce effects, or, more subtly that if events are caused physically then they cannot be caused by events in virtual machines. Some researchers have inferred that human mental events and processes, such as weighing up alternatives, and taking decisions cannot have any consequences: they are mere epiphenomena. There is no space here for a full rebuttal of this view, but the key idea is that causation is not like some kind of fluid or physical force that flows from one thing to another. Rather, the concept of X causing Y is a very subtle and complex concept that needs to be analysed in terms of whether Y would or would not have happened in various conditions if X did or did not occur, or if X had or had not occurred, or if X does or does not occur in the future. The truth or falsity of those 'counterfactual conditional' statements depends in complex ways on the truth or falsity of various laws of nature, which we attempt to express in our explanatory theories.

However, in addition to their role in true and false statements counterfactual conditionals may also play a role in instructions, intentions, or motives. Thus 'Had he not cooperated I would have gone ahead anyway' may be not so much a retrospective *prediction* as an *expression of resolve*. Plans, intentions and strategies involve causation as much as predictions and explanations do. These too can play a causal role in virtual machines.

Real Causation in Virtual Machines

Suppose one of your files disappears. A software engineer may conclude (after thorough investigation) that one of your actions activated a bug in a running program, which led another part of the program to remove that file. These are events and processes in a virtual machine. This diagnosis could lead the programmer to make a change in the operating system or file management system that alters its future behaviour. Sometimes this can be done by altering software rules without restarting the machine – rather like telling a person how to do better.

Typically, the causal connections discovered and altered by software engineers do not require any action by electronic engineers or physicists to alter the physical machine, and most of the people with expert knowledge about the hardware would not even understand how the bug had occurred and how it was fixed.

Moreover, virtual machine components can constantly change their mapping onto hardware, so the change made by the software engineer need not have any specific physical location when the software is running.

The Mind-Brain Identity Defence

Some philosophers defend the thesis that 'only *physical* events can be causes' by claiming that the objects, events and processes in virtual machines actually *are* physical objects, events and processes viewed in an abstract way. This is a 'Mind-brain identity theory', or a 'virtual-physical machine identity theory', sometimes presented as a 'dual-aspect' theory. A detailed rebuttal of this thesis requires showing (a) that it is based on a bad theory of causation, (b) that the claim of identity being used here is either vacuous or false because the actual relations between contents of virtual machines and contents of physical machines are quite unlike other cases where we talk of identity (e.g. because the virtual to physical mapping constantly changes even within the same machine) (c) that the concepts used in describing virtual machine phenomena (e.g. 'software bug', 'failure to notice', 'preferring X to Y', etc.) are too different from those of the physical sciences to be capable of referring to the same things. More obviously, those virtual machine concepts are not *definable* using only concepts of the physical sciences.

This is a very compressed defence of the claim that virtual machine events can have causal powers. There are more detailed discussions on the Birmingham Cogaff web site.

Psychotherapy as Virtual Machine Debugging

This design-based approach using the concept of a 'virtual machine' can, in principle, justify techniques that deal with a subset[32] of human mental problems by manipulating virtual machines instead of manipulating brains using chemicals, electric shocks, etc. If a virtual machine is suitably designed then it is possible to identify and in some cases repair, certain 'bugs', or 'dysfunctional' processes, by interacting with the running system. Such debugging is common in teaching.

Doing that well requires *deep, explanatory* theories of how normal mental virtual machines work. Some debugging techniques for minds have evolved through various kinds of social experimentation without such deep theories, and many of them work well. For instance, when a child gets the wrong results for long division we do not call in a brain surgeon, but check whether he has perhaps learnt the wrong rule or misunderstood one of the mathematical concepts involved. In such cases the lack of competence can be seen as a bug that can be fixed by talking, drawing diagrams, and giving simpler examples. It might be thought that this is unlike fixing bugs on computers because in the latter case the process has to be stopped, the new program compiled and the program restarted whereas humans learn and change without having to cease functioning. But some programs are run using interpreters or incremental compilers, which allow changes to be made to the running program *without* stopping and restarting the process with a recompiled program. Several AI programming languages have that kind of flexibility. More sophisticated software development tools can also plant mechanisms for interrogating and modifying running systems.

Of course, teaching someone how to improve his long division, and giving counselling to enable a patient to understand how he unintentionally causes family rows to escalate, differ in detail from the process of fixing a typical bug in a computer program. For example the teaching and therapy depend on kinds of self understanding that few computer programs have at present. But in future it will not be uncommon to find virtual machines, that, instead of being forcibly altered by a user editing rules, instead change themselves as a result of being given *advice* about how to behave. That's not even science fiction: it is not hard to achieve. This comparison should undermine three assumptions: (a) the belief that the only way to understand and fix problems with minds is to work in a totally bottom up way, namely understanding and modifying brains, (b) the belief that nothing we know about computers is relevant to understanding and repairing minds, and (c) the belief that machines can only be *programmed* to do things, not advised, inspired, or instructed or cajoled into doing them.

Motivation in Virtual Machines

So, states, events, and processes occurring in virtual machines can have causal influences which alter both other things in the virtual machine and also the physi-

---

[32] Don't expect it to remove brain-tumours, for instance!

cal behaviour of the system (e.g. physical events in memory, hard drives, internal interfaces, what is displayed on a screen, the sounds coming from speakers, or various attached motors, etc.). However, we still need to understand the variety of kinds of causation in virtual machines.

In very simple computer models nothing happens until a user gives a command, which can then trigger a cascade of processes. In more sophisticated cases the machine is designed so that it initiates various activities from time to time, e.g. checking for email, checking whether disks need defragmenting, checking whether current scheduling parameters need to be revised in order to improve processing performance. It can also be designed so that events initiated from outside trigger new internal processes. E.g. a user attempting to access a file can trigger a subprocess checking whether the user has the right to access that file. All of these cases require the designers of the virtual machine to anticipate kinds of things that might need appropriate checking or corrective actions to be performed.

But there is no difficulty in building a machine that acquires new competences while it is running, and also new conditions for exercising old competences. If the virtual machine architecture allows new goal-generators to be acquired, new strategies for evaluating and comparing goals, new values to be employed in such processes, then after some time, the machine may have goals, preferences, intentions, etc. that were not given to it by anyone else, and which, as remarked in [2], can only be described as its own. It would be, to that extent, an *autonomous* machine, even if the processes by which such motives and motive-generators were acquired involved being influenced by things said and done by other people, for instance teachers, heroic figures, and other role models. Indeed a machine might be designed specifically to derive motives, values, preferences, etc. partly on the basis of such influences, tempered by experimentation on the results of trying out such values. Is that not what happens to humans? This point was made long ago in Section 10.13 of [2]. So motivation in artificial virtual machines, including self-generated motivation, is not a problem in principle.

The Myth that Intelligence Requires Emotions

In recent years, especially following publication of Damasio's [8] and Picard's [9], much has been made of the alleged need to ensure that intelligent systems have emotions. I have argued elsewhere that the arguments are fallacious for example in [10], [11], [12], [13], [14], and [15]. Some of these claims are merely poorly expressed versions of Hume's unsurprising observation that without *motives* an intelligent system will have no reason to *do* anything: this is just a confusion between emotions and motives.

There is also a more subtle error. 26 years ago, [10] argued that intelligent machines need mechanisms of kinds that perform important functions and which *in addition* can sometimes generate emotional states and processes as *side effects* of their operation, if they lack sufficient processing power to work out what needs to be done. This does not imply that they *need* emotions, just as the fact that some operating systems need mechanisms that are sometimes capable of generating 'thrashing' behaviour does not imply that operating systems *need* thrashing behaviour. Desirable mechanisms can sometimes have undesirable effects. There can be particular sorts of situations where an 'alarm system' detects a putative need to override, modulate, freeze, or abort some other process, and where because of shortage of information or shortage of processing power, a rough and ready rule operates. It would be better if the situation could be fully evaluated and reasoned about (without any emotions), but if there is inadequate capacity to do that in the time available then it may be better to take the risk of false alarms, especially if the alarm system has been well trained, either by evolution or individual learning, and does not often get things wrong.

The Need for Therapy to Undo Bad Learning

However the danger in having such powerful subsystems that can change as a result of learning is that they may change in bad ways – as clearly happens in some humans. In some cases, it may be possible to undo bad changes by re-programming, e.g. through discussion, advice, teaching, re-training, therapy, etc. In other cases help may be available only when it is too late. Of course there are also cases where such systems go wrong because of physical damage, disease, malfunction, corruption, etc., which may or may not be reversible, e.g. chemical addictions.

2.1.3   *The Variety of Mental Virtual Machines*

There is not just one kind of mind. Insect minds are different from minds of birds and monkeys. All are different in many ways from adult human minds. Human minds are different at different stages of development: a newborn infant, a nine- month old crawler, a two-year old toddler, a four year old talker, a 50 year old professor of psychiatry. Apart from differences resulting from development and learning there are differences that can be caused by genetic deficiencies, and by brain malfunctions caused by disease or injury. What's more obvious is that the brains are different too. So we need to find a general way of talking about different minds, different brains, and the different kinds of relationships that hold between (a) the minds, i.e. the virtual machines, and (b) the brains, i.e. the physical machines in which they are implemented.

Even Virtual Machines Have Architectures

We are talking about a type of complex system with many concurrently active parts that work together more or less harmoniously most of the time but can sometimes come into conflict. These parts are organised in an information- processing architecture that maps onto brain mechanisms in complex, indirect ways that are not well understood. So we should ask questions like this if we wish to do deep science studying a particular kind of mind:

- What sorts of component parts make up the architecture of this sort of mind?

- What are their functions?
- For each such component, what difference would it make if it were modified or removed, or connected in a different way to other components?
- Which parts of the architecture are involved in various processes that are found to occur in the system as a whole?
- Which parts are connected to which others, and how do they interact?
- What kinds of information do the different parts acquire and use, and how do they obtain the information?
- How is the information represented? (It could be represented differently in different subsystems).
- Can the system extend the varieties of information contents that it can make use of (extend its ontology), and extend the forms of representation that it uses?
- What is the total architecture in which they function, and how is it made up of sub-architectures?
- How are the internal and external behaviours selected/controlled/modulated/coordinated?
- Can conflicts between subsystems arise, and if so how can they be detected, and how can they be resolved?
- What mechanisms in virtual machines make those processes possible, and how are they implemented in brains?
- In how many ways can the different virtual machines either individually, or through their interactions go wrong, or produce dysfunctional effects?
- How did this virtual machine evolve, and what does it have in common with evolutionary precursors and with other contemporary animal species?

Answering these, and similar, questions requires a long term investigation. One of the reasons why optimistic predictions regarding imminent successes of Artificial Intelligence (or more recently robotics) have repeatedly failed is that the phenomena we are trying to explain and to replicate, are far more complex than anyone imagines. I.e. the main reason is NOT that the wrong programming languages, or the wrong models of computation were used (e.g. symbolic vs. neural), or that the test implementations used simulations rather than real robots, but that what the researchers were trying to get their systems to do fell very far short of what they implied in their predictions and promises would be achieved, because they had not analysed the requirements adequately.

Moreover, it could turn out that *all* of the currently understood models of computation are inadequate, and entirely new kinds of virtual machine are needed, including machines that grow their own architecture, as suggested in [3].

2.1.4   What is Information?

There are many questions still to be answered about the concept of information used here. What is 'information'? What is an information-user? What is involved in understanding something as expressing a meaning or referring to something? Is there a distinction between things that merely manipulate symbolic structures and things that manipulate them while understanding them and while using the manipulation to derive new information? In how many different ways do organisms acquire, store, derive, combine, manipulate, transform and use information? How many of these are, or could be, replicated in non-biological machines? Is 'information' as important a concept for science as 'matter' and 'energy', or is it just a term that is bandied about by undisciplined thinkers and popularists? Can it be defined? Is information something that should be measurable as energy and mass are, or is it more like structure, which needs to be described not measured (e.g. the structure of this sentence, the structure of a molecule, the structure of an organism, the properties of a toroid)?

We currently have the paradoxical situation that philosophers who are good at conceptual analysis are badly informed about and often have false prejudices about computing systems, whereas many software engineers have a deep but unarticulated understanding, which they use very well when designing, implementing, testing, debugging, maintaining, virtual machines, but which they cannot articulate well because they have not been taught to do philosophy.

Information-Theory is not About Information!

The word 'information' as used (after Shannon) in so-called 'information theory' does not refer to what is normally meant by 'information', since Shannon's information is a purely syntactic property of something like a bit-string, or other structure that might be transmitted from a sender to a receiver using a mechanism with a fixed repertoire of possible messages. Having that sort of information does not, for example, allow something to be true or false, or to contradict something else. However, the more general concept of information, like 'mass', 'energy' and other deep concepts used in scientific theories, is not explicitly definable. That is to say, there is no informative way of writing down an explicit definition of the form 'X is Y' if X is such a concept. All you'll end up with is something circular, or potentially circular, when you expand it, e.g. 'information is meaning', 'information is semantic content', 'information is aboutness', 'information is what is expressed by something that refers', 'information is a difference that makes a difference' (Bateson), and so on.

But that does not mean either that the word is meaningless or that we cannot say anything useful about it. The same is true of 'energy'. It is sometimes defined in terms of 'work', but that eventually leads in circles. So how do we (and physicists) manage to understand the word 'energy'?

The answer was given above in Section 2.1.1: we understand the word 'energy' and related concepts, by understanding their role in a rich, deep, widely applicable theory (or collection of theories) in which many things can be said about energy, e.g. that in any bounded portion of the universe there is a scalar (one-dimensional), discontinuously variable amount of it, that its totality is conserved, that it can be transmitted in various ways, that it can be stored in various forms, that it can be dissipated, that it flows from objects of higher to objects of lower temperatures, that it can be used to produce forces that cause things to move or change their shape, etc. etc. As science progresses and we learn more about energy the concept becomes deeper and more complex. The same is happening with "information".

An Implicitly Defined Notion of "Information"

We understand the word 'information' insofar as we use it in a rich, deep, and widely applicable theory (or collection of theories) in which many things are said about information, e.g. that it is not conserved (I can give you information without losing any), that instead of having a scalar measure of quantity, items of information, may form a partial ordering of containment (information I2 is contained in I1 if I2 is derivable from I1), and can have a structure (e.g. there are replaceable parts of an item of information such that if those parts are replaced the information changes but not necessarily the structure), that two information items can share some parts (e.g. 'Fred hates Mary' and 'Mary hates Joe'), that it can be transmitted by various means from one location or object to another, that it can vary both discontinuously (e.g. adding an adjective or a parenthetical phrase to a sentence, like this) or continuously (e.g. visually obtained information about a moving physical object), that it can be stored in various forms, that it can influence processes of reasoning and decision making, that it can be extracted from other information, that it can be combined with other information to form new information, that it can be expressed in different syntactic forms, that it can be more or less precise, that it can express a question, an instruction, a putative matter of fact, and in the latter case it can be true or false, known by X, unknown by Y, while Z is uncertain about it, etc. etc.

Information is Relative to a User, or Potential User

Whereas energy and physical structures simply exist, whether used or not, information in a physical or virtual structure S is only information for a type of user. Thus a structure S refers to X or contains information about X *for a user of S, U.* The very same physical structure can contain different information for another user U', or refer to something different for U', as shown by ambiguous figures, and also written or spoken languages or notations that some people understand and others do not. This does not make information inherently subjective any more than mountains are subjective because different people are capable of climbing different mountains. (Indexicality is a special case, discussed below.) The information in S can be potentially usable by U even though U has never encountered S or anything with similar information content, for instance when U encounters a new sentence, diagram or picture for the first time. Even before any user encounters S, it is *potentially* usable as an information bearer. Often, however, the potential cannot be realised without U first learning a new language, or notation, or a new theory within which the information has a place, and which provides substantive ontology extension for U, as discussed in [16]. This may be required for growth in self-knowledge too.

A user with appropriate mechanisms has potential to derive infinitely many distinct items of information from a small structure, e.g. infinitely many theorems derivable from Peano's five axioms for arithmetic. Physically quite small objects can therefore have infinite information content, in combination with a reasoning mechanism, though limitations of the implementation (e.g. amount of memory available) may constrain what is actually derivable. It follows that physical structure does not constrain information content, unless a type of user is specified.

Information Processing in Virtual Machines

Because possible operations on information are much more complex and far more varied than operations on matter and energy, engineers discovered, as evolution had 'discovered' much earlier, that relatively unfettered information processing requires use of a virtual machine rather than a physical machine. E.g. digital electronic calculators can perform far more varied tasks than mechanical calculators using cog-wheels.

It seems to be a basic law that increasing usable information in a virtual machine by making implications explicit requires the physical implementation machine to use energy. Similarly (as suggested by Jackie Chappell), using greater information content requires more energy to be used: e.g. in storage, sorting and processing information. So biological species able to acquire and process vast amounts of information must be near the peak of a food pyramid, and therefore rare.

Causal and Correlational Theories of Meaning Are False

It is often thought that learning to understand S as referring to X, requires an empirical discovery that there is a causal relation between S and X, such as that occurrences of X always or often cause occurrences of S to come into existence, or such that the occurrence of X is shown empirically to be a reliable predictor of the occurrence of S. That this theory is false is shown by the fact that you have no reason to believe that occurrences of the word 'eruption' are correlated with or reliable predictors of eruptions, and moreover you can understand the phrase 'eruption that destroyed the earth 3000 years ago' even though it is impossible for any such correlation or causal link to exist, since the earth was not destroyed then. Further we can use concepts that refer to abstract entities whose existence is timeless, such as the number 99, or the shortest proof that 2 has no rational square root. So empirical correlations and causal influences are impossible in those cases.

Information Content Determined Partly by Context

It is sometimes thought that artificial minds would never be able to grasp context-sensitive information. For example, an information-bearing structure S can express different information, X, X', X'', for the same user U in different contexts, e.g. because S includes an explicit *indexical* element (e.g. 'this', 'here', 'you', 'now', or non-local variables in a computer program). Indexicality can make information incommunicable in the sense that the precise information content of one user cannot be transferred to another. (Frege, for example, showed that one user's use of the word "I" has a sense that another person is incapable of expressing.) A corollary is that information acquired by U at one time may not be fully interpretable by U at another time, because the context has changed, e.g. childhood 'memories'. In [17] it was argued that such indexicality accounts for the 'ineffability' of qualia. However this does not usually prevent the *intended function* of communication from being achieved. The goal of communication is not to replicate the sender's mental state, or information content, in the receiver, but to give the receiver information that is adequate for some purpose.

Many structures in perceptual systems change what information they represent as the context changes. Even if what is on your retina is unchanged after you rapidly turn your head 90 degrees in a room, the visual information will be taken to be about a different wall – with the same wallpaper as the first wall. Many examples can be found in [18].

Sometimes U takes S to express different meanings in different contexts because S includes a component whose semantic role is to express a higher order function which generates semantic content from the context, e.g. 'He ran after the smallest pony'. Which pony is the smallest pony can change as new ponies arrive or depart. More subtly what counts as a tall, big, heavy, or thin something or other can vary according to the range of heights, sizes, weights, thicknesses of examples in the current environment.

There are many more examples in natural language that lead to incorrect diagnosis of words as vague or ambiguous, when they actually express precise higher order functions, applied to sometimes implicit arguments, e.g. 'big', 'tall', 'efficient', 'heap', or 'better' (discussed in [19]).

Information Content Shared Between Users

Despite the above, it is sometimes possible for X to mean the same thing to different users U and U', and it is also possible for two users who never use the same information bearers (e.g. they talk different languages) to acquire and use the same information. This is why relativistic theories of truth are false: although I can believe that my house has burned down while my neighbour does not, one of us must be mistaken: it cannot be true for me that my house has burned down but not true for my neighbour. Truth is not 'for' anyone. Meaning depends on the user. Truth does not.

2.1.5   Information-Using Subsystems

An information-user can have parts that are information users. A part can have and use some information that other parts cannot access. When we ask 'Did X know that P?' it is not clear whether the answer must be 'yes' in cases where some part of X made use of the information that P. E.g. human posture control mechanisms use changes in optical flow. Does that mean that when you are walking around you know about optical flow even though you do not know that you know it? Your immune system and your digestive system and various metabolic processes use information and take decisions of many kinds. Does that mean that *you* have the information, that you know about the information, that you use the information? Some people might say yes others no, and some may say that it depends on whether you know that you are using the information.

Likewise there are different parts of our brains that evolved at different times that use different kinds of information (even information obtained via the same route, e.g. the retina or ear-drum, or haptic feedback). Some of them are evolutionarily old parts, shared with other species (e.g. posture control mechanisms), some newer and possibly some unique to humans, (e.g. human face recognition mechanisms, and mechanisms that can learn to read music and other notations).

A deep feature of at least the human architecture, and perhaps some others also, is that sensors and effectors providing interfaces to the environment can be shared between different subsystems, as described in [20]. E.g. your vision mechanisms can be shared between: posture control, visual servoing of manipulation actions, mechanisms involved in reading instructions, and affective mechanisms that appreciate aesthetic qualities of what is seen. Your walking mechanisms can be shared between a subsystem concerned with moving to the door and also a subsystem concerned with social communication, e.g. flirting by walking suggestively. We can describe such architectures as using 'multi-window' perception and 'multi-window' action, whereas current artificial systems mostly use only 'peephole' perception and 'peephole' action, where input and output streams from each sensor or two each effector are go along channels of restricted functionality.

Sometimes the sharing is concurrent and sometimes sequential. Conflict resolution mechanisms may be required when concurrent sharing is impossible. Some of the mechanisms that detect and resolve conflicts may be inaccessible to self-monitoring (discussed later), so that an individual may be unaware of important decisions being taken.

Much philosophical, psychological, and social theorising misguidedly treats humans as unitary information users, including Dennett's intentional stance and what Newell refers to as 'the Knowledge level'.

Just the Beginning of an Analysis of 'Information'

The analysis of the concept of 'information' presented here amounts to no more than a small fragment of the full theory of types of states, events, processes, functions, mechanisms and architectures that are possible in (virtual and physical) information-processing machines. I doubt that anyone has produced a clear, complete and definitive list of facts about information that constitute an implicit definition of how we (the current scientific community well- educated in mathematics, logic, psychology, neuroscience, biology, computer science, linguistics, social science, artificial intelligence, physics, cosmology, ...) currently understand the word 'information'.

E.g. there's a great deal still to be said about the molecular information processing involved in development of an individual from a fertilised egg or seed, and in the huge variety of metabolic processes including intrusion detection, damage repair, transport of materials and energy, and control by hormones and neurotransmitters. It may be that we shall one day find that far more of the brain's information processing is chemical than anyone dreams now is possible.

A more complete theory would provide a more complete implicit definition of the concept 'information' required for understanding natural and artificial systems (including far more sophisticated future artificial systems). A hundred years from now the theory may be very much more deep and complex, especially as information processing machines produced by evolution still seem to be orders of magnitude more complex than any that we so far understand.

2.1.6   Biological Information Processing

All living things, including plants and single-celled organisms, process information insofar as they use sensors to detect states of themselves or the environment and use that information either immediately or after further processing to select from a behavioural repertoire. The behaviour may be externally visible physical behaviour or internal processes.

While using information an organism normally also uses up stored energy (usually chemical energy), so that it also needs to use information to acquire more energy.

There are huge variations both between the kinds of information contents used by different organisms and between different ways in which information is acquired, stored, manipulated and used by organisms. The vast majority of organisms use only two kinds of information: (a) *genetic information* acquired during evolution, used in replication, physical development and maintenance of the individual, and (b) *transiently available* information used by online control systems. I call the latter 'implicit' information. There may also be some less transient implicit information produced by gradually adjusted adaptive mechanisms.

Since the vast majority of species are micro-organisms the vast majority of information-using organisms can use only implicitly represented information; that is to say they use only information that is available during the transient states of activation produced by information being acquired and used, and the information is represented only *transiently* in activation states of sensors, motors and intervening mechanisms, and also in parameters or weights modified by adaptive feedback mechanisms.

The short term transient implicit information in patterns of activation and the longer term implicit information in gradually changing adaptive mechanisms together suffice for most living things – most of which do not have brains! In some animals brains, add only more of the same. However, in humans and many other animals there additional kinds of information and information processing.

Brains are Needed for More Than Movement

Most organisms manage without brains. Why not all? One function is resolving conflicts between different parts responsible for decisions that could be incompatible, e.g. decisions to move one way to get to food or to move another way to avoid a sensed predator. This requires coordination, possibly based on dynamically changing priorities. A different kind of requirement, discussed later, is doing more complex processing of information, e.g. in order to acquire a better understanding of the environment, or to acquire something like a terrain map, or to plan extended sequences of actions. In organisms with many complex parts performing different functions it may also be necessary to coordinate internal changes, for example all the changes involved in reaching puberty in humans, or during pregnancy, where many internal changes and external behaviours need to be coordinated.

Lewis Wolpert wrote, in an *Observer* book review, March 24, 2002: *"First the only function of the brain from an evolutionary point of view is to control movement and so interaction with the environment. That is why plants do not have brains."* This often quoted, but grossly misleading, claim reflects a widespread current focus on research that models brains as sensory-motor control mechanisms, e.g. [21]. Even if the original function of brains was to control movement, much of what human brains do has nothing to do with control of movement – for example explaining what is observed, predicting future events, answering a question, and doing mathematics or philosophy.

So we see that information in organisms may be implicit and transient, implicit and enduring, explicit and capable of multiple uses, used only locally, used in controlling spatially distributed information processing, and may vary in kind of abstraction, in the ontology used, in the form of representation used and in the kinds of manipulation that are available.

Varieties of Explicit Information

Only special conditions bring about evolution of mechanisms that create, store and use *explicit*, that is *enduring, re-usable*, information structures that can be used in different ways (e.g. forming generalisations, making predictions, building terrain maps, forming motives, making plans, remembering what happened when and where, communicating to other individuals, etc.) As explained in [22], this requires new architectures involving mechanisms that are not so directly involved in the sensorimotor relationships and their control.

There may have been intermediate stages of evolution in which non-transient information was stored only in the environment, e.g. in chemical trails and landmarks used as sources of information about the presence of food or predators, or about the routes followed by con-specifics (pheromone trails), or about location relative to a nest or an enduring source of food. (Some insects can use land-marks so this capability probably evolved a long time ago.) This is possible only for animals in an environment with stable structures, unlike some marine environments.

Mechanisms that evolved to use external enduring information may have been precursors to mechanisms using internally stored explicit information. There probably were many different evolutionary transitions, adding extra functionality. In previous papers (e.g. [23]) colleagues and I have emphasised three main categories of competence requiring different sorts of architectural components, namely reactive, deliberative and meta-management capabilities, but there are many intermediate cases and different sorts of combinations of cases that need to be understood – not only for understanding how things work when everything functions normally, but also in order to understand the many ways things can go wrong, including both the consequences of physical malfunctions and also the consequences of dysfunctional processing of information in virtual machines.

Deliberative Mechanisms

A very small subset of organisms (and some machines) have a 'deliberative' information-processing capability insofar as they can construct a set of information structures of varying complexity (e.g. plans, predictions, theories), then compare their merits and de-merits in relation to some goal, and produce new information structures describing those tradeoffs, then select one of the alternatives and make use of it. That ability to represent and reason about hypotheticals was one of the first aspects of human processing modelled in AI subsystems, but it was probably one of the last to evolve, and only a very tiny subset of organisms have it in its richest form, described in [24].

It is now fashionable to contrast that early 'symbolic' AI work with *biologically inspired* AI work. But that is just silly, since all of the work was biologically inspired insofar as it was an attempt to model processes that occur in human beings. A human making a plan or proving a theorem is just as much a biological organism as a human running, jumping or catching a ball, even though more species share the capabilities required for the latter activities. Biology is much richer than some researchers seem to realise.

Meta-Semantic Competences

Another relatively rare biological information-processing capability involves the ability to refer to, reason about, or care about things that themselves contain, or use information. For instance when you discover what someone thinks, and when you worry about someone's motives you are using such meta-semantic competence. Humans can also apply this *meta-semantic* competence to themselves (though probably not at birth: the architecture needs time to develop). Having this ability requires a more complex architecture than merely being able to refer to things without semantic content (e.g. physical objects and processes). That's partly because having meta-semantic content involves representing things that are treated as true in only a hypothetical, encapsulated, way. The very same form of representation may be used for what an individual A believes is the case and for what A thinks another individual B believes is the case, but the functional roles of those two forms of representation will be different.[33] In particular, the former will be 'referentially transparent' and the latter 'referentially opaque', and the conditions for truth of the two beliefs are quite different – something young children do not learn in the first few years.

The ability to think and reason about the mental contents of another has much in common with the ability to think and reason about one's own mental contents as far as mechanisms and formalisms are concerned, though obviously different sensor mechanisms are involved, i.e. external and internal. However the evolutionary benefits are very different. It is not clear which evolved first. Probably both types of meta- semantic competence co-evolved, each helping to enrich the other.

Incidentally, nothing said here implies that any organism has *full* self-knowledge. On the contrary, self-knowledge will inevitably be associated with a bottleneck with limited capacity – a point that has relevance for psycho-analysis. Compare [25].

Perhaps a more widely-shared ability to formulate internal questions or goals was an evolutionary precursor to meta- semantic competence, since formulating a yes-no question ('Will it rain today?') requires an ability to represent propositions that are capable of being true or false without a commitment to their truth value. The same is true of having desires and intentions. So the ability to represent 'X thinks I will eat him' and to reason about the consequences of X's thinking that (e.g. X will run away) may have arisen as a modification of the ability to represent one's own goals ('I will eat X') that have not been achieved, plans that have not yet been carried out, tentative predictions that have not yet been tested, and questions that have not yet been answered.

Old and New in the Same Architecture

It is not always remembered that besides having such sophisticated (and possibly unique) capabilities for explicit manipulation of information, humans share many information processing capabilities with other species that lack the distinctively human capabilities. For instance many animals have excellent vision, excellent motor control, abilities to cooperate on certain tasks, hunting capabilities, the ability to avoid predators, nest-building capabilities, as well as the information-processing involved in control of bodily functions. The notion that somehow human cognition, or human conscious processes can be studied and replicated without taking any account of how these more general animal mechanisms work and how they interact with the distinctively human mechanisms, may lead both to a failure to understand humans and other animals and also to designs for robots that don't perform as required.

Some of the ideas developed with colleagues in Birmingham about the different components in such a multi-functional architecture are reported in the paper by Brigitte Lorenz and Etienne Barnard presented at this conference and will not be

---

[33] J. Barnden's ATT-META project has developed a way of making that distinction which is also related to the ability to think metaphorically.

developed here. From this standpoint it is almost always a mistake to ask questions like 'How do humans do X?' Instead we should ask 'How do different subsystems in humans do X?' (e.g. X = control actions, interpret visual input, learn, store information, react to interruptions, generate motives, resolve conflicts, etc.) And we should expect different answers not only for different subsystems, as in Trehub's [26], but also for humans at different stages of development, in different cultures, with and without brain damage, etc.

2.1.7   Pre-Linguistic Competences

It is often assumed that there is a massive discontinuity between human linguistic competence and other competences (e.g. as argued by Chomsky in [27]), though there are many who hotly dispute this (e.g. Jablonka and Lamb [28]). But people on both sides of the dispute make assumptions about human language that may need to be challenged if we are to understand what human minds are, and if we wish to produce working human-like artificial minds. In particular it is often assumed that the essential function of language is communication. But, as argued in [29], and more recently in [30] many of the features that make such communication possible (e.g. the ability to use varieties of information with rich and varied structures and with compositional semantics, and non-communicative abilities to check whether some state of affairs matches a description, to notice a gap in information, expressed in a question) are also requirements for forms of information that are involved in perception, planning, expressing questions, formulating goals, predicting, explaining and reasoning.

From that viewpoint, communication between individuals in a public medium was a *secondary* function of language which evolved only after a more basic kind of competence evolved – the ability to use an internal, non-communicative, language in mental states and processes such as perceiving, thinking, supposing, intending, desiring, planning, predicting, remembering, generalising, wanting information, etc. and in executing intentions or plans.

It is clear that human children who cannot yet talk, and many animals that do not use an external language can perceive, learn, think, anticipate, have goals, threaten, be puzzled, play games, carry out intentions and learn and use facts about causation. I know of no model of how any of that can be done without rich information processing capabilities of kinds that require the use of internal languages with compositional semantics. But I know of no detailed model of what those 'prelinguistic' languages are, how they evolved, how they develop, how they work, etc.

Fodor, in [31], postulated a 'language of thought' (LOT) which was supposed to be innate, available from birth and capable of expressing everything that ever needs to be said by any human being, but he left most questions about how it worked unanswered and was not concerned with non- human animals. Moreover he supposed that external languages are translated into the LOT, whereas there is no reason to believe such translation is necessary, just as compiling to machine code is not necessary for computer programs to run, if an interpreter is available.

Moreover if it were necessary, then substantive ontology extension during learning and development would be impossible.

Nobody is yet in a position to say what the prelinguistic languages do and do not share with human communicative languages. In particular, it may be the case that the main qualitative competences required for human language use already exist in these pre-linguistic competences in young children and other animals, and that the subsequent evolutionary developments related to human language were mainly concerned (a) with developing means of generating adequately articulated external behaviours to communicate their structures, (b) improving perception of such behaviours, and (c) extending internal mechanisms (e.g. short term memories) that are equally useful for sophisticated internal information processing (such as planning) and for communication with others.

If so, information (unconsciously) acquired, used and stored for later use in early childhood may have much richer structures and deeper semantic content than has previously been thought possible. Whether it includes the kind of content required to support psychoanalytic theories remains open.

Non-Auditory Communication

It is easier to achieve a large collection of perceptually discriminable signals by using independently movable fingers, hands, mouth, eyes, head than to do it all by modulating an acoustic signal – especially as that would interfere with breathing. So, if some animals already had hands for which they were learning to produce (and perceive, during controlled execution) large numbers of distinct manipulative competences concerned with obtaining and eating food, grooming, climbing, making nests, fighting, threatening, etc., then perhaps the first steps to communicative competence used structured movements, as in that involved producing and perceiving structured sign language, rather than vocal language.

There is much suggestive evidence, including the fact that humans find it almost impossible not to move hands, eyes, facial muscles, etc. when talking, even when talking on the phone to someone out of sight. More compelling is the ease with which deaf babies are able to learn a sign language, and the reported fact that some children with Down syndrome seem to learn to communicate more easily if sign language is used. Most compelling is the case of the Nicaraguan deaf children who invented their own highly sophisticated sign language, leaving their teacher far behind. See [32].

An implication of this is that the human ability to develop linguistic competence does not depend on learning a language that is already in use by others. Language learning then appears to be a process of collaborative, creative problem-solving, which may be constrained by the prior existence of a social language, but need not be.

In [33] Arbib proposed that action recognition and imitation was a precursor to the evolution of language, but the arguments are somewhat different, and do not share our hypothesis in [29] that the existence of a rich non-communicative language preceded the evolution of communicative language, though the commentary by Bridgeman [34] makes a similar point, emphasising pre-linguistic planning capabilities.

Once something like this form of structured communication had developed, enabling information that was previously only represented in internal languages to be shared between individuals via a public language, the enormous advantages of having such a communicative competence, both in solving problems that required collaboration and in accelerating cultural transmission, might have led to an unusually rapid process of selection for changes that enhanced the competence in various ways, e.g. developing brain regions specialised at storage of ever larger vocabularies and more complex rules for construction and interpretation of action sequences.

Another development might have been evolution of brain mechanisms to support construction and comparison of more complex symbolic structures, e.g. more deeply nested structures, and structures with more sub-structures.

The Nature of Linguistic Communication

If linguistic communication evolved from collaborative non- linguistic activities, controlled by sophisticated internal languages, this must change our view of human language. It is often thought that linguistic communication involves a process whereby an information structure in one individual gets encoded in some behaviour which is perceived and decoded by another individual who then constructs the same information internally. However, in collaborative, creative, problem solving the shared physical and task contexts provide much of the information, and need not be communicated. All that is needed is whatever suffices to produce desired results, without copying an information structure from one brain to another.

If a mother hands a child a piece of fruit, the child may thereby be triggered to form the goal of eating it, without the mother having had the goal of eating it. Moreover, the mother does not need to anticipate the precise movements of bringing the food to the mouth nor the precise chewing and swallowing movements produced by the child. Those details can be left to the child. More generally, context allows communication to be schematic rather than concrete and detailed.

This is obvious when A asks 'Where are my keys?' and B replies 'Look behind you?'. B may not know exactly where the keys are except that they are on the table behind A. A turns round and sees the keys, getting the precise information he wanted, as a result of B's communication, even though B did not have that information. Both have achieved their communicative goals, but not by transmitting a specific information structure from B to A.

So, many kinds of linguistic and non-linguistic communication involve communication of a *partial* or *schematic* structure, leaving gaps to be filled by the receiver using information that may or may not be available to the sender. In that case, many linguistic expressions may be best thought of as having a higher-order semantics, to be applied to more specific non-linguistic information as context demands, in collaborative problem solving rather than in a process of information transfer. (This is related to Grice's maxims of communication, and to current concerns with 'situatedness' in language understanding, perception and action.)

If the contents of linguistic communication *between* individuals make such use of the context then it may be equally true for thought processes *within* an individual. In that case, much of the information content of your mind is outside you. This may be one of the ideas people who emphasise the importance of 'embodiment' are getting at. However, everything said here could apply to minds in individuals with simulated bodies located in simulated environments. The important thing is what sorts of structures, processes and causal interactions occur within the individual's information processing architecture and how they are related to its environment. Whether that environment is physical or another virtual machine environment makes little or no difference to the kind of mind discussed here.

From Sign Language to Sound Language

Use of sign language has many problems. As has often been noted, it can work only when the sender's hands are free, and when the receiver can see the sender, thereby ruling out communication in many situations where visual attention must be used for another task, or where there are obstructions to vision, or no light is available. So perhaps our ancestors started replacing hand signals with sounds where possible. This could then lead in the usual way to a succession of evolutionary changes in which the vocal mechanisms, auditory mechanisms, and the neural control systems evolved along with physiological changes to the vocal tract, etc. But the ability to learn sign languages remains intact, even if not used by most people.

If all this is correct then it may have deep implications for clinical developmental psychology as well as for understanding precisely what the effects of various kinds of brain damage are.

2.1.8   Architectures That Grow Themselves

Much research in AI and robotics is concerned with what sort of architecture an intelligent system needs, and various rival architectures are proposed. If the architecture needs to grow itself, some of this research effort may have been wasted.

It was argued in [3] that the architecture should not be regarded as genetically determined, but as grown using both innate and acquired meta-competences influenced by the physical and cultural environment, as depicted in Fig. 2.1.1. So there may be considerable differences in how adult minds work, as a result of a different developmental trajectories caused by different environments. This might affect kinds of self- monitoring, kinds of self control, kinds of learning ability, and so on. Another implication is that attempts to quantify the influence of genes and environment in terms of percentages is completely pointless.

146

2.1.9   Conclusion: Machines in Ghosts

Although I was asked to report on developments in AI that might be relevant to psycho-analysis I have instead focused on some very general features of requirements for AI models rather than specific AI models, because I think that despite all the advances achieved in AI, the models are still very primitive compared with human minds and have a very long way to go before they can be directly relevant to working therapists dealing with real humans.

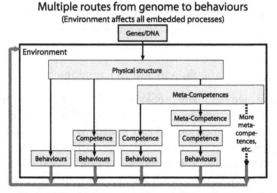

Fig. 2.1.1 This indicates some of the many and varied relationships between the genome and behaviours, produced at different stages of development after layers of competences and meta-competences have been built up [3].

Nevertheless, I hope it is now clear why a human-like machine, a human-like intelligent ghost, and indeed a human- like human must contain very specific kinds of information processing virtual machines. So our ability to understand human minds and ways they can go wrong, including acquiring false, beliefs, inappropriate motives, inadequate strategies, tendencies to be over-emotional, and worse, must be informed by deep knowledge about information-processing systems of the appropriate kinds.

The complexity of the machine, and the fact that self- monitoring mechanisms within it provide only a very limited subset of information about what is going on cause many individuals to start wondering what is really going on in them. Ignorance about information-processing mechanisms can make things seem so mysterious that some people invoke a special kind of stuff that is quite unlike physical matter. From there it is a short step to immortality, souls, etc., and the conceptual confusions discussed by Ryle in [35]. But if we don't go that far we may still come up with deep new ways of thinking about phenomena that were previously thought resistant to computer modelling. This could, among other things, lead to a revolution in psychoanalysis.

2.1.10 Acknowledgement

Thanks to Dean Petters, Jackie Chappell, Matthias Scheutz, Margaret Boden, members of the EU-Funded Cosy Project, and other collaborators and discussants over the years. Some of the ideas here were inspired by John McCarthy and Marvin Minsky, as well as Immanuel Kant [6]. Related papers and presentations are in the Birmingham CogAff and CoSy project web sites. Jackie Chappell and Gerhard Pratl made useful comments on an early draft.

References

[1]   D. C. Dennett, Brainstorms: Philosophical Essays on Mind and Psychology. Cambridge, MA: MIT Press, 1978.

[2]   A. Sloman, The Computer Revolution in Philosophy. Hassocks, Sussex: Harvester Press (and Humanities Press), 1978, http://www.cs.bham.ac.uk/research/cogaff/crp.

[3]   J. Chappell and A. Sloman, "Natural and artificial meta-configured altricial information-processing systems," International Journal of Unconventional Computing, 2007, http://www.cs.bham.ac.uk/research/projects/cosy/papers/#tr0609.

[4]   A. Sloman and J. Chappell, "The Altricial-Precocial Spectrum for Robots," in Proceedings IJCAI'05. Edinburgh: IJCAI, 2005, pp.1187–1192, http://www.cs.bham.ac.uk/research/cogaff/05.html#200502.

[5]   S. Harnad, "The Symbol Grounding Problem," Physica D, vol. 42, pp. 335–346, 1990.

[6]   I. Kant, Critique of Pure Reason. London: Macmillan, 1781, translated (1929) by Norman Kemp Smith.

[7]   A. Sloman, "The irrelevance of Turing machines to AI," in Computationalism: New Directions, M. Scheutz, Ed. Cambridge, MA: MIT Press, 2002, pp. 87–127, http://www.cs.bham.ac.uk/research/cogaff/00-02.html#77.

[8]   A. Damasio, Descartes' Error, Emotion Reason and the Human Brain. New York: Grosset/Putnam Books, 1994.

[9]   R. Picard, Affective Computing. Cambridge, Mass, London, England: MIT Press, 1997.

[10]   A. Sloman and M. Croucher, "Why robots will have emotions," in Proc 7th Int. Joint Conference on AI. Vancouver: IJCAI, 1981, pp. 197–202.

[11]   A. Sloman, "Towards a grammar of emotions," New Universities Quarterly, vol. 36, no. 3, pp. 230–238, 1982, http://www.cs.bham.ac.uk/research/cogaff/96-99.html#47.

[12]   A. Sloman, "Real time multiple motive-expert systems," in Proceedings Expert Systems 85, M. Merry, Ed. Cambridge University Press, 1985, pp. 213–224.

[13]   A. Sloman, "Damasio, Descartes, alarms and meta-management," in Proceedings International Conference on Systems, Man, and Cybernetics (SMC98), San Diego. IEEE, 1998, pp. 2652–7.

[14]   A. Sloman, R. Chrisley, and M. Scheutz, "The architectural basis of affective states and processes," in Who Needs Emotions?: The Brain Meets the Robot, M. Arbib and J.-M. Fellous, Eds. Oxford, New York: Oxford University Press, 2005, pp. 203–244, http://www.cs.bham.ac.uk/research/cogaff/03.html#200305.

[15]   A. Sloman, "Do machines, natural or artificial, really need emotions?" 2004, Invited talk: http://www.cs.bham.ac.uk/research/projects/cogaff/talks/#cafe04.

[16]   A. Sloman, 'Ontology extension' in evolution and in development, in animals and machines, University of Birmingham, UK, 2006, PDF presentation: http://www.cs.bham.ac.uk/research/projects/cosy/papers/#pr0604.

[17]   A. Sloman and R. Chrisley, "Virtual machines and consciousness," Journal of Consciousness Studies, vol. 10, no. 4-5, pp. 113–172, 2003.

[18]   A. Berthoz, The Brain's sense of movement, ser. Perspectives in Cognitive Science. London, UK: Harvard University Press, 2000.

[19]   A. Sloman, "How to derive "better" from "is"," American Phil. Quarterly, vol. 6, pp. 43–52, Jan 1969, http://www.cs.bham.ac.uk/research/cogaff/sloman.better.html.

[20]   A. Sloman, "The mind as a control system," in Philosophy and the Cognitive Sciences, C. Hookway and D. Peterson, Eds. Cambridge, UK: Cambridge University Press, 1993, pp. 69–110.

148

[21] M. Lungarella and O. Sporns, "Mapping information flow in sensorimotor networks," PLoS Computational Biolology, vol. 2, no. 10:e144, 2006, dOI: 10.1371/journal.pcbi.0020144.

[22] A. Sloman, "Sensorimotor vs objective contingencies," University of Bimringham, School of Computer Science, Tech. Rep. COSY-DP-0603, 2006, http://www.cs.bham.ac.uk/research/projects/cosy/papers/#dp0603.

[23] A. Sloman, "Architecture-based conceptions of mind," in In the Scope of Logic, Methodology, and Philosophy of Science (Vol II). Dordrecht: Kluwer, 2002, pp. 403–427, (Synthese Library Vol. 316).

[24] A. Sloman, "Requirements for a Fully Deliberative Architecture," University of Bimringham, School of Computer Science, Tech. Rep. COSY-DP-0604, 2006, http://www.cs.bham.ac.uk/research/projects/cosy/papers/#dp0604.

[25] M. Minsky, "Matter Mind and Models," in Semantic Information Processing, M. Minsky, Ed. Cambridge, Mass,: MIT Press,, 1968.

[26] A. Trehub, The Cognitive Brain. Cambridge, MA: MIT Press, 1991, Online http://www.people.umass.edu/trehub/.

[27] N. Chomsky, Aspects of the theory of syntax. Cambridge, Mass: MIT Press,, 1965.

[28] E. Jablonka and M. J. Lamb, Evolution in Four Dimensions: Genetic, Epigenetic, Behavioral, and Symbolic Variation in the History of Life. Cambridge MA: MIT Press, 2005.

[29] A. Sloman, "The primacy of non-communicative language," in The analysis of Meaning: Informatics 5 Proceedings ASLIB/BCS Conference, Oxford, March 1979, M. MacCafferty and K. Gray, Eds. London: Aslib, 1979, pp. 1–15, http://www.cs.bham.ac.uk/research/projects/cogaff/81- 95.html.

[30] A. Sloman and J. Chappell, "Computational Cognitive Epigenetics (Commentary on [28])," Behavioral and Brain Sciences, 2007.

[31] J. Fodor, The Language of Thought. Cambridge: Harvard University Press, 1975.

[32] L. Osborne, 'A Linguistic Big Bang', The New York Times, 24 October 1999, http://www.indiana.edu/ langacq/E105/Nicaragua.html. A five minute video including examples of the invented language is available at http://www.pbs.org/wgbh/evolution/library/07/2/l07204.html.

[33] M. A. Arbib, "From monkey-like action recognition to human language: An evolutionary framework for neurolinguistics," Behavioral and Brain Sciences, vol. 28, no. 2, pp. 105–124, 2005.

[34] B. Bridgeman, "Action planning supplements mirror systems in language evolution," Behavioral and Brain Sciences, vol. 28, no. 2, pp. 129–130., 2005.

[35] G. Ryle, The Concept of Mind. London: Hutchinson, 1949.Aaron Sloman

2.2　Simulating the Primal Affective Mentalities of the Mammalian Brain: A Fugue on the Emotional Feelings of Mental Life and Implications for AI-Robotics

Jaak Panksepp

In order to simulate the operations of the human mind we must consider both the genetic and epigenetic construction of the human brain. We must be clear about what is genetically fundamental and what is epigenetically derivative. Perhaps the most accurate simulations need to get the genetically-provided subsystems properly represented in Read only Memory (ROM) as key operating systems and to configure Random Access Memory Space (RAM) in such a way that developmental epigenetic programming can simulate the natural ecologically and culturally constrained developmental landscapes that transpire in neocortical maturation of the human child. The foundations of the higher mind are intimately linked to the genetically prescribed infrastructure of the lower, affective mind. Basic social emotional systems will need to have the greatest attention, in order to conceptualize how intersubjective dynamics and hence human language and thought gradually emerge through epigenetic programming of higher regions of the brain. In my estimation, it becomes critical to have clear visions of how emotional and other affectively valenced feelings are actually created in biological brains. The Affective Neuroscience approach envisions seven basic emotional operating systems, embedded within a core-self and self-related infor-mation processing infrastructure. The basic emotional systems may generate large-scale neurodynamics that provide attractor landscapes for harvesting information from the external world that is especially important for survival. In other words, a series of emotional-instinctual action circuits are built into the brain as large-scale network functions that may need to be described in terms of "state-spaces" that regulate "information-processing" algorithms. Thereby most cognitive mentality may epigenetically arise from a solid, genetically based foundation of emotional-affective processes. A persistent dilemma for computational modeling of mind is that, at present, the only experiencing minds that currently exist in the world are ultra-complex network-based brain capacities, based unambiguously on certain biophysical and neurochemical properties that are intimately linked to the dynamics of living bodies. Can such affective-emotional properties of biological brains be closely emulated in virtual machines? Can a simulation that does not possess any of the relevant biochemical/biophysical properties ever be substantively simulated on non-biological platforms? Only future work can tell. Such issues cannot be solved "in principle" at the present time. I will cover this topic in fugue fashion, repeating previously shared arguments by adding new dimensions of complexity in successive layers.

150

2.2.1   Preface

Originally, this paper was a commentary of Aaron Sloman's contribution "Machines in the ghost." Since Professor Sloman's final contribution to this volume was modified, in part, because of that critique, this paper has been recast to be a more general statement about my own views on the cardinal role of basic brain-mind processes, many of which are survival promoting emotional and motivational systems, on the developmental/epigenetic emergence of the mature human mind. Sloman's focus is largely on the higher cognitive aspects of human mind, while this essay may serve as a counterpoint to how those higher aspects actually emerge during development. In my estimation the higher cognitive aspects of mind may constitute the easier dimensions of the 'Hard Problem' of consciousness, while the harder foundational aspects require an intimate acquaintance with the biological nature of affective experiences which guide much of what the cognitive apparatus does. In other words, the developmental goal of much of higher mentation, much of it learned, is to provide strategies for most effectively achieving the 'comfort zones' of living and minimizing the many 'discomfort zones' that living in the real world entails. Many subtle comfort and discomfort zones reside in social interactions, where skills with emotionally based inter-subjective interactions are often more important than high cognitive intelligence in promoting reproductive success and happiness.

I agree with Sloman's view that high human intelligence may not require any concurrent emotional arousal (indeed, neocortex tends to inhibit primary-process emotionality). However, strong evidence from brain damage studies indicate that higher brain functions could not exist without emotional circuits providing a solid foundation of primal self-hood (Panksepp, 1998a,b) which can be linked to higher self-related information processing within the brain (Panksepp & Northoff, 2008). Emotional systems seem to lie at the heart of what it means to be human. Thus, I doubt if "emotional states" are simply "side effects" of certain operations in "intelligent machines" as Sloman phrased it. My evolutionary reading of brain-mind evolutions is that a series of basic emotional systems and fundamental forms of self-representation, concentrated within the deeper subcortical reaches of the mammalian brain, are essential value generating structures around which much of the epigenetically emergent cortical-cognitive apparatus revolves. If we damage those ancient biological-value encoding regions of the brain, mentality evaporates; if we damage the higher cognitive reaches of the brain, it does not (Merker, 2007; Panksepp, 1998a). It is a vastly more difficult enterprise to model the affective foundations of the human mind than the remarkable information-processing capacities which largely emerge epigenetically – through education and the cultural molding of developmental landscapes through lived experiences in the world. My goal here is to argue that a deep understanding of the subcortical tools for living and learning are the biggest challenge for any credible future simulation of the fuller complexities of mind. The cognitive aspects may be comparatively easy challenges since many follow the rules of propositional logic.

2.2.2   Prelude: How much of Adult Human Mind is Genetically Endowed?

Before we can simulate the true nature of human mentality on any artificial computational platform, we must be clear about the foundational aspects of mind that evolved through the long course of natural selection. I will argue that subsequent aspects, much of what might be called the cognitive mind, emerges through developmental landscapes – epigenetically programmed by experiences of living in the world. The first is more of a genetic question and the second largely a plasticity and epigenetic one. *Epigenesis* is the study of those semi-permanent, non-mutational changes in gene function (through DNA-chromatin methylation, acetylation, etc.) that can dramatically and permanently modify body and brain functions, and be passed down through generations with no modification of fundamental gene structure – with no change in the classic nucleotide pairings of DNA sequences. That is one of the main types of plasticity that helps program higher regions of the brain. The view I will advance here is that if one simply takes the final product we recognize as the adult human mind, with its very complex cognitive capacities, then one is dealing largely with the experientially refined aspects of mind. One may need to be clear that those higher mental complexities may largely arise from developmental processes rather than one's genetic heritage. Indeed, most higher aspects of mentation, including language and reasoning, may reflect epigenetic co-ordinations between lower and higher brain functions. They are dependent as much on cultural and ecological constraints on brain development as the permissive tools for mentation built into the fabric of the brain by evolution. Indeed, the failure to clearly envision the difference between the basic tools for living and the final results of living have led to much confusion in current evolutionary thinking about the nature of mind (Panksepp, 2007).

Evolutionary Psychologists, most of whom find their intellectual roots in social-personality psychology rather than neuroscience, are commonly all to ready to claim that a great deal of the cognitive sophistication of adult human brain functions – from cheater detection modules to language instincts – are contained in some pre-packaged albeit rudimentary modular form within our genetic heritage. They rarely consider the fact that much of what is unique about the human brain, namely our massively expanded neocortical space for the association of events and eventually ideas, anatomically resembles Random Access Memory (RAM) space more than pre-ordained cognitive operating systems. By failing to consider the nature of the brain machine, they may be proliferating error upon error, a stance that computationally oriented mind-simulators may be wise to avoid. Evolutionary Psychologists rarely provide even a vague approximation of how one might go from the underlying genetic information to the instantiation of mature cognitive brain functions. If they did, then they would have to pay as much attention to epigenetic developmental landscapes as modularity-mythology and just-so adaptive stories.

What is actually contained within the epistemological engravings of the genetically provided circuits of the brain? This is a momentous question not only for those who wish to simulate the human mind, but also for those who simply want to understand the basic neuropsychological functions of the human mind. Much of what Evolutionary Psychologists have asserted to be reasonable conjectures about the evolved aspects of human minds, are based on genetic mythologies with no biological or neuroscientific substance. If so, then those who wish to create artificial mentality need to be clear about neurobiological and neuropsychological approaches that are based on solid evidence-based ideas rather than mythology-based conjectures. My reading of the evidence is that most of the higher functions of the human brain may be largely programmed through environmental and cultural encounters with real world events rather than through any ancestral dictates. Admittedly, this is as extreme a position that one can currently take, but it is a didactic position I will develop here: At the outset, we must admit that the newborn human child shows no indication of evolved cognitive abilities of the kind Evolutionary Psychologists opine. If so, it remains possible that most higher aspects of human mind emerge epigenetically, guided by more primitive emotional and perceptual tools, most of which we share with all other mammals, operating on higher general-purpose association fields of neocortex. As far as we factually know at present, humans have few intrinsic cognitive abilities not possessed by other mammals. However, because of our massive general-purpose cerebral endowments, with abundant association fibers connecting different regions of neocortex, we acquire them throughout development.

If neuroscientists, neuropsychologists and neuro-psychoanalysts are going to provide anything really useful for those who wish to create artificial minds that simulate, with any accuracy, living human beings, then we must first work from the bottom up rather than from top down, especially if the 'top' is created epigenetically from certain foundational skills. In the mature system, the top-down controls begin to prevail, because of a life time full of self-related information processing (Panksepp & Northoff, 2008). Accordingly, my goal here is to identify and summarize some of the foundational tools for mentality that are truly part of our genetic heritage. In this realm, most neuroscientists recognize that basic attentional, motivational, emotional, sensorial/perceptual and memorial systems do exist as birthrights of all mammals. At the same time, we have little credence about genetic sources of higher cognitive and perceptual processes. Indeed, it remains possible, even likely for some of us, that most higher brain functions, from detailed visual perception to linguistic reasoning, are epigenetically created by experiences in the world.

2.2.3 Much of our Knowledge Exists in the World Before it Exists in our Brain.

Most of our knowledge about the world exists first in the world. It does not need to be encoded in our brains and minds through genetically based neuronal wiring. It is quite sufficient to have a plastic brain that can absorb information about relationships in the world through learning. This learning can include things as subtle as language and our capacity to think. If true, this could make the job of creating artificial intelligence much easier than if we first had to fathom currently unknown genetic algorithms. In this view, a more profound difficulty arises from the foundational processes, created by evolution, which may not follow any obvious rules of associationism or propositional logic. At the ground level, we are confronted by momentous questions. How was a core-self built into the brain? How were survival values encoded? How many survival values and associated emotional and motivational systems are genetically built into the brain? Are we yet up to that task? What now seems clear is that what was actually built into the brain through eons upon eons of natural selection, were a variety of fundamental survival processes that may be so far removed from traditional computational principles, that they may require a complete re-thinking of where we need to begin to construct the ground floor of mind. Also, considering that these processes in living brains are coded in a large diversity of neurochemistries, it would seem that mother-nature could not readily build precise survival solutions into the brain with non-chemical information processing codes. Where does this leave any form of computationalism that cannot simulate the complexities of living matter?

In my evidence-based estimation, what was built into the foundational aspects of the brain in every human and every mammalian species, aside from certain basic sensory/perceptual, attentional/learning and motor abilities, obvious to most observers, were several categories of more subtle mind tools, including: i) a core-self which has traditionally, and incorrectly, been envisioned as a "ghost in the machine," and ii) a set of value structures that we currently refer to by the generic labels of "emotions" and "motivations" which iii) allow for self-related information processing, with all of the above iv) embedded in circadian sleep-waking mechanisms that govern each organism's capacity to confront the world effectively. Here I will focus only on the first three of the subtle mind processes that may constitute the most difficult challenges for anyone wishing to simulate mind on non-biological computational platforms, with most effort devoted to the second category. I have no confidence that the natural reality of those processes can be computed with any existing procedures that artificial-intelligence has offered for our consideration.

It is possible that the three value generating processes noted above constitute the very foundations of consciousness that cannot be instantiated on non-biological platforms. However, my aim here is not to argue the improbability that those neurobiological, and genetically promoted, mind processes will ever be instantiated on artificially engineered devices, but to at least introduce them to the AI community as major challenges to consider in full neurobiological (as opposed to flow-diagram) depth. There is only space here, however, to make introductions rather than to summarize our admittedly preliminary understanding of such brain systems. See Panksepp (1998a, 2005a) for a more comprehensive discussion of the nature of the brain mechanisms of intrinsic biological values – the basic emotions and motivations. For a more in depth discussion of the core self and self-related processing, see Northoff, et al., (1996), Panksepp (1998a, b) and Panksepp and Northoff (2008). The ancient neural fabric for self-representation, laid out in the ancient visceral nervous system, may be a worthy target for initial modeling if anyone really wants to come close to simulating the biological nature of mind.

Before proceeding to a discussion of the genetically prescribed mental tools of brain, we need to briefly discuss how higher brain functions may emerge through epigenetic developmental landscapes. Once more, *epigenesis* covers changes in gene expression patterns that engender brain and bodily specializations that are honed by environmental interactions rather than information intrinsically contained in the genome. In mental development, we must distinguish what is functionally emergent from that which is genetically prescribed. For instance, to the best of our knowledge *neocortical* visual functions (Sur & Leamey, 2001), perhaps language also, are functionally emergent specializations of higher brain regions rather than genetically pre-ordained functions of the brain.

2.2.4 The Epigenetic Foundations of Mental Life

Only a decade ago, experts estimated that our genome contained approximately a 100,000 genes. Following recent advances in molecular biology, we must now ponder how the subtlety of the human mind can emerge from our vastly more modest hereditary storehouse of some 22 thousand genes. Much of the answer may arise from the fact that we have an enormous neocortex, repetitive tissue that evolved very rapidly in recent brain evolution to its present vast proportions in top predators, especially human beings. The most probable way this could have happened so rapidly is that our neocortex is more akin to the classic *tabula rasa* – the blank-slate of philosophers – than to a massively "modularized" evolutionary toolbox envisioned by too many Evolutionary Psychologists who rarely discuss or explicitly contemplate the true nature of our brains. Evolutionary psychology continues to thrive on "just-so stories" that have, at best, modest statistical support from behavioral studies, typically with little connection to either genetic or neuroscientific issues (Panksepp, 2007a). There is too much gratuitous talk about evolved cognitive "modules" by investigators who display little devotion to what real genetics can or cannot explain about the human mental apparatus at the present time. Those who wish to simulate human mind may be well advised not to make that same mistake, and not to believe, before credible data is provided, that much of the higher reaches of the human brain is pre-programmed by our genetic heritage.

At birth, the neocortex is largely a "blank slate" upon which experience writes the unique scripts of individual lives. The neocortex consists largely of endlessly repetitive "columnar" units of about three thousand neurons. These basic computational units of neocortex required no great genetic complexity to proliferate, at least once their basic structure had been pre-ordained by gene controlled neurodevelopmental landscapes. These cortical columns, with a touch of metaphor, resemble computer chip-like units that constitute random access memory (RAM) space of our digital computers. To the best of our knowledge, they do not contain read only memories (ROMs) of our ancestral past, but only the general-purpose associational space where higher brain states, concordant with the basic, subcortical survival needs, are programmed by experience. What we do inherit are lower brain functions, shared by all mammals, which are stable, instinctual, brain operating systems that allow a massive amount of complexity to be developmentally programmed into RAM space.

To the best of our current knowledge, modest as it is, we are genetically provided with only a series of primitive tools for living – namely, some kind of representation of the organism as an organism (a core-self) with a series of genetically-ordained attentional, emotional and motivational tools for creating distinct brain states, that monitor survival-related brain bodily states, all working with the basic capacity to learn about how to survive in the world – all the strategic details of which get stored away in neocortex. All those ancient, genetically-provided instinctual networks – those powerful subcortically concentrated emotional and motivational tools for living and learning – are remarkably similar (homologous) in all mammals. What makes the human mind unique is the capacity of higher brain regions to become refined as repositories for specific cognitively-based survival and thriving strategies. And here "cognition" refers to all that information processing that is based on inputs through the various exteroceptive sensory portals of our bodies. In contrast, the foundational issues are within-brain/body operating principles.

In this chapter I will largely focus on the emotional networks that are some of the most important evolutionarily provided tools for living in the world. Among them, we find various basic emotional urges: the desire to engage with the world (SEEKING), anger at being thwarted (RAGE), trepidation toward the scary things of the world (FEAR), and eventually blossoming sexuality (LUST), the desire to nurture the young (CARE) and always the awful feeling of being alone, without social support (PANIC). Perhaps the most important tool for the construction of higher social brain functions, is that joyous enthusiasm to engage PLAYfully with others. I use capitalizations to highlight that we are talking about complex brain networks, elaborating instinctual action tendencies accompanied by specific feelings (Panksepp, 1998a, 2005a,b) to guide the maturation of each and every child. It is noteworthy that there are distinct neuropeptidergic codes for these emotional functions, providing many novel targets for new psychiatric drug development (Panksepp & Harro, 2004). One could argue that the basic social tools – LUST, CARE, PANIC and PLAY – are among the most important ones to consider when we try to envision how the higher human social brain – the capacity for inter-subjective sharing and mirroring – perhaps language itself – are developmentally programmed into higher regions of the brain. If there are other epistemological engravings of cognitive strategies in neocortex, *determined by our genes* – from language instincts to mirror neurons – they remain to be empirically demonstrated.

For instance, the PLAY urge, which allows young organisms to learn about what they can or cannot do with each other, may be one of the main tools that can help construct our social attitudes in the initially empty executive and memorial spaces of the neocortex, along with learning motor and balance skills necessary for survival. From this vantage, the mature social brain arises from the utilization of the various evolutionarily provided emotional tools to program higher reaches of the brain, probably by epigenetic rules that are barely beginning to be explicated by neuroscience. If so, the lesson for artificial simulation of higher mental life would be to construct certain kinds of basic subcortical operating systems, with plenty of higher neocortex-type RAM space with which to interact, into the technological brains of robots. It is an open question whether that fabricated RAM space needs to be constructed in a certain way, perhaps best initially to resemble columnar structures of the neocortex, that can link up the kinds of emotional and motivational state spaces – perhaps simulating large-scale attractor-type network dynamics able to self-referentially integrate real-life confrontations with the world through self-referential organization of neocortex type neural nets. In other words, simulation of biological minds may initially need to expend more effort on the foundational issues rather than the higher cognitive competences that emerge from experience. Successful pursuit of that project may require us to envision how higher brain functions within that RAM space (reflecting the details and abilities of individual lives that can last a life-time) arise epigenetically from lower brain functions.

To reemphasize a key point: massive, multi-dimensional forms of plasticity mediate the emergence of higher brain function. For instance, in biological brains, epigenetic brain-mind landscapes arise largely from methylation of the chromatin supports surrounding genes. The resulting changes in the three dimensional environments around genes control how readily transcription factors can control the expression of specific genes. This is one of the ways bodily organs differentiate, allowing different cells to do different things. For our purposes, this is a beautiful exemplar, among many other types of plasticity, that is emerging as a major factor that controls how higher brain networks specialize and learn to handle specific life skills. These stable and far reaching brain changes, leading to cellular specializations, and thereby control how long-term neurobehavioral and neuropsychological changes in higher brain regions come to reflect the qualities of social and physical environments (Zzyf, McGowan, & Meaney, 2008).

From this vantage, much of higher brain development – neocortical maturation and programming – is regulated as much by environmental factors as "information" encoded in the genome. Thereby, many aspects of personality and temperament emerge through real-life social-environmental interactions that are guided by ROM-like emotional systems encoded more intimately within our genes. Thus, it might be fair to claim that the fully human, higher social brain can emerge only under the dictates of brain emotional systems, such as those that engender PLAY and the other basic emotions. If the higher human mind is created as much by social learning as ancestral genetic molding, then we have special responsibilities for promoting better, emotionally intelligent, child-rearing practices (Sunderland, 2006). This may also provide a unique vision for how we might simulate human mentation on non-biological computational platforms (the concern of this volume).

Evidence for the abundance of epigenetic processes in development is growing dramatically, including the life-long protective effects of maternal devotion (Meaney, 2001). Such findings have profound human psychiatric implications, and they encourage caution in assuming that higher cognitive functions of the human brain are genetically inherited A most compelling exemplar is available for the development of visual cortex: If those occipital cortical regions that developmentally acquire visual competence are surgically destroyed *in utero,* animals still develop into visually competent creatures. Thalamic visual outputs simply migrate to adjacent, initially general-purpose, neocortical regions allowing normal visual cortex to emerge developmentally (Sur & Leamey, 2001). In other words, genetically prescribed visual competence in subcortical circuits can establish visual competence apparently in any region of intact neocortex, at least early in development before neocortical specializations have emerged. The implication of this for understanding biological minds and constructing artificial minds are profound: There may be no need to program functional competence into higher brain regions. All that may be needed is a sufficiently complex and plastic neocortex-type functional platform, that can be programmed into functional competence (e.g., even perception) through subcortical brain operating systems that can guide learning and other forms of plasticity in RAM space. I expect the optimal architecture for the RAM chips would be best based on the "columnar" structure of mammalian neocortex.

2.2.5   The Genetic Foundations of Mental Life

It is a fact of human life that an enormous cornucopia of *feelings* are our birthright. These feelings come in two major varieties: A) Our ability to perceive the world through our various sensory portals – many of these are relatively neutral in terms of the intensity of feelings of goodness (pleasure) and badness (displeasure), even though they readily get linked, through classical conditioning, to worldly objects and events that can evoke strong affective feelings. Neuroscience has done a fine job explicating such relatively neutral information-providing "feelings" of the brain, and artificial intelligence (AI) is making strides in providing robots with various sensory-perceptual capacities. But more importantly for present purposes, there is another class, B) the strongly valenced affective feelings – the varieties of ways brains can feel good and bad – which can also come in a variety of flavors, including i) *sensory pleasures and displeasures*, especially tastes, smells and the varieties of touch (in principle, these should be among the easiest to simulate in virtual machines because they are tightly linked to exteroceptive sensory receptors of the body – e.g., for touch, taste and smell), ii) then there are the varieties of homeostatic *bodily feelings*, from hunger and thirst to fatigue and exhilaration, (easy to simulate superficially as intrinsic "motivational" processes, but hard to simulate deeply because they are characterized by affective feelings that remain poorly deciphered physiologically, even though we know that medial subcortical areas of the brain are most critical for their elaboration (Denton, 2006; Panksepp,

1998a), and finally, and perhaps most mysteriously, iii) there are the *emotional feelings*, that are strongly linked to the ways nervous systems were built during the long evolutionary journey that eventually led to brains like ours.

Emotional feelings are very hard to simulate veridically since emotions are much more than cognition-disrupting bodily "thrashings," as some have envisioned (including Prof. Sloman in the original ENF paper submitted for my commentary, with his indictment of most everyday semantic-emotional usage). In fact, brain emotional systems generate all-consuming affective feelings that are essential for guiding critical behavioral choices and hence the quality of life. Although I can readily appreciate superb analyses, such as Sloman's contribution to this volume, of how we should scientifically approach the functional organization of virtual machines, including the varieties of emotional and other affective processes. Unfortunately, those of us seeking to create a meaningful functional neuroscience really have few semantic alternatives if neuropsychological entities, such as primary-process emotions ranging from anger, fear, separation distress, lust, maternal care and play (Panksepp, 1998a), to the complexities of pain (Aydede, 2006; Eccleston & Crombez, 1999) and the large variety of cognitively related feelings arising from various hopes and regrets (Camille, et al., 2004; Zeelenberg, et al., 1998), actually govern how biological organisms organize their behavioral choices.

There is nothing logically or biologically incoherent about the assertion that affects are the primary brain states via which all humans and animals live their lives. Affects may be the intrinsic, evolutionarily provided value indicators in all mammals. In pursuing these issues scientifically, perhaps it is important to accept the fact that we must start with fuzzy categories, which will allow more rigorous neurophysiological definitions to emerge from the ongoing scientific work. These are not really conceptual issues, but matters of neural nature. In my estimation, we have absolutely no way to begin approaching these subtle brain organizational issues, if we discard everyday semantic usages right at the outset, even though we should aspire to ground such concepts eventually in specific brain functions. This is a project I have been pursuing for thirty some years (Panksepp, 1974, 1982, 1998a, 2005a). Although Sloman has written extensively about such matters from a functionalist-engineering perspective (his references 10-15), he seems to decisively side with the cognitive side of the mind-divide, envisioning emotions as simple "alarm" type behavioral interrupt systems that often have "bad" effects on behavioral and cognitive competence. In fact, the underlying networks have many non-deliberative properties, and may be necessary foundations for consciousness (Panksepp, 2007b).

Although the "alarm" function is certainly true for various negative emotions, especially fear, emotional affects also tell us much about what we are doing well in our behavioral choices, as in the joy of social play and positive social engagements that allow young organisms to learn about the opportunities and limits of their social worlds. Thus, we should not underestimate the nature and impact of affective processes in the governance of adaptive behaviors, even though Sloman may be correct in envisioning that some modern views of affect, such as Damasio's peripheralist sensorially-focussed, somatic-marker hypothesis are deeply flawed (Dunn, et al., 2006; Tomb, et al., 2002). However, it does not follow that the possibilities summarized here are equally flawed – namely that within-brain, internally generated emotional action tendencies, accompanied by raw emotional feelings, do control many behavioral strategies. In any event, in my estimation anyone who believes that the affective-experiential aspects of emotions are not all that relevant to the simulation of the real-life cognitive aspects of mental life have a flawed vision of what mental life is all about. Likewise, those who subscribe to "ruthless reductionism" – namely that only neuronal firing counts, while subjective experience does not – are probably misunderstanding and misrepresenting what the brain actually does. Since mentality is thoroughly neurobiological, large-scale neural network activities concurrently generating instinctual actions and internal affective experiences can developmentally begin to guide more subtle, cognitively molded, behavioral choices. Thereby, cognitively molded behavioral variability would continue to track hedonic gradients. If so, understanding how affect is generated within the brain is of foremost importance for simulating what real brains do.

Accordingly, I advocate the position that it is only when we neuroscientifically fathom the affective mechanisms of the brain will we be in a position to appreciate how organisms, through massive cortical endowments, eventually become cognitively capable of becoming aware of their place in nature, with the many life decisions they must make, which, in humans, motivates us to think about the deepest mysteries of experienced existence (as is often best highlighted in novels and other artistic undertakings). Such brain processes appear to lie at the foundation of mental livingness; affective experiences may be essential for all higher cognitive forms of consciousness (Merker, 2007; Panksepp, 1998a, b, 2007a, b). Although we may be able to effectively simulate cognitive processing on sophisticated information-processing platforms, it is debatable, perhaps even doubtful, that we can ever create artificial creatures that have the depth of psychological existence that our affective consciousness permits. To make progress on simulating the neurobiological mind, we should not minimize and caricatures the nature of emotions, as if they were little more than the occasional "thrashings" that disrupt well-oiled, adaptive cognitive decision-making machines. That does not appear to be the case, either as far as evidence from human experience or the functional organization of the animal brain is concerned. Affective feelings may be the energetic drivers of our practical rationality (which is not quite the same as our intelligence) – from the varieties of life choices we make to the counterfactual imagination which enriches all our desires, hopes and expectations, as well as our anxieties and worries (Byrne, 2005).

One of the most important aspects of mammalian emotional feelings, and other affects, is to generate *internal value codes* for certain life-challenging circumstances, so that psychological and behavioral responses can be coordinated toward adaptive ends (survival). It feels bad to feel fear, and delightful to be in a playful mood. Good feelings, which come in many varieties, tell us we are probably pursuing courses of action that enhance our lives; bad feelings, which are equally diverse, inform us our opportunities for a good life have diminished. Feelings come in many forms, and exist at multiple levels of complexity – primary, secondary and tertiary processes, which parse approximately into innate, and learned and thoughtful functions of the brain (going progressively from tighter genetic to more emergent epigenetic controls). The neural nature of affects has barely been deciphered by neuroscience and psychology – and the primary process forms remain most neglected (e.g., see critique by Barrett, 2006; and response by Panksepp, 2007c). These are the aspects of emotion that become critical for veridical simulations of key behavioral control processes in virtual machines, if they are to have any hope of achieving true mentality as opposed to sophisticated simulacra status. Such issues may not be as important for simulating behaviors of the simpler organisms (paramecia, etc.) that may be more deeply robotic, even though substantive knowledge at that level, where there are no neural homologies, must remain thin for the foreseeable future.

There are no assurances of how such living brain and virtual machine functions are best coordinated. Thus, we should close no doors in our attempt to decipher these processes neurobiologially, and then to determine if it is feasible to simulate them in substantive depth. In my estimation, the only working hypothesis that currently has the potential to yield scientific progress on these issues is the hypothesis that includes the recognition that emotional affects are isomorphic with the neurodynamic of instinctual emotional systems in action (Panksepp, 1998a, b; 2005a, b). Whether such intrinsic brain functions can be adequately modeled using computational information flow within virtual neuronal machines remains uncertain. Although I have little confidence that this can be achieved for the core affects (i.e., intrinsic values of complex animate system), I would briefly highlight the kinds of issues we may need to consider if the robotic dream, or nightmare, of manufactured feeling creatures, as the case may be, is to become reality. First, let me reaffirm that there is absolutely no principled reason that we cannot generate simulacra that are so psychologically compelling that most would be tempted to grant them more psychological sophistication than is actually contained in their charades. Indeed, in superficial form, that has been well achieved on celluloid for over a century. The ontological problem runs deeper, and depends on what is the most accurate view of the many layers of consciousness that exist in human, and probably other animal, brains (Velmans & Schneider, 2007). In my estimation, the most troublesome problem for mental life simulation in robotics and AI is the nature of affective consciousness – the way basic biological values – the ancestral voices of the genes – are instantiated by neural tissues (Panksepp, 2007a, b).

On first principles, it is realistic to assume that whatever consciousness is, at present it is only instantiated through carbon based molecules, where an enormous complexity of interacting biochemical systems are required to construct a platform for mental existence. This is no trivial matter, even though, and quite understandably, those generating simulacra on silicone based computational devices may wish to temporarily ignore such critical details. Whether mere information flow (as defined in classical Shannon-Weaver terms) is sufficient for instantiating mind will remain debatable for the foreseeable future. As a neurochemically oriented neuroscientist, I have no such convenient assumption to fall back on, for it is clear that the mammalian mind, especially the affective feeling parts, is, to the best of our knowledge, critically dependent on the specific molecules and complex state control circuits that constitute the living brain. These complex network states resemble energetic field dynamics more than digital-type computational information. In any event, we should not prematurely assume that everything in the brain-mind is informational, although a substantial part of cognitive perception and decision-making may well be, which would help explain why the cognitive side of the problem has attracted more focused attention than the affective side.

In sum, the hardest problem of consciousness may be how brain evolution eventually generated various feelings that are central to the experience of our existence. Since our best mechanistic understandings of processes, such as emotional feelings, has so far emerged from our understanding of the neuroanatomical and neurochemical basis of emotional states (Panksepp, 1986; 1998a, 2005a, 2007a, b) and since the major goal of modern biological psychiatry and perhaps neuro-psychoanalysis is to help re-achieve affective balance (i.e., by taking advantage of such neurochemical controls), it remains possible that the problem of mind cannot be solved without full devotion to such concerns. Many in AI are willing to make the traditional simplifying assumption that mind may be largely free of "platform constraints" but neuroscientists have no reason to have faith in such luxurious simplifications, which may turn out to be fool's gold. Thus, most of what I will have to say is a cautionary message that may need attention by anyone aspiring to compute the depth of mind using strictly computational, virtual machine "solutions." I will end with some ideas about "Turing Tests" that may need to be developed to evaluate for the existence of deep affective/emotional experience within AI and robotic life-simulation platforms.

2.2.6   Historical Interlude – the Biases of the Cognitive Revolution

The cognitive revolution emerged from the hope that the human mind resembled digital computer functions sufficiently well so that the general principles for mental operations would turn out to be largely computational issues independent of the specific platform on which the calculations were done. Indeed, at the very beginning, the fathers of the "cognitive revolution" decided that the issue of emotions and affective feelings was potentially so complex that it was wiser to put all that on a shelf, until the more solvable externally triggered perceptual and cognitive-thought processes were solved. The matter was really not much different in behavioral neuroscience, where learning and conditioning issues prevailed, and the nature of core emotionality was rarely discussed (for historical summary, see Panksepp, 2005a, b). This essentially left emotion research as a marginal player, only to arise, almost tsunami fashion, with the onset of the human brain-imaging revolution – an approach that has only given us some useful correlates and practically no knowledge about neural causality.

This sudden resurrection of a key problem of consciousness, has led to a pervasive vision of emotions that is so integrally linked to cognitive activities, that few remain who are willing to consider the apparent evolutionary fact that emotional operating systems of the mammalian brain evolved to a very sophisticated level by the time the simplest forms of learning (classical and instrumental conditioning) were preparing the way for the emergence of a cognitive mind that could dwell, in some depth, on problems of existence (Panksepp, 2007a). Only a few in the AI community, like Aaron Sloman (1982) and Rosalind Picard (1997), continued to conceptualize how emotions might be envisioned as either bodily commotions or key operating principles in virtual machines. Meanwhile, the few in neuroscience who sought to understand the primitive affective operating systems of the brain, were slowly constructing an image of brain organization where emotional and other basic affective abilities of the brain were the very foundation of consciousness (Panksepp, 2005a, 2007a) – perhaps the evolutionary substrates for primary-process mentality (Panksepp, 2007b).

The present evolutionary vision of brain organization highlights how mind may have emerged on the face of the earth. During the evolutionary development of the brain, raw affect (e.g., from the primal sting of pain onward) may have been critically important in how raw experience first emerged in brain tissues (Aydede, 2005), with potential constraints on all future mind developments. If one doubts the power of affect in human mental affairs, then one would have to explain the power of the torturer's tools. As Shakespeare understated in *Much Ado About Nothing* (Act V, Scene 1) "there was never yet a philosopher that could endure the toothache patiently." Clearly our affective weaknesses can overwhelm our cognitive strengths.

It is gratifying that revolutionaries such as Picard have sought to push the envelope of computational approaches to the emotional foundations of AI, with no presumption that the simulations will achieve any experiential reality. I think most agree that it would be quite easy to build various forms of "thrashing" into a computational beast. But that is not the point. The diverse feeling of fear and rage, desire and satisfaction, lust and love, loneliness and playful joy, and how they guide behavioral choices are the true challenges. The rest of this paper will try to flesh out how modern affective neuroscience is providing at least the vague outlines of ways to solve such neurophysiological-neurochemical dilemmas, yielding hopefully, conceptual approaches that may be useful for those interested in building truer forms of livingness into virtual machines.

In my estimation, a critical feature of such a synthesis must investigate of how extensive neural networks, whether biologic or computational, generate large-scale neurodyamics that are intimately related to the action readiness of living or robotic organisms. We must be able to envision how such networks mediate self-referential information processing (Northoff, et al., 2006), and how a nomothetic core-self that can developmentally lead to a variety of idiographic selves (Panksepp & Northoff, 2008) is constructed in the brain. Perhaps a clear biophysical understanding of resting "background states" within the brain (Greicius, et al., 2003) is a critically important stepping stone, providing a fundamental image of "self-representation" (Panksepp, 1998b), perhaps a necessary neural ingredient for the construction of various affective states of mind. Such a materialistic, monistic vision of affect has no more room for "ghosts in the machine" than a logical decision-making model of human mind. However, it insists that abundant room must be made in AI for a realistic confrontation with our emotional nature, its various joys and vicissitudes – perhaps even its deeply moral nature – than is contained in any mere cognitive decision making device, which has no intrinsic values, and may never have.

Can We Model Affective Brain Dynamics Computationally?

The proper answer to this is "we don't know." Our knowledge in this area remains woefully incomplete. Those who accept the central dogma of the neuron doctrine very seriously as the foundation of mind (i.e., the digital-type patterns of firings and non-firings) are bound to have more optimism than those who consider graded dentritic potentials and large scale network dynamics as being more important for mentality. Those who assume that practically all we need to know at the biophysical level about real minds lies in the computational domain, of on-off switches are ignoring a great deal of brain activity that mediates mental "information." The central dogma of neuroscience, the neuron doctrine, certainly envisions that neuronal firings do all the hard work of mind, that ultimately mind is a conglomeration of synaptic transmissions based on electrical and chemical signals, passed on from one neuron to another – as all-or-none electrical currents (action potentials) flowing through complex neuronal networks. Through the intermediaries of graded currents, activated by synaptic neurotransmitters, mind is gradually constructed. The simple-minded neurone-doctrine view of brain function, which is currently the easiest brain model to apply in AI/robotics, under-represents what biological brains really do. "State spaces" require more complex network dynamics.

Although the neuron-doctrine sounds like a fine materialistic starting-assumption, partly true, it has also lead to a pervasive vision in mind science that can be called "ruthless reductionism" – the assumption that only neuronal firing (e.g., information processing) counts, and mental experience therefore is redundant with such processes, and ultimately does not need to be conceptualized as a causal entity in what brains do. However, there is every reason to believe the feelings of the brain do have causal efficacy. When we injure a limb and avoid using it, the pain is preventing us from inflicting further injury and hence delaying restoration of function. Just consider that everyone born with congential pain insensitivity, dies prematurely (Brand & Yancey, 1997). If affects guard and guide our lives, then they should also do so within virtual machines that have any chance of approximating biological mental life – unless, of course, one is simply interested in generating ever more sophisticated and compelling simulacra (a worthy project in itself). In this way we must leave abundant room for psychology to have causal efficacy in the control of behavior, at least among the real (the non-virtual) beasts that roam the earth. To me the evidence seems overwhelming that experienced psychological processes control what organisms do, and that such states are not simply properties of brains that can be envisioned comprehensively through the lens of the "neuron doctrine." Such experienced conditions of 'mind flesh' arise from the dynamics of various "neuronal networks" working in conjunction with various non-neuronal (e.g., glial) properties of the brain. In sum, from a network doctrine point of view, there is abundant room for emergence of new functions – psychological dimensions – that have causal efficacy in the control of behavior.

As soon as one takes a network-view to the brain, then one is confronted by the dilemma that living brains may contain many biophysical complexities that are not well described in computational terms. There are complex glial supports for network activities, protein-protein, non-synaptic, paracrine interactions. A prime example is the functional role of nonsynaptic, gaseous neuroactive substances such as nitric oxide in the brain, which provides non-precise diffusional control over neural network activities. In short, there is a biophysical jungle of complexity that allows large-scale network functions that concurrently generate instinctual behaviors and affective feelings within the brain, establishing the possibility of various epigenetic landscapes within higher regions of the brain. We cannot forget or ignore these dimensions of living brains, as if they were just some kind of background noise or non-functional window-dressing for the more discrete "informational" neuronal firing aspects that some assume do all the hard work in brain functions. One ignores such biophysical aspects of mind at their peril. The widespread assumption that "information-processing" will suffice to solve most of the hard problems in this area is risky at best. Energetic emotional "states" may be more than informational functions of the brain, they may establish dynamic attractor landscapes that ensnare information from the environment into very complex network functions (Alcaro, et al., 2007).

What might be the lessons for AI? Because such network dynamics generate poorly-understood attractor envelopes for various types of behavioral actions (e.g., instinctual emotional responses), a critical ingredient for establishing the foundations of mind, it may be necessary to construct a coherent, wide-scale organismic representation, a coherent embodiment, as an enactive foundation for any virtual machine that could be expected to have any chance of experiencing feelings. These large scale emotional networks, still incompletely understood, have been identified by use of localized electrical stimulation of the brain (ESB). As noted, I decided to capitalize vernacular emotional systems designators. I presently claim there is sufficient evidence for the existence of at least seven such executive systems: These include SEEKING, FEAR, RAGE, LUST, CARE, PANIC and PLAY. The capitalizations highlight that these systems are only partly understood, and also to avoid part-whole confusions (mereological fallacies), thereby avoiding misleading every-day vernacular usages that Sloman correctly highlights as potentially problematic. These specific neural networks generating categories of emotional response tendencies are the referents. These systems have many functional properties (Table 1), providing the first neural definition of an emotional system. This usage also implicitly acknowledges that experimental science can only clarify functional brain parts, and never the coherent wholes that psychologists are often interested in contemplating. Of course, robotic simulations also aspire to create functional wholes – this is the nature of the engineering enterprise. More than any other scientific approach, neuroscience provides a true understanding of the functional parts that need to be simulated by AI.

Aside from what I believe are outdated peripheralist theories of emotions (e.g., see Dunn, et al's., 2006 critique of James-Lange type hypotheses), there are practically no coherent neuroscientific statements about how emotional feelings are created in the brain. Practically the only hypothesis, generated in the modern era, is that raw affect is a property of the large-scale networks that generate emotional-instinctual action tendencies (Panksepp, 2005a). This idea is premised on the robust fact that when ESB is used to evoke coherent emotional action tendencies, animals never appear to be neutral about such stimulation (Panksepp, 1982, 1998a, 2005). They readily learn to turn on stimulation that provokes emotional actions that are affectively positive (e.g., SEEKING, LUST, CARE and PLAY). They readily avoid brain arousals characteristic of FEAR, RAGE and PANIC. To the extent that this proposition is correct – that raw affect is a property of these kinds of psychobehavioral brain dynamics – it provides investigators ways to study emotional feelings in some biophysical and neurochemical detail (Panksepp, 1998a, 2005a). It also restores long discarded energy-metaphors (certain kinds of bodily force fields) to the forefront of thinking about the nature of emotional feelings and other affects (Ciompi & Panksepp, 2004).

My working premise is that there are networks for all of the basic emotions. They all interact with each other to some extent at various levels of the neuroaxis, including the highest levels of cognitive processing. Most prominently, it looks like the SEEKING urge may be recruited by the other emotional systems. It is required for everything the animal does; one could conceptualize it in psychoanalytic terms as the main source of libidinal energy. For a full coverage of the details that need to be considered in modeling this system, see Alcaro, et al. (2007) and Panksepp and Moskal (2007).

Further, all may be partly dependent on large-scale within-brain neurodynamic representations of the body, especially the visceral-autonomic aspects, situated diffusely throughout the trans-hypothalamic and medial-thalamic courses of the various emotional systems. This coherent bodily representation, labeled the core-SELF (Panksepp, 1998a, b), may be a shared substratum for the coherent representation of distinct emotional dynamics. This ancient somatic-skeletomuscular and visceral-autonomic continuum may constitute a genetically dictated neuro-symbolic representation of the virtual emotional body, one that is intimately linked, guiding the major instinctual-emotional expressions of the peripherally body. Various neuropeptides are envisioned as the specific regulators that can send this core-SELF into various types of basic emotional dynamics.

To the extent that the various resonances and attractor envelopes of this core-SELF may be isomorphic with raw emotional feelings, it would seem that AI needs to struggle with ways it might instantiate these foundational mechanisms in order to begin building feelings into virtual machines and robots. The cognitive layerings of mind may need to sit on this solid foundation of affect-based organismic coherence. Thus, a principled distinction may need to be made between primary-process affective consciousness, and secondary-process and tertiary-process cognitive consciousness (i.e., learning and thought related cognitive processes).

2.2.7 Neuro-Conceptual Parsing of Affective and Cognitive Consciousness.

From the perspective of affective neuroscience, the core affective/emotional systems of the brain (very largely sub-neocortical, and limbic/paleocortical) can be distinguished from the more recently evolved higher cognitive structures (thalamic-neocortical axis) on the basis of various criteria outlined previously (Panksepp, 2003a) and summarized in Table I. These distinct dynamics interact presumably to generate the all-pervasive emotion cognition interactions that occur in intact organisms.

Affective processes/values (Internal Brain States)	Cognitive processes (External Information)
more subcortical	more neocortical
less computational (analog)	more computational (digital)
intentions in action	intentions to act
action to perception	perception to action
neuromodulator codes (e.g., more neuropeptidergic)	neurotransmitter codes (e.g., more glutamatergic, etc.)

Table I: Distinct characteristics of emotional/affective and cognitive/informational systems of the brain

These distinctions coax us to consider emotional brain functions in quite different non-traditional ways (e.g., as just another form of information-processing), and recognize that core emotional systems operate in holistic dynamic ways (i.e., networks of neurons yield energetic brain and bodily *field-dynamics*) as opposed to discrete information processing algorithms. Such distinctions provide novel ways of looking at those pervasive emotion-cognition interactions that are currently of great interest to psychologists and other brain/mind scientists.

As discussed initially in Panksepp (2003a), at the most fundamental (primary-process) level, the core of affective/emotional processes may be more deeply holistic and organic (i.e., their status cannot be understood without a "network doctrine" as well as molecular coding approaches to how distinct global neurodynamics are generated). In contrast, cognitions are more deeply informational (i.e., their status may be dramatically clarified by information-only approaches). Thus, the governance of mind and behavior is a multi-tiered, hierarchical process with several distinct types of mechanisms – some more instinctual and built into the system by evolution, and the other more flexible and deliberative, relying heavily on learning mechanisms.

State functions vs. channel functions. Marcel Mesulam (2000) highlighted that some aspects of the brain operate via discrete information channels (e.g., sensory-perceptual processes) while others operate more globally and control wide swaths of brain activity (e.g., obvious examples are the biogenic amine transmitters, which act as brain-wide "spritzers"). These amines include norepinephrine, dopamine and serotonin. They regulate neuronal arousability in global ways. At present, those systems represent the best understood global state controls, but they are rather non-specific in terms of discrete emotions. In line with such a distinction, we envision emotional operating systems to consist of large ensembles of neurons working together, under the "symphonic" control of neuropeptides to produce coherent organic *pressures for action, thought and feelings* (from such a conception, an "energy" dimension for the core emotional state control systems is apt, and is supported by various affective neuroscience research strategies, especially neuropeptide regulation of such global processes - Panksepp, 1998a; Panksepp & Harro, 2004). In other words, many neuropeptides operate almost like "orchestral conductors" to arouse the integrated network dynamics of discrete emotions and motivations. This distinction can help us conceptually differentiate brain processes that produce highly resolved perceptual qualia from aspects that are less sensorially distinct, holistic-energetic, and in the category of raw feels. The global *state-patterns* elaborated by such brain networks may generate an essential psychoneural context for perceptual consciousness - establishing a solid organic grounding for specific cognitive mental activities linked to discrete information-processing channels.

Computational vs. non-computational forms of consciousness. The computational view claims that channel-functions, since they are dependent on the coding of neuronal firing patterns in anatomically delimited channels, may be instantiated using arbitrary (platform-independent) symbol-manipulating models of mind. On the other hand, the more organically instantiated forms of affective consciousness, although also dependent on neuronal systems, are not computational in the same sense. These systems depend on extensive networks in which the patterns of neuronal firings do not convey discreet information, but rather ensembles of neurons develop analog pressures within the brain/mind, creating certain types of holistic feelings and actions. We believe that these global state envelopes, most evident in the instinctual emotional urges of animals, are critically dependent on not only a variety of generalized multi-functional chemistries (e.g., amino acids and biogenic amines) but also emotion and motivation specific neurochemistries (e.g., endogenous pleasure opioids, cholecystokinin, corticotropin releasing factor, dynorphins, orexin, neurotensin, substance P, etc.,) that are emerging as promising targets for new psychiatric drug development (Panksepp & Harro, 2004).

Intentions-in-action vs. *intentions-to-act*. During mind-brain evolution, the state-control systems of the brain help to establish embodied instinctual behavioral patterns along with internally experienced affective states (Panksepp, 1998a,b). These instinctual arousals constitute ancient psycho-behavioral controls allowing fundamental forms of cognitive activity and intentionality to emerge as an intrinsic part of the action apparatus. This, we believe, is what John Searle (1983) was referring to in his classic distinction between *intention in action* (e.g., hitting when angry; freezing when scared) and *intentions to act* (e.g., jumping off a diving board; turning on a car). It is only with a more differentiated sensory-perceptual cognitive apparatus, such as that which emerged with higher cortical encephalization, that certain organisms can operate in a virtual reality of cognitive-type activities, and thereby select and generate more deliberative behavioral choices (i.e., intentions to act) based upon the nuances of their perceptual fields.

Action-to-perception processes vs. *perception-to-action processes*. This distinction overlaps with the previous one. It assumes that the affective/emotional state-control systems help direct and focus specific attentional-perceptual fields. Only with the emergence of cognitive neural mechanisms capable of resolving highly detailed perceptual fields did the possibility emerge of buffering decision-making by higher executive processes (largely through the working-memory capacities of frontal lobes). Thus, sensorial awareness was transformed into perceptual guidance devices, permitting higher-order cognitive actions by organisms (yielding eventually the widely accepted perception-to-action processes, which are closely related to intentions-to-act processes). If ancient action-to-perception processes are fundamental for affective experience and primary-process intentionality (i.e., intentions-in-action), then we may be better able to generate some general principles whereby emotion-cognition interactions transpire in the brain (i.e., indeed, perceptions and cognitions may be strongly linked to the field dynamics of emotional systems).

Neurochemical codes vs. *general glutamatergic computations*. Neuroscientists have long recognized that a distinction needs to be made between the rapidly acting neurotransmitters that directly generate action potentials (with glutamate being the prime example of an excitatory transmitter), and those neuromodulatory influences which bias how effectively the rapidly acting transmitters operate (with the abundant neuropeptides being prime examples of neuromodulators that may regulate emotionally and motivationally specific state-variables in widely-ramifying neural networks). In other words, there are distinct neurochemical codes for the core states of the nervous system, and a better understanding of these codes will allow us to better conceptualize how *state* functions become dynamic governors of *channel* functions, with the former weighing more heavily in generating distinct forms of affective consciousness and the latter being more important in resolving cognitive details. For instance, there will be neuropeptides that are able to conduct a neurotransmitter orchestra into new global field-dynamics (Panksepp, 2005a;

Panksepp & Harro, 2004). In this view, one can conceptualize how affective and cognitive processes are remarkably interpenetrant, while at the same time recognizing that the conceptual distinction can help guide both neuroscientific and psychosocial inquiries.

2.2.8   Evolution of the Brain

In sum, mammalian brains reflect layers of neuronal evolution. Although some still hold to the view that such layering does not exist, with there being simply a unified pulsating organ of mind. This is a fantasy. The simple fact that one can eliminate the whole neocortex of most mammals at birth, and still have very sophisticated emotional creatures (e.g., ones that play almost normally; see Panksepp, et al., 1994), highlights a key dimension of this layering. Organisimic emotional and motivational coherence is unambiguously dependent on the lower reaches of the brain – at a level where animals are not very bright cognitively. Does the evolutionary layering of brain and mind need to be simulated in any virtual machine that has a chance of simulating existing living minds? I suspect that an understanding of such layering will be very useful for realistic simulations. If so, the first order of business is to try to construct primitive raw affects, and emotional values and behaviors into the system. This is a tall order. It is clear that our higher cognitive abilities, emerge from higher neocortical brain regions, but unless those processes are grounded in biological-type values, it may be that any realistic simulation of the living process will simply not be possible.

There are also good reasons to argue that all cognitive decision-making is tethered to the nature of our biological values. Those who do not spend much effort on coordinating these levels, but move rapidly toward information-processing solutions of cognitive processes, will have virtual machine that remain ungrounded in biological-survival realities. Of course, superficial "motivations" are easy to build into information-processing machines. But are simulations of simple energy-hunger and lubrication-thirst type mechanisms, easily modeled, really dealing adequately with the foundational role of homeostasis in all aspects of living and decision-making? I don't think so. Homeostasis penetrates the living brain at all levels, and it is very clear from neuroscience, that the highest parts of the brain simply have no capacity for consciousness without the lower, cognitively less able parts of the brain. Thus, as a first-order simplification, it may be worth envisioning the lower regions of the brain as being constituted of various genetically constructed ROMs, with the higher regions being more general-purpose computational spaces with remarkable similarities across organisms to RAMs; cortical columns, the highly repetitive infrastructure of neocortices, do resemble memory chips. From a genetic perspective, it is easy to add more, allowing more sophistication for cognitive virtual machines. Within the living brain, the intervening basal ganglia processes integrate information from the ROMs and RAMs, generating new *habit structures* or Programmable Read Only Memories (PROMs).

In any event, the lower brain is full of attentional, emotional and motivational state functions that are foundational for all cognitive processing. The same amount of damage in subcortical regions of the brain is much more devastating for continued mentation than comparable damage to higher regions of the brain where the details of cognitive activities are computed. Indeed, no thought ever leaves the brain to guide the body unless it has been repeatedly massaged by the primitive processes of brain regions such as the basal ganglia. Is it not important to understand such processes if we are seeking to simulate "real" mental life? The bottom line of mind organization is that value structures, emotional, homeostatic and sensory affects, were built into the brain very early in brain evolution, providing apparently, a raw, non-reflective, phenomenal consciousness. A good start for the most universal of emotional systems, the general purpose SEEKING system, is our recent updating in Alcaro, et al., (2007).

These lower regions interact massively with higher brain functions, and they are absolutely essential for those higher functions to operate in living organisms. It is a reasonable data-based conjecture that the lower brain functions are made of different kinds of neuronal processes than the cognitions and decision making processes, and may be well simulated by information-processing metaphors. Can those ancient mind-brain functions be neglected in any attempt at a "real" simulation of mental life? Although many in the AI community have remained silent about such issue, they can only do so if they prioritize information-processing as the core aspect of mind. In my estimation, the most important question in all of mind science, necessary for progress at other levels, is a clearer delineation of how the more ancient and foundational regions of brains create feelings. If you don't have the foundations engineered correctly, the rest may be fatally flawed.

2.2.9   Different Emotion Theories

This is not to say that my views on emotions and affects are the only ones, or the correct ones. I would simply claim they incorporate more of the existing neuroscience evidence than other current ideas.

At present, beside the affective neuroscience view summarized above, there are two other major brain views. The classic behavioristic view, very much alive, is that affects are created by reinforcement contingencies, and only reach consciousness (in this case, the term seems to be used synonymously with "awareness") when re-represented in language (Rolls, 2005). According to this view, animals have no feelings, and deaf and mute philosophers feel no aching teeth or empty stomachs. In my estimation, this is just a continuation of the methodological triumph, and intellectual tragedy, foisted on psychology by never-"mind" behaviorism – where emotions were mere fictional causes of behavior. This view was empty at its core because there was no clear conception of the raw capacities of organisms derived from evolutionary descent. Organisms were simply envisioned as information-processing devices, a view that may have also contributed to the rise of modern cognitive computationalism.

Damasio (1994, 2003), has advanced a modernized James-Lange type of peripheral feedback theory of emotions, where bodily commotions are read-out of feeling by the higher reaches of the brain (i.e., the somatic marker hypothesis).

This view has never been consistent with the mass of evidence concerning basic emotional circuits of the mammalian brain, revealed through brain research from Walter Cannon and Walter Hess (1957) onward, but it again reflects how modern psychology can get totally magnetized by a half-truth, while almost neglecting the robust and detailed neuroscience data from our fellow mammals (Panksepp, 1998a). The somatic-marker view, based on correlative psychophysiological recordings (e.g., skin conductance) has been negated by abundant human data (e.g., most recently by Hinson, et al., 2006). Indeed, one of the very best attempts to image human emotional affects (happiness, sadness, fear and anger) strongly indicated that such human feelings emerge from subcortical brain system already explicated through cross-species affective neuroscience research (Damasio, et al., 2000). One of the remarkable features of higher (cortical) and lower (subcortical) brains, is that they seem to be reciprocally related. High neocortical activity, especially of frontal executive regions, can suppress the influence of primary-process subcortical emotional systems; when one feels emotions intensely, this seems only possible when there is diminished neocortical information processing (Liotti & Panksepp, 2004; Panksepp, 2003).

The cognitivists have done little yet to clarify the nature of affect in the brain. Most of their brain-imaging results are more relevant for the information-processing changes that accompany emotional arousal. Their animal behavioral neuroscience expatriates, such as LeDoux (1996) have been explicit about the supposed epiphenomenal status of emotional feelings – mere "icing on the cake" (p. 302) – as they have proceeded to magnificently clarify how the amygdala participates in fear learning and perceptions of anxious stimuli. This important work has little relevance for the understanding the primal nature of affective values in brain organization. Their denial of affect promotes a ruthless reductionism which leaves no room for those network functions of the brain that help generate primary-process experiential states, ranging from the exhilaration of joy to the agony of pain. The quest to understand the biology of the neural "soul" (the core SELF) is a worthy enterprise (Panksepp, 1998b; Northoff, et al., 2006); it will require more attention to the ancient emotional functions that we share with the other mammals than the remarkable cognitive abilities that are uniquely human. We are inheritors of mental structures that appear to be constituted of much more than "information processing". Raw phenomenal feelings of various forms of goodness and badness arise from very ancient regions of the brain. From this vantage, an understanding of higher cognitive processes is the easier part of the "hard problem."

Do affective mind processes need to be understood neurobiologically before we can attempt to simulate true mentality in robots? I think so. And a fundamental principle may be that the higher reaches of the brain (subserving cognitive activities) may be more readily computed with available, platform-independent algorithms, than the lower substrates which would require types of non-linear dynamics, strongly linked to the power of brain molecules to create large-scale state-spaces, than simply modulating the frequencies of action potentials. In other words, the subcortial reticular circuits are much more complex than the repetitive columnar (RAM-like) *general-purpose* learning structures of neocortex upon which experiences write life-stories. As noted, even the visual cortex is not genetically devoted for vision, for if it is eliminate *in utero*, before birth, a fine visual cortex emerges epigenetically because subcortical visual information eagerly searches for a home in which individual perceptions are readily computed (Sur & Leamey, 2001). In short, much of what the cortex learns, perhaps even the capacity for language, is constrained by subcortical function.

Unless there is a major conceptual breakthrough in AI, this may be a lasting "brick wall" in attempts to generate credible simulations of mind on traditional silicone platforms. How much do we still need to know about the mammalian brain before we can make *real*, as opposed to surface, progress on the affective issues? An enormous amount! Cognitive functions may remain easier to compute for the foreseeable future. Those functions are much closer to traditional logical and algorithmic problems, presumably readily solved by a brain-free cognitive science with little help from neuroscience. That is not the case when it comes to the affective foundations for living. Those mysteries cannot be solved with flow-diagrams. They require a fuller understanding of the neurobiologically generated large-network neurodynamics that generate raw affective experience. How might one compute real pain? Such problems need to be fully confronted by the AI community. I suspect AI cannot make any more progress on these issues than is being made, indeed first needs to be made, at the neuroscientific level.

2.2.10   The Problem of Other Minds

I will not spend much time on this critical issue, except to say that obviously there is a diversity of minds on the face of the earth. Many are less sophisticated than ours, and hence potentially easier to model. Others are more sophisticated in many poorly understood ways. For instance, new brain regions provide new functional capacities.

Dolphins and other *cetacea* have massive paralimbic cortices in brain regions where our anterior cingulate cortex is situated (Marino, et al., 2007). Cingulate cortex in humans and other mammals is devoted strongly to the generation of social emotions, ranging from feelings of separation-distress to maternal care, providing presumably one important substrate for empathy (Panksepp, 1998a). What might a creature with much more neural tissue in this brain region (which receives input from subcortical emotional systems) be capable of in the social realm? Might they be able to read the emotions of others directly, using their sonar to read the internal autonomic activities of other dolphins? Even humans? We simply do not know, but such issues need to be always kept in mind when one contemplates simulating the emotional and cognitive abilities of living creatures that actually live on the earth.

In sum, the project to simulate mental life depends critically on whether one is aiming to have excellent simulacra or the real thing? It is currently impossible to be definitive about what the real thing is, but simulations need to have a clear vision of the true nature of mental apparatus. Perhaps psychoanalysis provided a more realistic vision than either the behaviorism or cognitivism of the past century, not to mention Evolutionary Psychological modularism of the past decades. Perhaps affective neuroscience has now provided the most realistic image of those aspects of the subcortical mind shared by all mammals.

2.2.11   Postlude – Emotional Turning Tests

There are currently many logical and reasonable analyses of how *cognitive* virtual machines need to be constructed. However, most of those visions need to struggle for a realistic confrontation with primary-process emotions and other affects – the real value systems – of organisms. These mind-brain functions appear to be foundational for the existence of a higher cognitive apparatus. There seems to be no way to finesse this neurobiological reality – a confrontation with those intrinsic, ancient evolutionary brain circuit functions responsible for the natural-instinctual behavior patterns, those intrinsic rough-and-ready tools for living, that evolution constructed firmly into our brains and minds. Affective feelings arise from such natural emotional systems. These aspects of mind cannot be ignored or trivialized if we wish to simulate the human mental apparatus. Our affective experiences give meaning (i.e., intrinsic values) to life. We can encapsulate the whole dilemma, of such mental functions, for future work on AI and robotics by asking: *How would you compute the feeling of an orgasm?*

Our orgasms evolved long before our ability to negotiate the social-cognitive terrain in complex information-processing ways. All normal humans value high-quality orgasms. To our knowledge no computer-based virtual machine or robot does, and it is hard, but perhaps not impossible for some, to imagine how to build such real experiences into man-made computational devices. I have no idea how one might compute an orgasm, nor have I heard anyone put a workable idea on the table. What could ever inform us that a computer was capable of experienced feelings? It would be easy to make them shake-and-shiver after a certain amount of thwarting or artificial skin stimulation, but would that be the real thing? How would one know? In order to evaluate for the presence of an affectively resilient mind in a robot, we would need new kinds of Turing tests that would convince us of the ability of robots to report human-like affective experiences (Panksepp, 2000).

In any event, all of us can report high-quality feelings with little trouble; no one has any problems discriminating exquisite orgasms from the spasms of nausea. Most of us can also easily distinguish fine wine from swill. Most of us can luxuriate in a fine foot or body massage. These are the kinds of endpoints we need to construct in our robots before we should have any confidence in their livingness. Does this mean that we have to build sensitive tongues, feet and bodies into our robots and see if they can discriminate the above experiences as well as humans?

Those are the kinds of processes we need to evaluate in any credible Turing Test we might devise for the existence of an affective life (Panksepp, 2000). Simulation of cognitive decision making capacities will not be enough. Probably more of our cognitive activity is unconscious than our affective life. If we want to

174

challenge ourselves with the problem of simulating mental life within a robot, we really have no option but to come to terms with the affective nature of human and animal mentality. Without that, all may be smoke and mirrors, and an endless flow of simulacra. Of course, it could be that after a certain amount of information processing, a robot would have feelings that are unique to robots, but those matters are bound to remain arcane forever.

At the same time, let me not be coy. I suspect many of us might treasure time spent with a truly bright and sympathetic, good-looking, robot that had been well programmed to be interested in us as a people, that ("who"?) knew how to alleviate our loneliness and to share in our interests and sense of human companionship. This is a worthy project even if it is improbable, in principle, to build real affects into virtual machines.

Because of the biochemical complexity of the platforms for our lives, perhaps the dream of mechanical life will forever remain a dream. Still, there is much room for innovation – after all a feeling is *just* an incredibly complex physiochemical dynamic within the brain. What if we consider bio-robotics, perhaps using pluripotent stem cells created by nuclear transfer to generate cloned neuronal tissues, perhaps even tissue that has the capacity to specialize into a fully organized brain? What if we cultured such brains in the imaginative philosopher's vats, to adult proportions? What would be the mental competencies of such organs of mind? Would they be able to have experiences and skills, without ever having had bodies with which to explore the world, without ever having had rich environments with which to interact? Maybe the path to "success" will be first achieved through "biobrains" wired into machine bodies that have reciprocating sensors to feed information about their inner and outer states back to the master routines instantiated in real neural tissues? Overall, such projects may be more technologically feasible than the more far-fetched computational simulation of the affective-experiential richness of mind within a silicone platform. If someone where able to succeed, even partly, we would be well advised to contemplate aspects of such enterprises that have not received sufficient attention here – cyborg simulacra as embodiments of evil, should continue to haunt us as long as we continue to explore such radical intellectual possibilities.

References

Alcaro, A., Huber, R. & Panksepp, J. (2007). Beheavioral functions of the mesolimbic dopamienrgic system: An Affective Neuroethological perspective. *Brain Research Reviews,* In Press.

Aydede, M. (ed.) (2006). *Pain: New Essays on Its Nature & the Methodology of Its Study.* M. Aydede (ed.) The MIT Press, Cambridge, MA.

Barrett, L.F. (2006). Are emotions natural kinds? *Perspectives on Psychological Science,* 1, 28-58.

Brand, P. & Yancey, P. (1997). *The Gift of Pain.* Zonervan Publishing House, Grand Rapids, MI.

Byrne, R.M.J. (2005). *The Rational Imagination.* MIT Press, Cambridge, Mass.

Cannon, W.B. (1927). The James-Lange theory of emotions: A critical examination and an alternative theory. *American Journal of Psychology,* 39, 106-124.

Camille, N. Coricelli, G. Sallet, J. Pradat-Diehl, P., Duhamel, J. & Sirigu, A. (2004). The involvement of the orbitofrontal cortex in the experience of regret. Science, 304(5674), 1167-1170.

Ciompi, L. & Panksepp, J. (2004). Energetic effects of emotions on cognitions – complementary psychobiological and psychosocial finding. In R. Ellis & N. Newton (eds). *Consciousness & Emotions, Vol. 1. (*pp. 23-55), John Benjamins, Amsterdam.

Damasio, A.R. (1994*). Descartes' Error,* New York: G.P. Putnam's Sons.

Damasio, A.R. (2003). *Looking for Spinoza.* Orlando, Fl: Harcourt Brace.

Damasio, A.R., Grabowski, T.J., Bechara, A., Damasio, H., Ponto, L.L.B., Parvizi, J. & Hichwa, R.D. (2000). Subcortical and cortical brain activity during the feeling of self-generated emotions. *Nature Neuroscience,* 3: 1049-1056.

Denton, D. (2006). *The primordial emotions: the dawning of consciusness.* New York: Oxford University Press.

Dunn, B. D., Dalgleish, T., & Lawrence, A.D. (2006). The somatic marker hypothesis: A critical evaluation. Neuroscience and Biobehavioral Reviews, 30(2): 239-271.

Eccleston, C. & Crombez, G. (1999). Pain demands attention: a cognitive-affective model of the interruptive function of pain. Psychological Bulletin, 125(3): 356-366

Greicius, M. D. Krasnow, B., Reiss, A. L., Menon, V. (2003). Functional connectivity in the resting brain: A network analysis of the default mode hypothesis. Proceedings of the National Academy of Science, 100, 253-258.

Hess, W.R. (1957). *The Functional Organization of the Diencephalon.* Grune & Stratton: New York.

Hinson, J.M., Whitney, P., Holben, H., & Wirick, A.K. (2006). Affective biasing of choices in gambling task decision making. *Cognitive, Affective & Behavioral Neurosciences*, 6, 190-200.

LeDoux, J.E. (1996). *The Emotional Brain,* New York: Simon & Schuster.

Liotti, M., & Panksepp, J. (2004). On the neural nature of human emotions and implications for biological psychiatry. In Panksepp J (ed) *Textbook of Biological Psychiatry,* pp. 33-74. Wiley, New York.

Marino, L., Connor, R.C., Fordyce,R.E., Herman, L.M., Hof, P.R., Lefebvre, L., Lusseau, D., McCowan, B., Nimchinsky, E.A., Pack, A.A., Rendell, L., Reidenberg, J.S., Reiss, D., Uhen, M.D., Van der Gucht, E., & Whitehead, H. (2007). Cetaceans have complex brains for complex cognition. *PloS Biology,* 5, 966-972.

Meaney, M. J. (2001). Maternal care, gene expression, and the transmission of individual differences in stress reactivity across generations. *Annual Review of Neuroscience,* 24, 1161-1192.

Merker, B. (2007). Consciousness without a cerebral cortex: a challenge for neuroscience and medicine. *Behavioral and Brain Sciences,* In Press.

Mesulam, M. (2000). *Principles of behavior and cognitive neurology.* 2nd ed. New York: Oxford University Press.

Northoff, G., Heinzel, A., de Greck, M., Bermpohl, F. & Panksepp, J. (2006). Our brain and its self – the central role of cortical midline structures, *NeuroImage,* 15, 440-457.

Panksepp, J. (1974). Hypothalamic regulation on energy balance and feeding behavior. *Federation Proceedings,* 33, 1150-1165.

Panksepp, J. (1982). Toward a general psychobiological theory of emotions. *The Behavioral and Brain Sciences,* 5, 407-467.

Panksepp, J. (1986a). The anatomy of emotions. In R. Plutchik (Ed.) *Emotion: Theory, Research and Experience Vol. III. Biological Foundations of Emotions* (pp. 91-124). Orlando: Academic Press.

Panksepp, J. (1998a). *Affective neuroscience: The foundations of human and animal emotions.* New York: Oxford University Press.

Panksepp, J. (1998b). The periconscious substrates of consciousness: Affective states and the evolutionary origins of the SELF. *Journal of Consciousness Studies,* 5: 566-582.

Panksepp J. (2000). Affective consciousness and the instinctual motor system: the neural sources of sadness and joy. In: Ellis R, Newton N, eds. *The Caldron of Consciousness: Motivation, Affect and Self-organization, Advances in Consciousness Research.* (pp. 27-54), Amsterdam: John Benjamins Pub. Co.

Panksepp, J. (2003). At the interface of affective, behavioral and cognitive neurosciences: Decoding the emotional feelings of the brain. *Brain and Cognition. 52,* 4-14.

Panksepp, J. (2005a). Affective consciousness: Core emotional feelings in animals and humans. *Consciousness & Cognition,* 14, 19-69.

Panksepp, J. (2005b). On the embodied neural nature of core emotional affects. *Journal of Consciousness Studies,* 12, 161-187.

Panksepp, J. (2007a). The neuroevolutionary and neuroaffective psychobiology of the prosocial brain. In: R.I.M. Dunbar and L. Barrett (eds.) *The Oxford Handbook of Evolutionary Psychology.* (pp. 1455-162). Oxford University Press, Oxford, UK.

Panksepp, J. (2007b). Affective consciousness. In: M. Velmans and S.Schneider (eds.) *The Blackwell Companion to Consciousness.* (pp. 114-129). Blackwell Publishing, Oxford, UK.

Panksepp, J. (2007c). Neurologizing the psychology of affects: How appraisal-based constructivism and basic emotion theory can co-exist. *Perspectives on Psychological Science, 2,* In press.

Panksepp, J. & Harro, J. (2004). Future prospects in psychopharmacology. In. J. Panksepp (ed) *Textbook of Biological Psychiatry,* (pp. 627-660) Wiley:Hoboken, NJ.

Panksepp, J., Moskal, J. (2007). Dopamine, and SEEKING: Subcortical "reward" systems and appetitive urges. In Handbook of approach and avoidance motivation, (pp. 281-296) A. Elliot, ed. Lawrence Erlbaum Associates, Mahwah, NJ, in press.

Panksepp, J., Normansell, L. A., Cox, J.F. and Siviy, S. (1994). Effects of neonatal decortication on the social play of juvenile rats. *Physiology & Behavior, 56,* 429-443.

Panksepp, J. & Northoff, G. (2008). The trans–species core self: The emergence of active cultural and neuro-ecological agents through self related processing within subcortical-cortical midline networks. *Consciousness & Cognition,* In press.

Picard, R. W. (1997). Affective computing, Cambridge, Mass. : MIT Press.

Rolls, E.T. (2005). *Emotions explained.* Oxford, UK: Oxford University Press.

Searle, J.R. (1983). *Intentionality: An essay in the philosophy of mind.* New York: Cambridge Univ. Press.

Sloman, A. (1982). Towards a grammar of emotions. *New Universities Quarterly, 36,* 230-238.

Sunderland, M. (2006). *The science of parenting.* London: DK Publishing Inc.

Sur, M. & Leamey, C.A. (2001). Development and plasticity of cortical areas and networks. *Nature Reviews Neuroscience, 2,* 251-262.

Tomb, I, Hauser, M., Deldin, P. & Caramazza, A. (2002). Do somatic markers mediate decisions on the gambling task? Nature Neuroscience, 5, 1103-1104.

Velmans, M. & Schneider, S. (eds.) (2007). *The Blackwell Companion to Consciousness.* Blackwell Publishing, Oxford, UK.

Zeelenberg, M. , van Dijk, W., van der Plight, J. Manstead, A., van Empelen, P. Reinderman, D. (1998). Emotional reactions to the outcomes of decisions: The role of counterfactual thought in the experience of regret and disappointment. Organizational Behavior and Human Decision Processes, 75, 117-141.

Zzyf, M., McGowan, P. & Meaney, M.J. (2008). The social environment and the epigenome. *Environmental and Molecular Mutagenesis.* 49, 46-60.

2.3   Discussion Chaired by Authors

In the following, the most important questions and comments of the discussion after the paper presentations of Aaron Sloman and Jaak Panksepp at the ENF 2007 are summarized. In this context, "Q" is used for any type of question or comment, usually from the auditorium, while "A" stands for "answer", usually from the chairs.

Q1:   J. Panksepp argued that it is not possible to simulate hormones in a computer. But as hormones are actuated, it should very well be possible, simply by saying that neurons are one information part, and hormones another.

A1:   This is true. What is missing is the powerful energy concept built into the computer as an organismic foundation. Additionally, there is not only one state but various state shifts, and large networks. It is an enormous challenge to build these in. And considering the biochemical complexity of life, ultimately this might only be a simulacrum.

Q2:   Where is the difference between Panksepp's attributing affective states to his rats and Sloman's attributing affective states to his bubbles?

A2.1:   It is predictive validity. Human medicine has always relied on studying the physiology and the biochemistry of animals. As soon as it is possible to understand these chemistry, it will be possible to make predictions to human affective experience. But computers will have to prove that they can make qualitatively similar predictions.

A2.2:   One difference is that it is possible to know what is going on inside the computer. The second one is that it really is very simple, so there is no actual intention to say that this is happiness or sadness as in humans.

178

3 Session 3

3.1 Cognitive and Affective Automation: Machines Using the Psychoanalytic Model of the Human Mind

Peter Palensky, Brigitte Palensky (née Lorenz) and Andrea Clarici[34]

Automation systems of the future are envisioned as dealing with massive amounts of potentially unreliable data and expected to distill reliable information and decisions out of that. Classical mathematics - filters or correlations - and classical artificial intelligence (AI) – such as artificial neural networks or intelligent agents - can substantially help to deal with this problem. There are, however, examples in nature that are capable of focusing on just the right thing in just the right moment, even when the system is overloaded with useless other information: higher living creatures, especially humans.

It is the goal of this paper to follow a bionic path towards an intelligent automation system. The chosen approach tries to combine the latest findings of neurology, psychoanalysis and computer engineering to create an artifact, capable of perceiving, storing, remembering, evaluating and recognizing situations and scenarios, consisting of an arbitrary number of sensor information.

3.1.1 Introduction

Automation is the art of designing and running a technical system that – on behalf or instead of somebody – optimizes a certain process. This abstract definition holds for all applications and technologies so far. Initially, mechanical automata like automatic looms were introduced to perform repetitive tasks. With advanced mechanics and, later, pneumatic, hydraulic, and electromechanical components, the complexity and flexibility of such machines grew, essentially contributing to the industrial revolution.

Soon, the working of such machines was decomposed into the well known INPUT, PROCESSING, and OUTPUT chain (Fig. 3.1.1). From the 1950s on, the materialistic view of such systems was changed into a more abstract one where raw materials and finished goods were replaced by input data and output data. Processing matter gradually was supported by, or predominantly turned into, processing information. The link between the world of information and the real world was then – and is now – established via sensors and actuators.

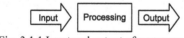

Fig. 3.1.1 Input and output of a process

---

[34] The authors would like to thank the Consorzio per lo Sviluppo Internazionale dell'Universita di Trieste, and the Foundation Kathleen Foreman-Casali (Trieste, Italy), the Fondo Trieste (Provincial Government of Trieste) and Georg Fodor and Dorothee Dietrich (Psychoanalysts, Vienna) and the entire ARS-PA team for their generous support in this research.

The advent of computers is inherently connected with the change of focus from processing matter to processing information. Computers were – up to the 1940s – specialized machines, dedicatedly constructed to perform *one* particular task. All data and relevant information was converted and coded into numbers (typically binary ones) that the computer performed its operations on (in the form of physical processes). John von Neumann and Alan Turing introduced a revolutionary view of such machines: Not only INPUT and OUTPUT are described as numbers, but *also the rules of PROCESSING*. This resulted in the concept of machines capable of performing every (algorithmic) job that exists (given unlimited time and memory). Such machines are decomposed into the following two components:

- a storage – the memory – that holds the input data and the rules of processing (the computer program) in the form of numbers, and
- a universal machine that neutrally executes on the given input what is described by the program to produce the desired output.

Every contemporary computer is a universal machine in the above sense, and it is not possible to build a computing machine which is more powerful than the simple universal machine described by Turing. A consequence out of the above conception is that any given (universal) computer can emulate any other computer. Technically speaking, this means that an Intel-based personal computer (PC) can fully emulate a Motorola-based Apple computer, and vice versa. Here, "fully" means precisely 100%! Every single aspect of one machine can be imitated by another one. Take the case of a Java virtual machine.[35]. Here, emulation is a permanent situation, and the structure of the real machine is completely hidden and unknown. In sum, a universal machine cannot only run arbitrary programs, it can also run other universal machines.

For traditional automation tasks, this universal computing power is not really necessary. The control module of a welding robot performs a repetitive task; if a bit smarter, it recognizes twisted workpieces with sensors or cameras. Most automation systems execute some repetitive sequence of tasks, and react on sensory inputs and events. For such systems, dedicated computers and specialized languages were developed, easing the specification of what to do, where, and when.

Flexible and universal systems are not always successful. For describing what the system should do, one fundamental thing is necessary: a model of the process and its solution. Describing the task of welding is a simple 3D-problem, controlling the temperature in a car engine leads to a model based on differential equations, etc. Typically, it is possible to describe a sufficient model of the process, but sometimes it is not. Some jobs are simply too complex to be described in a consistent and clear-cut model based on some formal language. Such processes range from image recognition up to climate predictions. This is one situation where engineers apply methods of artificial intelligence.

---

[35] An artificially designed computer, typically implemented in software and run on a physical computer

180

An artificial neural network (ANN) for instance can be trained to recognize certain implicit patterns. This recognition might be way too complex to be described in a model, but the ANN is capable of doing it. After a certain training phase, it apparently "guesses" the right answer pretty often. The particular problem of pattern recognition is solved in this case, but the problem is neither understood nor solved in a general manner. A nice-to-have thing in this context would be to add given heuristics to the ANN. If it is for instance obvious that a blue spot on an image always means something defined, it would be good to share this knowledge with the ANN. But since we do not know what the internal structures (the weights and connections of the artificial neurons) mean, we do not know where to inject this knowledge. Such a hybrid heuristic/blind design of an ANN is surely possible but is often pretty inelegant.

Note, that the function of a neural network can be imitated by a universal machine. Even if the concept of an ANN is not sufficient to solve our future problems, this gives hope that our universal machines can implement and execute everything that we might invent.

In future, automation systems will increasingly deal with complex and fuzzy environments where no consistent model of the mechanisms at work is possible. There might be rules on what is good and what is bad, but the designers of such systems can never oversee and describe everything. Important aspects might be completely unknown, hidden correlations can lead to unexpected behavior, and some rules might even change over time: things that a static program, however smart it is, cannot cope with. All this is spiced with a flood of input data, where no heuristic program can tell what is important and what is noise.

Think of an evacuation system for a big railway station. It is virtually impossible to consider all situations that might happen in order to define a consistent decision tree for an automatic evacuation system. Currently it is always necessary that people support such a system by judging the situation, by making the right decisions. An automatic system would need millions of sensors to determine who is where and to be sure which door to close and which to open in case of fire. Additionally, sensors and network infrastructure might be damaged in such an event, making reliable information even harder to get. A traditional computer program, even if enhanced by the very useful tools of artificial intelligence, cannot solve this problem in a satisfactory way. This article discusses, that the desired "intelligence" very much depends on "soft things" like e.g. emotions. A claim that was also described in [1] by proposing affective architecture for everyday machines like computer servers. These affective machines would be anxious if something is going wrong (or if it looks like this could happen soon). A machine calling for help before breaking, is obviously better, the question is how does a machine figure out the right level of "fear"?

The following list summarizes the troubles that future applications are supposed to be confronted with:
- Complex processes: no explicit model
- Dynamic processes: changing characteristics

- Data sources: inaccurate or even wrong pieces of data are included
- Large amounts of data: relevant aspects are hidden
- Meaning of data: chunks of information have to be combined to get meaningful knowledge

Applications where some, if not all, of these aspects turn up can be found for example in traffic management, vehicle control, industrial automation, complex process industry, building automation, public security, or disaster management.

The bionic approach to meet the above challenges is to take a natural system, capable of managing such complex tasks, as a hint on how to design a machine to do the same. The subject of our studies is nothing less prominent than the human mind.

The human mind – and this certainly applies to many higher animals as well – is capable of filtering relevant information, focusing on important details, estimating risks, anticipating results, and recognizing complex situations. Machines have replaced humans only in very simple jobs. Those jobs that require flexibility, comprehension, or creativity are still the domain of people. Often it is necessary to have people's judgment on-site, even in such unfriendly places as a collapsed coal mine 10 km underground, or a broken nuclear power station. Here, a smart machine, on the right place, making the right decisions to prevent a disaster, would help a lot.

Humans can deal with complex contexts, the question is: How do they do it? The mind is equipped with a number of "functional blocks", serving as memory, as fast lane for information, or as inhibition for others, making it a very flexible and universal machine. This machine can survive in a complex and dynamic environment that it does not fully understand. It is the goal of this paper to name these functional elements, to describe them, and to start a discussion on how they can be implemented in a technical system.

3.1.2 Related Work

In [2], Turkle describes the way of psychology. In the first half of the 20th century, the black-box methodology of the behaviorists was dominating. Then, from the 1960s on, psychological terminology was strongly influenced by a new discipline: artificial intelligence (AI). For computer scientists it was straightforward to postulate "inner states" and "competing and conflicting inner agents" – a language, psychologists had not dared to use for quite some time. This development made it again scientifically possible for them to look for the inner structure of the human mind, thereby allowing to overcome the black-box methodology.

Among the most prominent elements supposed to make up the inner structure of an "intelligent" being or artifact are memories, and mechanisms for decisions. In AI, it is distinguished between an

- explicit (i.e. symbolic), and an
- implicit (i.e. emergent)

representation of these ingredients. Traditional computer science naturally uses explicit algorithms and data structures. Creating an intelligent machine in this way, we could call it intelligent design: The creator knows how to do it. Opposed to this is emergent design. Here, functionality and data are distributed over a connectionist platform rather than localized in information-bearing chunks, providing only a *potential* for intelligence.

As described by Turkle, both of these two entirely different design methods would allow for some psychoanalytic interpretation [2]. For example, changing one single bit in symbolic design often leads to dramatic consequences when turning "yes" to "no", or "left" to "right". Such accidentally changed bits might be compared to Freudian slips. On the other hand, a different "feature" of the human mind can be found in emergent design: graceful degradation. Here, changing one bit leads to a blurry image, but not to the complete destruction of the image.

Similar questions and issues as in neurology, psychology, and/or psychoanalysis not only turn up when designing an intelligent computer, but already when designing an ordinary computer architecture. Two extreme examples of computer architectures are the von-Neumann architecture and the connection machine [3]. While the former, the popular von-Neumann machine, is clearly structured and explicit, the latter is of an emergent nature. Both, however, suffer from a semantic gap: High level data structures and functions are not recognizable or locatable on the hardware level. This is a striking similarity with the relationship between psychological concepts and neurons. Another example where similarities appear is connected with the role of time. Data might be seen as a (static) mental image or memory, while data processing reminds of thinking. It is this data processing, where traditional computer programs and the human mind differ greatly.

In a technical metaphor due to Minsky, mind is seen as a society of conflicting agents (censor agents, recognition agents, anger agents, etc.) [2], [4]. These agents are considered – as the name suggests – as pretty autonomous entities, competing and inhibiting each other. While the "agents" of the human mind are a result of evolution, stemming from different evolutionary eras of mankind, software agents [5] are typically designed by the programmer. There is, however, no reason against using evolutionary mechanisms in the design of such a multi-agent system (MAS). Although traditional programs do not even use the multi-agent paradigm, the design of truly intelligent machines will supposedly require such a conflicting structure.

Software agents in their original sense were not seen as competing entities, but rather as collaborative individuals: If a problem is decomposable or of distributed nature, let an army of cooperating software agents solve it in parallel [6]. Some MAS consist of a community of specialized agents [7], each of them an expert for a certain problem type. In the same manner, a MAS of expert systems or multiple expert ANNs can be used as an arsenal of problem-solving specialists.

The presented work decomposes the complexity of the human mind into a number of cooperating and conflicting functional modules, interacting and exchanging information. Thereby, a main source of inspiration comes from (a modern view on) psychoanalysis. To our knowledge, psychoanalysis has not been so far seriously considered in the Artificial Intelligence community, maybe with the exception of the work of A. Buller [8], [9].

3.1.3   Technical Problems

Once the model and all its interfaces are described, one could expect it an easy job to put this into software. Technically implementing psychological models, however, has a number of pitfalls. Some of them are a barrier to understanding and accepting such systems, others make the system designer question the entire idea.

Problem No. 1: Biologic Terminology

Parts and pieces of a technical implementation might use biologic metaphors. If the individual functional parts of a system use some sort of inter-process communication, the designer is tempted to speak of "nerves" or "exchange of hormones". We believe it admissible to use biologic metaphors, however, one should be aware that they do not suggest that the technical implementation works in the same way as the biological counterpart.

Problem No. 2: Anthropomorphic Aesthetics

Calling anthropomorphic hardware like KISMET [10] an "emotional" machine just because it expresses emotions is considered to be inappropriate, since the only emotion then lies in the eye of the observer. Unless a machine does not *have* emotions to organize its behavior, solely producing emotional expressions stays only a superficial feature, not exploiting the real functional power of the concept of emotion as described by biology and psychology.

Problem No. 3: Design versus Evolution

From Lady Ada Lovelace, the famous and curious character in the history of computer science, there is the saying: "Computers can only do what you tell them to do". Thus, she did not expect machines to be capable of learning by themselves. In our case, the question is how soon this learning should start:
  (a) During the design phase: Evolutionary design.
  (b) During an initializing training phase: The system is adapted to its specific environment.
  (c) During runtime: A learning machine.
  (d) Never: The designer is capable of creating an intelligent machine from scratch.

In case (a), we let the system grow, evolve and develop. It can for instance learn (by natural selection or some other mechanism) which types of memories it needs, or how many and which kinds of "drives" for achieving its tasks. Following this path, an intelligent car will very likely not develop the same mental structures as a human: It has other senses, needs, goals, etc. The presented project has chosen to take the human mind as blueprint, because in this case a lot of design hints are available, coming from objective observation, *and* from subjective introspection. Once this works, the ultimate goal would be to head for alternative architectures, maybe completely differently designed ones, or just differently initialized and parameterized ones. Case (d) is hypothetical: A dynamic environment does not respect a static solution. So the element of learning will come into effect, maybe in the very beginning, certainly during runtime. Again, the bionic design methodology leads to important questions, one being raised by Lurija [11]. In a simplified version: The perception of small children is believed to solely rely on the flow of input data, at the same time, this data contributes to the formation of the semantic memory. Grown-ups, on the other hand, usually do not directly process input data, they let it only "come" to their semantic and episodic memory where it is compared with already stored mental images. These images, in turn, form the basis for all further processing done by the other parts of the mind. From an implementation point of view, the question is: What factors determine when and how to switch (see Fig. 3.1.2)?

A technical system is generally developed by going through the following phases:

- Specification of requirements
- Design
- Implementation
- Test

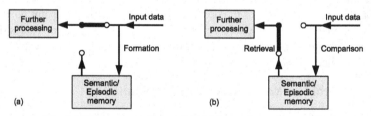

Fig. 3.1.2 Data flow of a small child (a), and a grown-up (b)

Problem No. 4: Design decisions

In the design phase, it is decided how a desired functionality should be realized. Sometimes there are multiple ways of implementation, sometimes there may be features not feasible to implement at all (although they might be easily described).

An example shall show, how design decisions can influence the nature and the performance of a system. Beside many other things – various types of memory have to be implemented when building a machine mimicking the human mind. Today, neurology can tell how nature does this. A Hebbian Neuron [12] learns (i.e. remembers and associates) by reinforcement, its architecture being based on feedback. Feedback is common to many (technical) elements that show memory-like behavior (e.g. flip-flops, integrators). Thus, if we want to implement Hebbian learning in a machine we have several entirely different ways to do this:

(a) Usage of hardware artificial neural networks (ANNs) to re-implement a Hebbian feedback

(b) Usage of software ANNs to re-implement a Hebbian feedback

(c) Usage of an ordinary digital memory cell plus some software-encoded timely behavior

All three ways of implementation can produce the behavior described in the specification, the level of abstraction, however, being different as can be. The level of abstraction, though, does make a difference in how easy it is to understand the system: An ANN is usually much more difficult to understand than a more explicit solution.

Problem No. 5: Implicit Functionality

Biological matter has much more side effects than what is its primary function. Modeling a gene by an integer number, a neural network by some matrix multiplication, or a semantic memory by a database might cover their main function, but leaves the side effects away. A side effect might be infinitesimally small at first, but this could change with time, or with a change of the size of the system.

Problem No. 6: Every Problem to be Solved by a Computer Must be Translated Into Algorithms and Numerical Data

Computer programs are algorithms, derived by decomposing complex problems into small steps, expressed via a formal language, and coded and stored as numerical data.

Originally, computer architecture was always tightly coupled to the nature of the given problem and to the available technology. The advent of the universal machines, based on digital microchips made an end to problem-oriented architecture: These machines are, by the very way they are defined, suitable to every problem type (posed as an algorithm to be performed on data). It may be speculated that the requirements of the presented cognitive architecture for autonomous systems might let this safe haven appear a bit limited.

It is clear that the model itself has to be described with traditional methods. Every functionality, every part and all their relations are described via data structures, numbers, modules, interfaces, mathematics, databases, and other familiar building blocks of IT. The choice of description languages is, as with every complex design, essential.

Although the model will be described with traditional methods, it may be the case that traditional programming languages will not allow a 100% successful implementation. Most programming languages still assume the von-Neumann architecture, an actually very easy and straight forward way of constructing a universal machine, its most remarkable trait being the clear distinction of data and program. However, we identify two problems with this architecture:

• Semantic gap
• Sequential architecture

The term "semantic gap" describes the fact that the programmer might use high level languages (HLLs) to express her problem and her solution, while the actual machine just understands a few, very simple low-level commands. Everything, described in a HLL, must be translated – by a compiler – into the above mentioned primitives. The processor has no chance to "understand" the given program, it just blindly executes the primitives. These primitives show no obvious structure, they are far away from the original problem: There is a gap. The problem of this gap is that the executing processor cannot recognize important structures and relations, what it would have to do in order to optimize its operation. Previously elegant structure is coded as a bulk of neutral primitives, and the executed primitives virtually cannot be understood by an observer (e.g. penetrating a computer with electric probes would not reveal the meaning of MP3 decoding).

The second disadvantage of the von-Neumann architecture is its sequential nature. Usually, these machines do not exploit potential parallelism in the problem or the solution. Modern computers do apply parallel aspects, either in the world of primitive macro instructions (out-of-order execution, VLIW, etc.) or – if supported by the programmer – via parallel programming languages. The latter is a step into the right direction, but the architecture of our choice would go way further.

As said in the beginning, the traditional way of implementing automata is to decompose its entire behavior into a set of algorithms and data structures. Computer science is entirely based upon these two elements. Virtually all contemporary automata are implemented in this way. There are, however, fundamental differences in how to get to such an implementation. Before being decomposed into unreadable bits and bytes, there is often a surprisingly easy and natural model about what is going on.

Fuzzy logic, and especially fuzzy control, is a perfect example for this. Once a problem is "fuzzificated" it is easy to handle, everything gets simple, and a solution is easy to find. Once a solution is found, it can be "de-fuzzificated" to affect reality. In the end, what you get as implementation of such a fuzzy control is a formula or a computer program that is just another mathematical transformation, indistinguishable from any traditional control transformation. The difference lies in the model. The fuzzy model is easier to comprehend and to handle, although it finally leads again to a (mathematical) algorithm. Another example is object-oriented design (OOD)[36]. As our world consists of objects and their relations, it is just natural for programmers to represent these things in an appropriate manner. OOD programs are typically well structured and less error-prone. But once compiled, they are again sequential code blocks, like any other program.

We face a similar problem, implementing psychological models for autonomous machines: In the end, everything is just a program. The difference comes with the used model. A module, responsible of deciding something, can be implemented as a rule-based decision tree, or as a set of dynamic, contradicting and

---

[36] A maybe even better example might be agent-oriented design.

conflicting "censors" [13], but finally both are just a program. The question is much more: Which model offers

- the desired functionality with
- the easiest description,
- the clearest structure,
- the best interfaces to the environment,
- the best adaptive power, and
- the highest implementation potential.

The implementation potential very much depends on the available technologies. It is a big step going away from von-Neumann to dataflow architectures, and an even larger one to leave digital computers at all and to use for instance analogue electronic ANNs [14], [15], or optic ANNs [16], [17]. This decision has to be done one day, for now we focus on the remaining requirements.

3.1.4 Psychoanalytic Principles and Models

The construction of an artificial mind designed on the basis of a psychoanalytic model should take as its starting point the definition of the basic mental processes and of the ways the psychic apparatus functions, in accordance with the specific features of the investigation of the functioning of the mind known as psychoanalysis.

The problem of the relationship between the mind's structural, psychological, anatomical, and functional components has always fascinated both neuroscientists (see [18], [19]) and psychoanalysts, starting with Freud himself [20]. The study of the memory, the "reminiscences", of his first hysterical patients, the nature of their amnesia, and the part played by psychopathogenic factors marked the starting point of the psychoanalytic exploration of the mind [21].

The aim of structuring the mind along the lines of a psychoanalytical model immediately comes up against a series of problems of various kinds. The first of these is that to date there is no standard psychoanalytical model shared by all psychoanalysts covering the functioning of the mind. As is well known, even Freud provided at least three different models of the functioning of the mind during the course of his scientific career and these models were based on three very different frames of reference for the psychic apparatus.

A. Psychoanalytic Principles

In view of the myriad psychoanalytical characterizations of the functioning of the psychic apparatus, what are the *unique characteristics* of psychoanalysis, namely the elements shared by the various schools of psychoanalysis which make it a single scientific discipline?

(1) The first unique characteristic we should mention is that psychoanalysis is concerned with the subjective life (an individual's inner world), with those experiences which cannot be rendered objective since they cannot be shared and which are unique to each individual. Other scientific disciplines concern-

ing the mind deal only with the objectively verifiable reality involving the outer world. However, the outer world can only be perceived subjectively through our senses.

(2) The second special feature of psychoanalytical thought is the focusing of *interest on the unconscious*, in other words the mental mechanisms lying beneath the threshold of consciousness. Many consider psychoanalysis as a general psychology, since it is concerned with and observes psychic activity dealing both with our perceptions and actions in the external world and the subjective inner world of feelings and emotions. Psychoanalysis posits the existence of a mental apparatus which basically functions largely unconsciously. However, unconscious functionings are not random but causal (brought about by psychic determinism, i.e., Freud brilliantly discovered that our unconscious acts are driven by motivational factors which do have a profound meaning to the agent), and that the mind is governed and balanced thanks to homeostatic principles, to achieve and maintain as much as possible intrapsychic psychological adaptation. This adaptation of an individual's functions is designed necessarily to achieve the greatest possible level of well being (in other words, with minimum levels of anxiety and conflict), and safety (in other words, the greatest possible stabilization of psychic functioning), so as to ensure the self-preservation of the individual and his social relations [22].

(3) The third exclusive feature which in our view is characteristic of psychoanalysis is its use of the term unconscious, which psychoanalysts consider to be a dynamic unconscious. The distinctive aspect of the dynamic unconscious derives from the fact that in this "zone" of the mind a particular mental experience, as well as the degree of affectivity associated with it, is diverted from consciousness by an equal and opposite force, which is thus of an inhibitory nature. These forces form the psychological defences whose purpose is to promote the functioning of the individual, "immunizing" him or her from disturbing impulses which would be destabilizing or even unbearable if brought to a conscious level, thus putting at risk the person's cognitive and affective processes. This concept of a play between "forces and counter-forces" distinguishes the dynamic unconscious from, for example, the Cognitive Unconscious, in other words, the world of the mechanisms and automatic connections present in all the sub-processes of the major cognitive functions (such as attention, perception, memory, language, etc.).

(4) Most interestingly, psychoanalysis is not a mono-personal psychology, but is instead based on a multi-personal model of the mind (normally bi-personal, or pluri-personal in the case of group psychoanalysis). In present-day psychoanalysis, the basic unit considered is no longer the individual functioning in its specific subjective reality, but rather the individual in relation to someone else; thus if the relationship represents the minimum unit of measurement in the psychoanalytical study of a person, the parameter measuring the change in time and space of a subjective relationship is provided by the con-

cept of wish. A subject's relationship with someone other than himself thus acts as a template for all surface derivatives (perceptions, behavior, symptoms, etc.). So, to understand the nature of specific human behavior one has to regard it as being derived from a relational process made up of at least four components:

(a) A representation of the subject in the relationship under consideration (meaning a particular image of the Self, of ourselves as agents urged on by a specific need, requirement or motive to an action thus aimed at a purpose);

(b) The need, in other words, the motive force (Freud would say the drive) leading a subject to seek something which could help it solve the problem arising from the sudden urge;

(c) The object of the relationship (in other words, a representation of an inanimate object, a thing, for instance food or a living object, or the internal representation of a person who is important in his affections) which can fulfill the motivating urge; and finally,

(d) The expected reaction by the object assigned or chosen to meet the need expressed.

The components (b) and (d) (the subjective need and the objective reaction expected by the subject) make up the fantasy of the wish (of which the individual may or may not be aware; see Fig. 3.1.3). Depending on how the fantasy of wish is imagined (largely unconsciously), a successful *wishfulfillment* will be achieved through re-experiencing the gratifying experience desired (*identity of perception*), or its frustration.

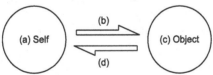

Fig. 3.1.3 The components of a relationship, the minimal unit of measurement in the psychoanalytical study: (a) a representation of the Self; (b) a need (the motive force); (c) an object of the relationship; (d) the expected reaction of the object thought to meet the need. Changes of a relationship are measured by wishes. The components (b) and (d) make up the fantasy of the wish.

(5) The final element shared by most psychoanalysts is the concept of the inseparability of the cognitive and affective components. This means that if we did not have a system of attributing emotional and motivational value, our cognitive representations (the internal images of inanimate or animate objects we have in our minds, or the words – the symbolic correlates of those images – that make up our thoughts) would lose all meaning; basically, we perceive in the world only those things to which we attach affective value. On the other hand, although it is true that if it were not for the emotions we would not be aware of the world which surrounds us, it is equally true that if

we did not have some kind of cognitive representation, our emotions would be blind, like Oedipus in Sophocles' tragedy. It could be said, taking up this example, that a mental representation (an image of an object, a person, a scene, or a word) which is not accorded affective value would dissolve and could not continue to exist as a subjective experience.

B. The Three Models of Sigmund Freud

Before describing a more recent psychoanalytic model [23], [24], some remarks regarding Freud's three starting theoretical models will be made. Those are (1) the *affective trauma framework*, (2) the *topographic framework*, and (3) the *structural framework*. The contemporary structural model presented afterward is directly stemming from Freud's ingenious theorizations. Generally, we believe that contemporary structural models (in respect to contemporary representational models based on the content of the mind) match best the findings of cognitive psychology and neuroscience, and may also be of practical use in pointing to the problems arising when using a psychodynamic model of the mind in order to conceptualize an intelligent device.

The Affective Trauma Model: Freud developed the first framework, concerning affective trauma, during a period not easily pinpointed between 1880 and 1897. Extremely briefly, it claims that as a result of a *traumatic event*, the mind can be subject to a dissociation generating a conscious and an unconscious psychic compartment. So, what differentiates this model from those Freud was to develop later is that a trauma (a real event experienced by the subject) leads to the dissociation of the mind into two "zones". This dissociation occurs thanks to a psychic mechanism (he called *repression*) designed to defend against expected pain which would be experienced at a conscious level as a result of the potential stimulus caused by the memory of a traumatic experience. At this time Freud therefore clearly had in mind a real event stored in the form of a *memory* of a traumatic experience, in other words composed of an unbearably powerful external stimulus, and is therefore destabilizing for the person's very integrity. At this time, Freud believed that a trauma, like for example abuse experienced during childhood, or mistreatment, serious deprivation, or a death in the family which cannot be overcome, basically something occurring outside the individual, was the *primary impetus* leading to this division of the mind into conscious and unconscious compartments (Fig. 3.1.4).

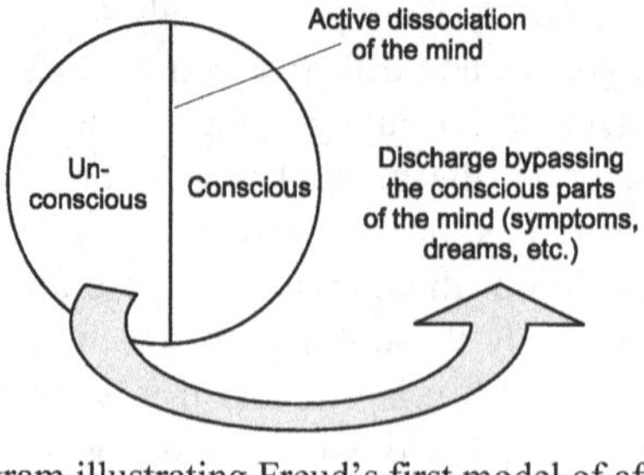

Fig. 3.1.4 Diagram illustrating Freud's first model of affective trauma

Freud honed his psychoanalytical technique over the following years and began to listen to his neurotic male patients and hysterical female patients, trying to comprehend the deep meaning of their statements: he soon discovered that an external trauma was not necessary for repression to occur (involving the shifting of the memory and the affective burden of the experience in question into the unconscious compartment), all that was required was a thought or a fantasy in the mind of the child or individual to cause the mind to split.

The Topographic Model: In Freud's second framework, during the child's mental development the prime generators of the dynamic unconscious are no longer the real traumatic events but above all intra-psychic events. From 1897 Freud therefore changed his thinking and therapeutic technique in important ways. Now Freud places greater emphasis on *the inner world*: The experiences that can act as potentially disturbing influences on subjective experience do not come from outside the individual so much as from within him. What puts in motion the need in the subject to protect consciousness with a special mechanism of defence (for example, repression) are the psychic representations of what he imagines and thinks, and his flights of fancy, rather than something that has happened in concrete reality. The unconscious is thus generated less by outer stimuli than by inner ones. This second Freudian framework (known as the *topographic model*) therefore includes a second great innovation in the understanding of the division and functioning of the mind: the unconscious no longer appears as the product of repression, in other words as a by-product of a dissociative event, affecting only someone who has been traumatized, perhaps at a critical moment in his development, but becomes a universal *site (topos),* present in every human being. Freud's second reference framework divides the mind up into three "zones", which at this stage Freud calls "systems", separated by two contact barriers, the first and second censorship (Fig. 3.1.5). The first compartment, referred to as the Conscious system, contains everything we are aware of in our minds at any given moment: Freud always associated this particular function with the processes of perception and all those processes which today we call cognitive (attention, memory, routine abilities, language, etc.). The second compartment, called the Preconscious system, contains all those mental representations that can be recalled to awareness from our memory by means of deliberate effort. This is a part of our mind we need not be constantly aware of and is therefore *descriptively unconscious*.

The third zone, called the Unconscious system, is the realm of the repressed. Freud sees it here as being the container of everything which in some moment of our lives we have perceived, imagined or experienced and which, since it is overwhelming our current consciousness, has been displaced beyond the initial contact barrier of the first censorship, that of repression. According to Freud, this censorship is also behind childhood amnesia and becomes entrenched in the child at around the age of four or five. "Beneath" this censorship can be found those early childhood experiences which according to Freud can no longer be accepted by the older child, still less by the adolescent or adult individual. But none-the-less, they continue to exert their influence through the affective charge contained in the

192

event or original thought, which finds its way as a surface behavioural by-product (for example, a symptom, or a dream). The Unconscious system, which Freud describes as "a boiling cauldron" (metaphorically, emitting spurts of affects toward the surface) which consciousness has to deal with, is also known as the *dynamic unconscious*, since it is supposed to contain *mental representations and motivating forces* which are restrained, inhibited and controlled by countervailing dynamic mechanisms.

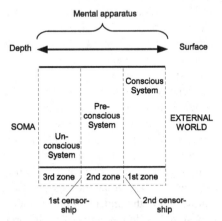

Fig. 3.1.5 The mental apparatus' "systems" in topographical relation to one another

The protective devices are the *psychological defences* [25]. A barrier, the second censorship, also exists between the Conscious and Preconscious system. As we shall see below, Freud saw this barrier as fine tuning what has access to consciousness and in contemporary Neo-Freudian psychoanalysis this censorship plays an even more central role.

The Structural Model. The last part of the concept of the psychic apparatus proposed by Freud includes a thoroughgoing revolution in the topographical model and the introduction of the *structural model*. Freud has two reasons for reviewing his model of the functioning of the psychic apparatus: (1) the central role of anxiety, as a signal putting into motion the mind's defensive needs [26], and (2) the idea that the homeostatic balance of the mind is kept constantly unstable by the presence of *psychological conflict*. Each individual's inner representational world is thus like a *gyroscope*, constantly part of re-establishing a precarious psychic balance, and is always at the mercy of a variety of motivational urges coming by turns from the outer world and from other unconscious compartments colliding against one another, causing anxiety and at times destabilizing conscious functioning and behavior.

In this third model of the mind, known as the "structural model", Freud highlights various psychic "entities", which he calls Id, Ego, and Super-Ego (Fig. 3.1.6). These psychic structures are in constant contact with each other and with the stimuli coming from outside, and when the interaction turns into conflict, the anxiety signal is activated. This signal catches the attention of the Ego and informs it of the presence and degree of mental destabilization being experienced, particularly with regard to an optimal range of *well-being* and *safety* (in other words, the essential and indispensable parameters for the individual's self-preservation, see [22]).

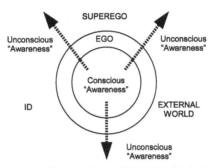

Fig. 3.1.6 A representation in diagram form of Freud's third, "structural" framework. The direction of the arrows indicates the theoretical possibility of having various degrees of awareness, ranging from stable consciousness to unawareness, which is never complete, according to the theory of identity of perception, a theory first put forward by Freud in [27] and further developed by Sandler in [23].

Freud's third model of the functioning of the psychic apparatus thus has at its core the structure he called the Ego. The Ego now has the governing role, and is the site of homeostatic regulation with regard to the instinctive drives of the Id. The Ego also deals with the requests of another structure which Freud calls the *Super-Ego*, a structural and mainly unconscious part of the individual's personality which develops in early childhood and contains moral rules, which often act to judge, inhibit and punish, but which are occasionally also a benevolent guide and the source of narcissistic approval. *"In relation to the Ego*, [the Super-Ego] *acts as a judge, like an external critic who observes the Ego, creating ideals for it* [and comparing it with them]" [28]. It is as if the Super-Ego were an inner voice telling us, *"You're like this ... but you ought to be like that ..."*, and if the alternatives being compared are too distant from one another, the result will be frustration and conflict within the person, occasioning an anxiety signal of varying intensity. The Ego thus stands at the hub of the Freudian structural model and is today mainly defined according to the functions it fulfills: On the one hand, the Ego is the *executive agent*, urging, planning and organizing voluntary action, on the other, it is the agent *controlling and inhibiting the senses*, both with respect to the outer world (perception) and the inner world, particularly with regard to wishes and instinctive and motivational urges. Finally, we have the Freudian structure par excellence, the Id, whose properties in Freud's third and final model (compare [29]) are not too different from those in the topographic Unconscious system in the second model. As a structure, the Id retains the special characteristics that Freud had ascribed to the Unconscious system in the second model: the Id remains the boiling container of experiences, thoughts, and fantasies that are repressed because they are unbearable to the conscious mind. According to [30], the Id's repressed experiences remain locked in an area not subject to the laws of logic (for example, there is no room for denial or for mutual contradiction, in which two phenomena that cancel each other out can coexist), or to the laws of time (the Unconscious system is atemporal), in which the passage of time is entirely unacknowledged. According to Freud, the atemporality operating in the Id explains the enduring influence of the repressed experience on observed subjective behavior.

C. The Model of Joseph Sandler

At the end of this section, a contemporary multidimensional model of the mind that is compatible with a structural approach shall be illustrated. In psychoanalysis, "*structure*" is a term which is very close to "*construction*" and to "*reconstructions*": In this analogy, what is structured can be constructed or re-constructed (see [31]) – maybe also artificially.

For anyone approaching the study of the components of human intelligence, or for that matter artificial intelligence, it is considered by us not to be enough just to answer the question, "*What is the functional structure of human intelligence?*". The proper question ought to be rather, "*How have these functions come to exist and how have they developed?*", thus, following in the footsteps of a developmental psychologist as he tries to understand a child's development.

The following outline of a recent psychoanalytical model developed by Joseph and Ann Marie Sandler [23] underlines the centrality of the concept of memory and repression in Freudian theory [32]. Of the great variety of clinical theories and practices now present in psychoanalysis, this model seems to be one of the most up-to-date, and one of the most closely adherent to the latest hard neuroscientific data in the matter of mental processes. We here describe a particular psychoanalytical model developed by Joseph Sandler [23], which involves dividing the mind into three compartments, *the Past Unconscious, the Present Unconscious,* and *the Conscious* part of the mind. In our opinion, Joseph Sandler's explanations are among the most convincing clinical ones of the nature of memory and repression [24]. They are useful in the practice of psychoanalysis, and at the same time, and strictly from a psychoanalytical point of view, they match up well and are consistent with data derived from the neurosciences. This model is actually a development of Freud's third model but added with the modern knowledge coming from neuroscience and cognitive psychology, so really better suitable for a practical application. Sandler was primarily an experimental psychologist before becoming president of the International Society of Psychoanalysis and his model has roots in rigorous scientific methodology while remaining very adherent to Freud.

The benchmark here, as mentioned above, is a model in which the mind is subdivided again in three main structures (Fig. 3.1.7, [33]), as in Freud's third model, but the inner functioning of each compartment is further defined and clearly differentiated. It is a model solidly based in the psychoanalytic Freudian tradition but also with strong connections to structural (reductionist) but also developmental empirical observations. In order to understand this model, it is useful to bear in mind, on the one hand, *the mental development of the child*, which is based on processes that are always the result of the working-out of genetic factors, and, on the other hand, the child's learning from the environment.

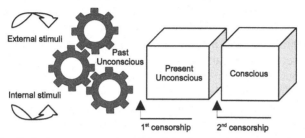

Fig. 3.1.7 Sandler's three-structured model of the function of the mind [33]: The first Section, representing the Past Unconscious, is not strictly-speaking a container but rather the site of consolidated devices (a gearing-chain), constituted of a series of implicit memories or repeated experiences which became procedures, psychological defenses, and automatic functions in the psychic apparatus. These procedural memories are modeled on the child's first basic experiences (up to the 4th–5th year) and influence all subsequent mental processes. The second Section, representing the Present Unconscious, contains unconscious or duly modified mental representations, in order to provide access to consciousness (the Conscious as the third Section). The contact barrier given by Freud's second censorship thus becomes the actual site of repression, rather than the one along the border from the Past Unconscious to the Present Unconscious (originally by Freud, from the dynamic unconscious proper or Id to the Preconscious).

A concept that should be taken into consideration in trying to understand this recent model of the functioning of the psychic apparatus is expressed by the term "experiential dimension". So, there is a first distinction to be made between *experiential context* and *non-experiential context*. The latter corresponds to the portion of the mind most easily pictured as being made up of devices, in other words, viewed as a machine or a set of functions. The non-experiential context thus covers anatomy and somatic physiology (especially the central nervous system) and fully corresponds to automatic responses of an organic or psychic nature. The non-experiential context also brings into play all aspects of a psychic nature, which are learned in the first years of life and become the templates for later relationships in life, psychological defences, or personal character traits.

For Sandler, the non-experiential mental context constitutes the first box and could be visualized as being made up of numerous molds or templates in which each template represents a functional device, a relational framework, memorized after being repeated a considerable number of times by the subject with another living or inanimate object in the world. Furthermore, every more recent template, or relational framework, tends to inhibit the earlier ones [34]. In [35], the whole of this first context largely inspired by relational experiences in early childhood has been called the past unconscious. It is not possible to re-experience parts of the past unconscious (nor to have memories of it) since it is intrinsically non-representational and non-experimental, but the workings of the past unconscious influence every single act in the life of an individual. The past unconscious is the part of the mind stemming from the more procedural and habit-like learning processes of the infant that appear to develop over the first three years of life, when the cerebral structures housing autobiographic memory (memories concerning ourselves set in a time frame and spatial context) are not yet fully formed. The past unconscious may be correlated to the cognitive concept of implicit memory. The structures for the autobiographic memories develop subsequently (from the third year of life onwards) in the human brain and they are known to be localized mainly in the hippocampus: This memory system forms the more narrative part of an individual's history and can be equated to the concept of *explicit memory* in cognitive psychology.

It is clear from the foregoing that in Sandler's structural model, the past unconscious no longer simply corresponds to the concept of the dynamic unconscious as it is described by Freud when he speaks of the system Unconscious in his second model[37]. However, the contradictions between Sandler's model and Freud's are only superficial. They concern the origin and the extension of Freud's dynamic unconscious, discussed in the following two paragraphs.

Concerning the origin, both Freud and Sandler ascribe a critical role to early childhood (up to school age). According to [29], the establishment of the dynamic unconscious, is owed mainly to repression, and connected to the overcoming of the Oedipus Complex: Freud uses the analogy with the Greek tragedy to underline the important stability the child acquires from his primary identifications with his/her parental models. Freud thus believed that childhood amnesia was caused by psychological reasons, mainly repression reinforcing the child's feelings and identifications toward its parents (which become unconsciously part of himself). Nowadays, while we continue to recognize the central role of the Oedipus Complex in identifications, in character formation, and in shaping personality development, contemporary psychoanalysis no longer seeks the root of childhood amnesia only in these dynamic psychological processes. Although, on the one hand, childhood amnesia coincides with the stabilization of parental identification (and thus of the Super-Ego), it would seem to be "caused" in reality by the biological development of the brain. Indeed, in early childhood implicit-type learning methods appear to predominate and coincide with what Sandler called the *Past Unconscious*. Brain structures concerned with conscious and autobiographical learning (located in the hippocampus within the temporal lobes) gradually form after

---

[37] For comparison between the psychoanalytic models, remember also that the Id of Freud's structural model can be basically equated to the system Unconscious of his second topographical model.

the third year. As a result of this critical phase in the brain organic development, the child acquires the capacity to produce stable mental representations out of his lived reality, constructing his own stories or fantasies bit by bit, and evolving more and more a sophisticated process: the capacity to think.

It is after this critical phase, that according to Sandler's model the second section of mental space develops, named the *Present Unconscious*: Since it contains mental representations of the world, that are manipulated and transformable according to the unconscious homeostasis of the subject, this portion of the mind can be still matched with the *dynamic unconscious* as described by Freud. On the other hand, the Present Unconscious also bears a certain similarity to the concept of the Preconscious System in Freud's second framework. However, the second section in Sandler's model (the Present Unconscious) differs from the Preconscious in that not all the mental representations there can be recalled by means of an effort of recollection or attention (since, after all, they are to all intents and purposes, unconscious mental representations). Although the representations in this section are mainly unconscious, they can be visualized in internal images, recalled in words or complex memories in the constructions of a psychoanalytic treatment. The Present Unconscious contains thus significant mental representations to the subject and fits into an experiential environment in the mind. Even so, these experiences (perceptions, memories, feelings, etc.) have varying degrees of access to consciousness, depending on the depth at which the so-called second censorship (the defensive or "countervailing" component which keeps a given experience beneath the threshold of consciousness in an active and dynamic manner) operates. In the Present Unconscious, every memory, fantasy, daydream, unconscious fantasy, dream or psychopathological symptom, is generated or molded in the first instance by the templates of the *Past Unconscious*.

As noted above, we recognize the present world (with our portion of consciousness which constitutes the third section in Sandler's model, the *Conscious*) on the strength of what we have learned implicitly in our *Past Unconscious* (engraved on our now consolidated memory, becoming a procedure, a motor habit, or an automatic act of perception). Once generated in this way on the "die" of a past experience, a given mental representation is modified in its passage through the *Present Unconscious* from the depths to the surface and changes according to the nature of the defences and dynamic qualities of the second censorship. In reality, the structuring effect of the templates of the *Past Unconscious* acts on every more recent psychological experience as if on a series of Russian dolls, each layer displaying differences of varying degrees of importance in the way the subject relates with the world around him.

3.1.5   A Neuro-Psychoanalytically-Inspired Cognitive Architecture

Within the ARS project [36] of the Institute of Computer Technology of the Vienna University of Technology, a new cognitive architecture for automation systems and autonomous agents has been introduced [37]. The architecture is sometimes referred to as ARS-PA (Artificial Recognition System-PsychoAnalysis) model. It is a comprehensive architecture, combining low-level, behavior-oriented forms of decision making with higher-level, more reason-oriented forms of behavior generation and selection. The integrating factor for the combination of the different modules of the architecture is the psychoanalytic view on the human mind. More specifically, the architecture is (loosely) based on Freud's third and final model, his "structural" model of the psychic apparatus consisting of the Id, the Ego, and the Super-Ego. The cognitive architecture can also be brought in correspondence with the model of Sandler (and maybe even better) which is no wonder since Sandler's model somehow is a contemporary version of Freud's third model. In the previous section, it has already been tried to outline how the models can be mapped onto each other. The reason why, in the following, it is still preferred to refer to the labels Id/Ego/Super-Ego rather than to *Past Unconscious/Present Unconscious/Conscious* is only because the question of machine consciousness is highly controversial. So, trying to technically implement a psychoanalytic model of the mind (as any other model) which is based on a distinction between various forms or stages of consciousness may spur unnecessary disputes, diverting attention from the actual functional contents of the cognitive architecture. In [13, p. 98], M. Solms describes the changing position of Freud on the necessity to base a model of the mind on consciousness: "In 1923, Freud recognized that the rational, reality-constrained, executive part of the mind is not necessarily conscious, or not even necessarily capable of becoming conscious. Consciousness, for Freud, was therefore not a fundamental organizing principle of the functional architecture of the mind." From that time onward, Freud assigned properties previously attributed to the system *Conscious-Preconscious* to the Ego whereby, however, only a small part of the Ego's activities was thought to be conscious. Instead, the Ego's new core capacity was *inhibition* which Freud considered to be the basis of all the rational, and executive competences of the Ego.

From a technical standpoint, the following two reasons were important for choosing the Id/Ego/Super-Ego model as basis for the psychoanalytically inspired cognitive architecture[38]:

- First, it is a structural model. Consciousness seems to be rather an emerging property of dynamic processes than a specific structure; however, when creating a technical system, in the beginning, a structure has to specified on which the – of course also to be specified – processes can be arranged.
- Second, the Id/Ego/Super-Ego model has already been able to be related (not in every detail, but regarding the main aspects) with neuroscientific and biological findings [38], [39]. These findings give important clues for the construction of the technical system.

The chosen psychoanalytic model will be complemented with established findings from neurology, neurobiology, ethology, and similar sciences, as long as there are no contradictions. This is necessary because psychoanalytic models, as

---

[38] Both arguments would also, and especially, apply to Sandler's model. However, as explained, we did not want to put the term "consciousness" on such a prominent place.

derived top-down from analyzing a very complex but already existing structure, naturally lack a lot of information about details which *do* have to be specified when artificially constructing and synthesizing an intelligent system bottom-up. Now, that it has turned out that neuroscientific findings can validate some of Freud's most central assumptions, the gap between the often antagonistic fields of neuroscience and psychoanalysis has narrowed considerably. This has manifested itself in the foundation of the International Neuro-Psychoanalysis Society[39]. For the technical design, the strategy is to proceed in both directions, bottom-up and top-down, in order to keep white spots in the cognitive architecture as small as possible. However, it is not intended to copy the brain on a structural level. The goal is to create an artificial system that aspires functional equivalence.

Apart from the chosen basic psychoanalytic model, a lot of further psychoanalytic principles and insights will be incorporated in the technical design, shaping the structure as well as the processes of the architecture. Those principles will be emphasized when they apply in the course of discussing the neuro-psychoanalytically inspired cognitive architecture. The architecture itself is a general one. It introduces various modules and how they are connected, thereby acting as a template upon which a variety of mechanisms and algorithms can be arranged, existing ones as well as still to be developed ones. Of course, the psychoanalytic principles that are taken as basis must not be violated. They distinguish the presented architectural approach from other suggested general architectures by providing very important and helpful constraints which guide the design of future elaborations of the architecture in a more specific way than which is for example the case for the architecture described in [40]. In the following, the neuro-psychoanalytically inspired cognitive architecture will be presented and discussed.

A. Inner and Outer World

A fundamental aspect of the proposed architecture is that the mind is rooted in the body. Consequently, there are two kinds of signals which have to be processed, internal (bodily) signals, and external (world) signals (Fig. 3.1.8). Alike, actions do not only influence the outside world, but also internal states. This conception is compatible with psychoanalysis as well as neuroscience.

In contrast to some of his followers, Freud wanted to establish psychoanalysis as a genuine natural science. He was a convinced naturalist and believed that there was a connection between the psyche and the body. He thought that what he described as "the Unconscious" from a psychoanalytic point of view, could also be described as bodily processes from an objective perspective. However, he did not consider both descriptions as identical, just as referring to the same thing *in itself*. To experience a thing *in itself*, according to Kant, we need the intermediation of the "perceptual apparatus". Freud now postulated that we actually do not just have an outwardly directed "perceptual apparatus", but also an inwardly directed one, the "psychic apparatus". It perceives the processes within ourselves. What psy-

---

[39] http://www.neuro-psa.org.uk

choanalysis does is to describe the working of this second perceptual apparatus from a top-down, subjective (introspective) point of view. Of course, the doings of the psychic apparatus can also be described objectively, at least insofar as one manages to correlate them with neurophysiological processes and structures.

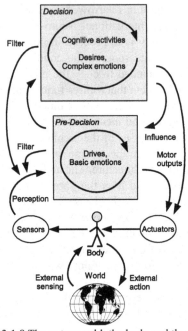

Fig. 3.1.8 The outer world, the body and the mind

During the last years, various neuroscientists have stressed that our subjective, psychological inner world is inherently rooted in the monitoring of our body, that is, in the way we *feel* ourselves [13], [41], [42]. In their theories, the linkage of signals from the self, representing bodily needs and regulatory elements (value systems), with signals from the surrounding outer world is critical for consciousness. As consciousness, in turn, obviously makes organisms more intelligent, the integration of external affordances, internal needs, and appropriate actions can be viewed as key element of intelligent behavior, especially if the integration, in connection with learning and evaluative memories, goes beyond "the current present" (compare [41]). Given what just has been said, it is strange that artificial intelligence has completely neglected all aspects related to the exact embodiment of their artifacts for such a long time.

This has changed with the upcoming of embodied cognitive science in which tradition the presented work can be viewed. Many of the basic assumptions and insights of embodied cognitive science are shared, however, the focus of the new approach is much more higher-level, and not so focused on the control of sensori-motor capabilities.

B. Interaction and Feedback

In recent years, people have realized that the sole purpose of perception – and in fact the origin of intelligence – is the guidance of action [13, p. 30]. Organisms cannot just lay back and take their time, they have to fulfill their bodily needs. Thereby, the particular form of how organisms are embodied – for example whether they have legs or fins – prescribes the kind of actions they can take. Moreover, as there is always feedback from the environment, which is again perceived, organisms are not just acting on, but interacting with the environment. For a long time, in the artificial intelligence community, system/environment interactions have not received the attention they deserve although they are the ultimate source for shaping an organism's cognitive capabilities. This is because feedback, in the form of *sensorimotor or emotional experiences*, serves as the basis for the formation of the cognitive categories and concepts a system has. Think of the well-known saying: "If the only tool you have is a hammer, then everything looks like a nail."

In the simplest case, there are just hard-wired rules between perception and action. These rules can be viewed as external regularities having been internalized during the course of evolution. With time, the incorporated processes between perception and action have become more and more complex [43]. For example, first there were solely unconditioned reactions, later on, by opening simple control loops, conditioned reactions came into being, the latter possessing a higher degree of flexibility and context-sensitivity. Mechanisms like reflexes, drives, instincts, conditional associations, or emotions are all members of a list that features distinguished levels of homeostatic control having been reached by evolution. They are all "ways to think" as M. Minsky argues in his latest book [44]. Finally, on top of the list, there are the more "intellectual sorts of thinking", like reasoning, drawing inferences, or planning.

It is clear that improvement in homeostatic control has to go hand in hand with better kinds of representations. So far, no one can exactly tell how complex relationships (especially symbolic ones) are represented in the (human) brain. It is known, that humans possess various kinds of memory, and that, during an individual's lifetime, the memory structures of the brain get gradually functional and more and more "filled" with contents. However, remembering is not just an act of passive retrieval of some fixed set of data, it is an active construction process [45], a great deal of which happens unconsciously.

In psychoanalysis, some kinds of interactions are given particular attention, for example those between an infant and his caretaker, or those between a therapist and her patient. Thereby, a process called 'transference' has been identified as important element. It means that in the course of interacting with each other, unconscious dispositions are not just passively remembered, but re-experienced.

202

C. General Architectural Scheme

In the following, the cognitive architecture, technically implementing the neuro-psychoanalytic view of the human mind will be described. Basically, it consists of two main blocks, named *Pre-Decision* and *Decision* unit (see Fig. 3.1.9). Each of these blocks is housing various functional modules which, in turn, again contain further sub-structures and processes. There are also different kinds of memory systems depicted as container-shaped blocks). Finally, the various functional modules and the memory units are connected by information and/or control flows.

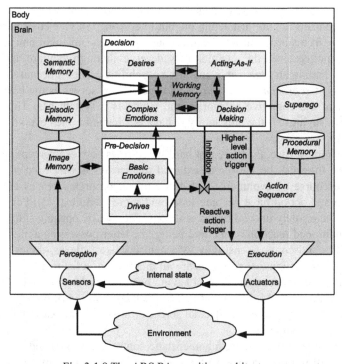

Fig. 3.1.9 The ARS PA cognitive architecture.

The modules of the architecture certainly do not anatomically correspond to structures in the brain, and, most probably, most of the functions bundled within a module are in reality much more "spread out" over the brain (in space and time) and much more intertwined with each other. The presented architecture is understood as a first attempt of a technical implementation, a feasible starting scheme that shall be improved iteratively. Note, that the general scheme of the architecture already shows some of the key ideas of the neuropsychoanalytic picture that have been implemented.

- First, the architecture includes the "body" of the autonomous system and the environment as elements that are integrated in the form of feedback loops.
- Second, it combines low-level and high-level mechanisms, summarized respectively in the two main functional blocks of the architecture, the Pre-Decision unit and the Decision unit, the first corresponding roughly to the psychoanalytic Id, and the latter to the psychoanalytic Ego. Each of these units contains several sub-structures and processes. The various processes within and across the smaller and bigger modules of the architecture give rise to different "levels of cognitive reasoning", e.g. reactive, deliberative, reflective, self-reflective. Generally, low-level responses are strongly, although not completely, pre-defined. Therefore, they are quick, but not always accurate. They provide the system with a basic mode of functioning in terms of in-built goals and behavioral responses. High-level processes take more time, but produce more distinguished forms of cognitive competences.
- Third, throughout the architecture, emotions, in combination with drives and desires, are used as important integrative and evaluative elements.

Freud did not use the term emotions but spoke of affects. He saw them as subjectively experienced manifestations of underlying physiological changes, the latter, being connected by him with his concept of drives. He formulated that drives originate from "within the body" in response to "demands placed on the mind in consequence of its connection with the body" [46, p. 214]. Technically speaking, emotions enable information processing systems to learn values along with the information they acquire. Therefore, a very important element of the architecture is the introduction of an episodic memory that contains emotionally evaluated previous experiences.

D. External and Internal Perception

The *Perception* module includes filtering and symbolization processes of – internal and external – sensory data. External, like internal, stimuli are continuously streaming into the system. The enormous amount of information has to be reduced, filtering out relevant pieces of information. At the end, there is, on the one hand, the perception of external objects and situations, and, on the other hand, the perception of internal bodily states.

In the part of the *Perception* module that handles the external perception, external stimuli run through a symbolization process that consists of several levels which are hierarchically organized [47]. Thereby, "symbols" (chunks of information) on the lower levels represent fragments of perceptual information (perceptual primitives), like edges, brightness, etc. Symbols on the higher levels contain more condensed information and represent more complex features. On the lower levels, different kinds of sensory values (optic, acoustic, etc.) are processed separately. The higher the level, the more the symbols are derived from the association of various channels. The perception process results in assemblies of symbols that are referred to as "images" (Fig. 3.1.10) and which represent objects, or situations (or parts of both) of the outside world. To reduce the computational load, the whole perception process only calculates changes. Moreover, throughout all levels, to recognize objects, situations, and sequences of situations – referred to as "scenarios" or "episodes" – the interpretation process makes heavily use of already memorized symbols and images, stored in the various types of memory of the system (Fig. 3.1.11).

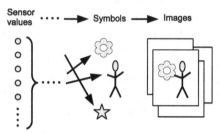

Fig. 3.1.10 Sensor values are, in steps, condensed to symbols which can in turn be grouped in images.

According to today's neuroscience, during childhood, our episodic and semantic memories are gradually built up [13]. As a consequence, the perceptual process is more and more governed by deeply encoded and abstract knowledge derived from learning experiences. Therefore, human perception is not unbiased and neutral but shaped by previous memories. We see, what we expect to see, and we are surprised or fail to notice when our expectations are contradicted. Moreover, perception is even more biased by the fact that input and evaluation of data are not arbitrary but organized as an active *screening process* searching for features that are required by the present state of body and mind. This constitutes a top-down influence from the higher-level processes of the brain on the perception process ("focus of attention").

Parallel to the perception of external stimuli, the *Perception* module also perceives internal stimuli. The part of the *Perception* module dedicated to the internal perception watches over the "bodily" needs of the autonomous system which are represented by internal variables. Each of these variables manages an essential resource of the autonomous system that has to be kept within a certain range, for example the energy level.

E. Memory Systems

The heavy use of predefined "mental images" as templates to shape, and also to speed up, cognitive processes is a central feature of the presented neuropsychoanalytic approach. Perception is not possible without memories. In fact, perception is actually a process of re-combining current inputs with previous experiences (projected as expectations onto the world) [13].

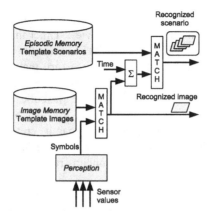

Fig. 3.1.11 Scenarios are ordered sequences of images.

Thereby, the role of memories is to organize previous experiences into recognizable chunks of knowledge. Thus, the processes of the perception units could be viewed as processes that take place within the perceptual memory structures, reorganizing their contents. Actually, the same is also true for all the other cognitive operations. However, depending on the cognitive level (low or high), the current goal or task (moving, listening, planning, etc.), and the current stage of the decision making process (more on the perception or more on the action side), different types of memories are involved. In general, one and the same experience or situation is stored in more than one type of memory – although in different forms. *Image Memory* is the most basic form of memory of the architecture. It is mainly associated with the perception process, and contains representations of symbols, objects, and situations. (More precisely, it contains representations of events rather than situations; thereby, events are defined as significant happenings occurring during a situation). *Image Memory* is a kind of passive, partly implicit, short-term memory supporting the perception process. It may be comparable to what is referred to as immediate memory by M. Solms [13, p. 144]).

In contrast, the *Working Memory* of the Decision unit is conceptualized as active, explicit kind of short-term memory. It supports the higher-level cognitive operations by holding goalspecific information and streamlining the information flow to the cognitive processes. As has already been mentioned, the perceptual processes can receive top-down attentional information from the Decision unit to search for task-relevant sensory data. This top-down information takes its origin in the *Working Memory*.

Semantic Memory contains facts and rules about the world, for example the physical rules of the environment, object categories, how objects relate to each other, propositions, etc. There is a strong connection between *Image Memory* and *Semantic Memory*. Perceptual processes, like other cognitive processes, are reliant upon knowledge from *Semantic Memory*. In the architecture, semantic knowledge describes possible associations between images.

The content of the *Episodic Memory* is based on individual autobiographic events and remembered from a first-person view. Other than semantic memories, episodic memories do not apply generally. When remembered, episodic memories should be re-experienced. Therefore, it is necessary that episodic memories include emotional ratings. Both kinds of memories, semantic and episodic ones, should be explicitly accessible. This requires a symbolic, localist representation. Semantic knowledge can come from *Episodic Memory* by a process called abstraction. Additionally, semantic memory units can be created by perceptual algorithms, or by explicit learning of facts and rules. The *Super-Ego* is viewed as a special part of the *Episodic Memory*. It contains rules for socially acceptable behavior.

Procedural Memory is a third type of long-term memory apart from *Semantic* and *Episodic Memory*. It is a kind of bodily memory, holding information necessary for the execution of behaviors, for example knowledge how to run, or how to drive a car. Normally, performing routine behavior is a largely implicit process.

F. Low-Level Decision Making – Pre-Decision Unit

Freud's central (at his time very controversial) statement was that most processes that determine our everyday feelings, wishes, and thoughts occur unconsciously. However, unconsciously motivated activities of the brain are not random, and they serve to ensure the individual's well-being (compare V). In this respect, drives and affects play an important role. Freud conceptualized drives as boundary phenomenon, signalling the mind the needs of the body [46]. Affects, possessing physiological as well as psychic aspects (they can be subjectively *felt*), were ascribed a similar intermediate role. Concerning their origin, Freud suspected affects as (inborn or learned) adaptations to prototypical environmental situations our ancestors had to deal with [48].

In Freud's structural model, unconscious, instinctual forces are enclosed in the Id which is governed by the "pleasure principle" (the seeking of pleasure). Today, J. Panksepp, one of the leading researchers of the brain organization and chemistry related to affective behavior, has identified at least four basic instinctual/emotional circuits shared by all mammals [49]. They are the SEEKING system (associated with a reward system called LUST system), the FEAR system, the ANGER/RAGE system, and the PANIC/LOSS system (responsible for social bonding). Additionally, a PLAY system is also investigated. The SEEKING/LUST system resembles Freud's concept of the libido drive. It is considered as pleasure-seeking system providing for the self-preservation of the body in the widest sense, thereby motivating most of our goal-directed interactions with the world in the first place. Pleasure creates an urge for repetition of the pleasurable behavior while frustration leads to avoidance. This is an important ingredient for learning.

In the cognitive architecture, the *Pre-Decision* unit can be roughly compared with the Id. It consists of the *Drives* and the *Basic Emotions* module. The *Drives* module gets input from the part of the *Perception* module handling internal perception. This input is connected with "bodily" needs, that is, technically speaking, resources of the system. In case that one of the internal resources is about to leave its range of well-being, this information is signified by the Internal Perception module to the Drives unit which, in turn, raises the intensity of a corresponding drive, for example hunger in the case of low energy. Provided that a certain threshold is passed, an *action tendency* to correct the impending imbalance is invoked, for the above example, to search for food. The output of the *Drives* module is forwarded to the *Basic Emotions* module where it meets with the current external perception. The task of the *Basic Emotions* module is to filter out stereotype situations, and to provide them with a first, rough evaluation. Stereotype situations are the ones that can be relatively easily recognized. As they are (by definition) often occurring, they usually possess some characteristic stimuli the system "knows" about (if only implicitly, that is unconsciously). After having been recognized, an impulse for a (greatly) predefined behavioral response is almost immediately evoked, without much further processing. On the whole, basic emotions enable agents to switch between various modes of behavior based on the perception of simple, but still characteristic external or internal stimuli. This helps the autonomous system to limit the set of potential actions among which to choose, and to focus its attention by narrowing the set of possible perceptions. The system starts to actively look for special features of the environment while suppressing others.

Each basic emotion is connected with a specific kind of behavioral tendency. Depending on the intensity of the basic emotion – and the strength of a potentially existing inhibitory signal from the Decision unit (Fig. 3.1.12) – the evoked behavioral tendency is either directly sent to the action module for execution, or, it is transmitted – together with the basic emotion values – to the *Decision* unit where it influences all further mental processing.

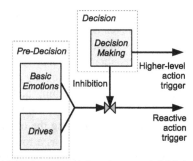

Fig. 3.1.12 Higher mental functions can inhibit lower ones.

Panksepp's basic emotions mentioned above are adapted to the needs of mammals in a natural environment. If not aiming at emulating human psychological or cognitive performance, they can be modified, depending on the demands of the technical application. Still, exactly the same basic emotions, when technically interpreted, can be very valuable for the design of any intelligent autonomous system. The SEEKING system is connected with the management of essential resources. The resources may be the system's own resources, but possibly also the resources of someone else the system has to take care of, for example those of handicapped or elderly people in the case of cognitive assistants. The task of the FEAR system is to identify potentially dangerous situations that require immediate reactions, thus, it plays the role of an alarm system. Again, the danger can be for the system itself or maybe someone else. Depending on the application and context, the associated reaction may be fleeing, hiding, or just being cautious. The ANGER/RAGE system leads to behavior where the autonomous system "impulsively" tries to defend its resources, or to remove obstacles preventing it from reaching a goal. It can do so by selecting a potentially appropriate reaction from a (small) set of possible ones. Additionally, it can "energize" (parts of) the system such that it momentarily increases the resources dedicated to the solution of the problem. The PANIC/LOSS system is important for the establishment of social interactions. This may be needed for automation or robotic systems that have to coordinate their behavior as an ensemble. It also plays a role for learning based on imitation.

The SEEKING system with its exploratory character – and, even more so, a dedicated PLAY system – substantially contribute to various forms of learning that go beyond simple reinforcement mechanisms. As outlined in [50], the key feature may be constant experimentation with external and internal actions during which re-usable chunks of information can be learned about possible combinations between external situations, potential behaviors, and internal states. This has to be supported by a memory that puts all these elements together, and a reward system that labels combinations as "good" or "bad", depending on whether they bring about pleasure or frustration. Freud wrote: "The raising of [...] tensions is in general felt as *unpleasure* and *their lowering as pleasure*. It is probable, however, that what is felt as pleasure or unpleasure is not the absolute height of this tension but something in the rhythm of the changes of them" [51]. Tensions can come from unsatisfied needs (bodily as well as higher ones like desires or goals), and also from emotional arousals (insofar as those are connected with potential future states of pleasure or frustration). To hope for something good to happen would establish a positive tension, and to fear that something bad will happen a negative one. Learning about a novel object, for example, could consist of a sequence of exploratory actions, like maybe, first, approaching the object carefully, then, putting it to the mouth, and finally, trying to use it as a tool. From the outcomes of these trials (whether they charge or discharge any of the system's tensions), several chunks of information about the novel object are created and stored, for example categorizing it as non-fear-inducing, bad-tasting object that can be used to crack nuts.

G. High-Level Decision Making – Decision Unit

Unless the *Pre-Decision* unit has not already "fired" – that is, initiated the execution of an action – the more exhaustive evaluation and decision making processes of the *Decision* unit are run through. At this stage, higher-level cognitive processes like reasoning or planning come into play. Still, there are also affective components (motivations and emotions) to attribute values to cognitive representations. In the presented cognitive architecture, these affective components are handled within the Desires and *Complex Emotions* unit. On the whole, the *Decision* unit can be referred to as Ego. All the functions fulfilled by the Ego are located there. [40]

The central task of the *Decision* unit is to associate desires, (basic and complex) emotions, and thoughts (which, in technical terms, are the elements of the *Acting-As-If* module and the *Working Memory*) such that the release of an "intelligent" action command follows. The whole construction corresponds to the agreed upon principle of psychoanalysis that meaning is achieved out of a combination of cognitive and affective components. In his psychodynamic theory, Freud described mental life as a kind of continuous battle between conflicting psychological forces such as wishes, fears, and intentions. The resolution of the conflicts, in one way or another, leads to compromises among competing motives. Thereby, a great deal of the involved mental processes occur unconsciously, and the conscious part of the mind has only limited access to the unconscious parts. In this respect, the role of emotions seems to be to inform the higher cognitive apparatus how world events relate to intrinsic needs.

Becoming aware of environmental changes or a changing status of the available internal resources is a necessary prerequisite for the control of the more hardwired, impulsive action tendencies of the *Pre-Decision/Id*. The arousal of these impulses cannot be prevented by the *Decision/Ego*, but their execution can be inhibited. Also, an original, primitive action tendency can be substituted by a more complex behavioral pattern.

The desires of the model are of similar nature as the drives. Like them, they are the motivational source of actions, but now, more sophisticated and more enduring actions. In general, desires activate behavior that aims at improving the subject-world relationship. A fulfilled desire discharges one or more tensions, the latter being related to the subject's (internal) needs, but also to the tasks it has to carry out. A desire is the wish to re-experience a once pleasurable (satisfying) situation.

Therefore the data structure of a desire has to involve a need, an object of desire, and a representation of the self and its expectations concerning the interaction with the object and the fulfillment of the desire. A desire can be initiated by a

---

[40] The mapping Pre-Decison unit with Id, and Decision unit with Ego is not 100% correct in a strict psychoanalytic sense. According to such, at least parts of the functions fulfilled by the *Desires* and *Complex Emotions* module should also be accounted to the Id. However, this is more a question of where to put which label and not necessarily a substantial one. Technically, everything works as described. When relating the presented architecture with Sandler's model, the *Pre-Decision* unit could be mapped with the *Past Unconscious*, the *Desires* and *Complex Emotions* module with the *Present Unconscious*, and the *Working Memory* (and especially the contents of the *Acting-As-If* module) with the *Conscious* part of the model.

drive (that is, a homeostatic, resource-related need), a memory (that is, a mental image occurring maybe in the course of a planning or association process), or an emotion that establishes a tension. The evocation of memories because of some aspect of similarity with what is currently perceived is the most basic process of all the processes of the *Decision* unit.

Like in nature, the objects of desire are not innate, but learned. When a tension is discharged, or an emotion of high intensity induced, the making of a new memory entry is triggered. This is a learning process that combines an external object or situation, a discharged or otherwise emotionally evaluated need, and a previously carried out action. Higher order learning processes can also be triggered like such which transfer recurring episodic memories as abstract schemes ("rules") into the *Semantic Memory*. Expectations about how to achieve the satisfaction of a desire ("wishfulfillment") take the form of *action plans* expanding the desire into goals and sub-goals (Fig. 3.1.13). For this expansion, especially experiences set on early in life, under the guidance of a "caretaker", become templates. The mechanism is connected with imitation learning. Influencing factors for imitation are the similarity of situations, the success of the observed behavior, and the status of the observed individual whose behavior is taken as a template. By this, out of a smaller set of innate motivational value systems, various desires are created. In parallel, also under the auspices of influential and respected others, (moral) standards and ideals acting as a counterpart to wishful drives and desires are formed. In the architecture, these normative elements are contained in the Super-Ego

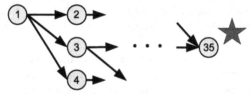

Fig. 3.1.13 Overview of the graph metaphor. The star represents the end state, the fulfillment of a desire. Nodes are states, edges are transitions between the states.

An activated desire directs the focus of attention to perceptions that are related to the fulfillment of the desire. There is also a focus on desire-relevant actions. Both filtering processes can be supported with information from the *Semantic Memory* if there is cooperation with the planning (*Acting-As-If*) module of the cognitive system. Other attributes of a desire include its intensity or tension (contributing to its priority), and the time it takes until a desire is given up. The latter is associated with an assessment of the degree of success (or failure) of a desire, which, in turn, is accompanied by a whole spectrum of emotional evaluations (ranging from hope to resignation).

The *Complex Emotions* module is responsible for all emotional mechanisms that go beyond the simple ones contained in the Basic Emotions module. Humans can greatly shorten, prolong, or otherwise modify the more hardwired emotional tendencies [49, p. 34]. With an increased complexity of situations the individual finds itself in, a broader emotional spectrum is necessary to handle them. An evolutionary elaboration has brought about a progressive differentiation in emotional control mechanisms from an inaccurate, global evaluation to a more accurate, local one, manifesting itself in emotions that possess more specific appraisals of situations, and more specific behavioral responses. Generally, it is assumed that such more elaborated emotions, like shame, envy, mercy, or hate, arise from interactions of the more basic systems with higher brain functions. Thus, a full emotional response includes rapid and unconscious processes, as well as slow, deliberative responses – up to conscious, verbal reflections on an emotionally challenging situation. In the architecture, the mechanisms of the *Complex Emotions* module deal with derivations of the basic emotions (labelled "complex emotions"), but also with the "pure" basic emotions themselves. This means that, for example, the (basic) emotion anger can not only be elicited or modified at the level of the *Pre-Decision* unit, but also by appraisals at the level of the *Decision* unit. In this case, however, anger is probably associated with a tendency to switch to a new method of problem solving rather than with an impulse to perform a simple attack behavior. Input to the *Complex Emotions* module is coming from the *Episodic Memory*, the *Super-Ego*, the *Pre-Decision* module, and the *Desires* module.

Functionally, emotions are control signals. There are two major lines along which complex emotions contribute to a more sophisticated control of behavior:

- First, some complex emotions extend present awareness in time. This class of emotions produces evaluations of expectancies that judge (future) possibilities.
- Another class of complex emotions establishes social hierarchies, contributing to rules for social coordination. For this class of emotions, the expressive aspects of the emotions using the face or the body are of particular importance.

The first class of complex emotions consists of affective reactions that evaluate success or failure of a desire, either prospectively or retrospectively. Thus, these emotions occur *when processing a desire* (Fig. 3.1.14). Prospective emotions evaluate possibilities that are still to happen. The spectrum ranges from confidence and hope on the positive end, over stages of doubt and fear, to hopelessness and desperation on the negative end. Retrospective emotions evaluate the finalization of a desire. The spectrum includes joy, satisfaction, disappointment, sadness, and distress.[41] All the mentioned emotions can be viewed as emotional states consisting of combinations of pleasure and frustration (joy and sadness) in various degrees of intensity. All of them are based on an assessment of the probability that the desire with which they are connected will be satisfied. This assessment is based on previous experiences related to the desire in question, and on the time how long the desire has already been unsatisfied.

---

[41] In the case of a "negative desire", that is, the wish that something bad does not happen, the joy of "wishfulfillment" would be referred to as "relief".

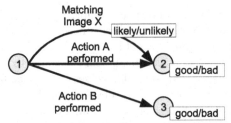

Fig. 3.1.14 Transitions in the desire graph. The transition of one state to another state can either happen (passively) by recognizing the next valid image or (actively) by performing a certain action.

The second class of complex emotions evaluates actions that are of social relevance. With these emotions the psychoanalytic principle that the individual has always to be seen in relation to someone else is acknowledged. Functionally, these emotions foster cooperative behavior. Influencing factors are whether actions are self-initiated or other-initiated, whether they are blame-worthy or praise-worthy. The emotions themselves can again be positive or negative. This leads to emotions such as shame, and pride when judging own actions; or reproach, and respect and admiration when judging the actions of others. All the above emotions are connected with social norms which are stored in the Super-Ego. In the case of humans, the norms, together with the corresponding emotions, are "learned" during infancy, for example, to feel pleasure when being judged positively by others. In the case of a technical system, an analog process might be a training or parametrization phase of an application.

Apart from the processing of a desire, the emotionally colored memories of the *Super-Ego* and the *Episodic Memory* (which the former is a part of) are the main source for the elicitation of a complex emotion. (Episodic) memories are not neutrally stored and remembered. Memories similar to the currently perceived image have associated emotions which are elicited when remembered, meaning that remembering is almost immediately followed by re-experiencing the associated emotion, including its physiological effects. Hence, emotions can be triggered by mental images just like by perceived images. With a given object or type of situation, possibly more than one kind of emotion can be associated.

Emotions have several effects on the decision making process of an autonomous system (some of them having already been addressed when discussing the *Basic Emotions* module). Most particularly, emotions contribute to the following tasks:

- generally, to inform the self about important changes in the body and the environment,
- to establish an alarm system,
- to establish social coherence,
- to support the pursuit of goals, or their abortion in time in case they turn out to be destructive,

- to set up new goals or desires in order to fix previous behavior that has turned out to be not so successful,
- to support learning of how to classify objects or other individuals, thereby influencing future interactions with them, for example increasing or decreasing the probability of the occurrence of such interactions,
- to support focus on relevant perceptions as well as relevant actions,
- to emphasize crucial events such that they can be memorized without having to be experienced repeatedly (which could be already destructive).

The stored experiences in the *Episodic Memory* include associated emotional values. Memories with high emotional values are remembered more easily and more often [p. 276] [52]. In general, the processes at the level of the *Decision* unit are characterized by an increased interaction with the memory systems (especially the *Episodic* and the *Semantic Memory*) compared to the processes at the *Pre-Decision* level. The memory interactions are also of improved quality, meaning that memories can now be accessed explicitly. This is particularly crucial for planning and inference capabilities, but also for the competence of performing reflections on affective states. In fact, it is widely assumed that direct accessibility together with explicit manipulability of memories is *the* essential feature of conscious processing.

The *Working Memory* is the place where the momentarily most salient external perceptions, desires, emotions, and thoughts come together (Fig. 3.1.15). Thus, affective contributions are combined with cognitive contributions to create "a feeling about something". The ultimate goal of this coupling is the release of an action command.

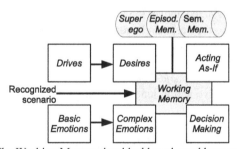

Fig. 3.1.15 The Working Memory is a blackboard, used by several modules.

The contents of the *Working Memory* changes on a moment to moment basis. The question now is which chunks of information populate the *Working Memory*. First, it contains the current external perception. This is already a construction where sensory inputs are complemented by similar memories. Second, these memories may have activated associated emotions which can also have gained entrance to the Working Memory in case they were above some threshold. Third, the currently most salient desires and tasks (the ones which produce the highest tensions) are the starting point to *actively* trigger the formation of a plan how to satisfy the specific desire or task. The building of (longer) plans is done by the *Acting-As-If* module which makes up the central part of the *Working Memory*.

The *Acting-As-If* module produces "thoughts". The name of the module indicates that thinking is viewed as a kind of acting, only that there is no execution of the invoked motor commands. Scripts for typical situations are given by episodic memories. They are used to expand desires or tasks into goals and sub-goals, and to evaluate the expected consequences of the planned actions. The evaluation process is done at every stage of the expansion. Both processes, expansion and evaluation, are supported by knowledge from the *Semantic Memory*, and by inference capabilities, whereby semantic facts and inference rules especially contribute to judge the feasibility and probability of potential actions or events. Thus, the evaluation consists of logically derived contributions from the *Acting-As-If* module (the seat of planning and reasoning), and of affective contributions coming from previous experiences stored in the *Episodic Memory*.

Apart from advancing the fulfillment of desires, another task of the *Acting-As-If* module is to constantly experiment with actions, their costs, and benefits in an *anticipatory* way in order to speed up cognitive reactions by narrowing the focus of attention (for perceptions as well as actions). For this purpose, to implement some kind of empathy, enabling the system to guess the emotions and intentions of others, might turn out to be very beneficial. Such capability could be realized, quite similar to planning, by initializing the scripts that are observed within the system itself. The quality of the plans derived by the system depends crucially on the length of the sequences of acts, or thoughts that can be projected ahead. For this competence, the usage of a symbolic language with syntactic rules is considered to be necessary.

Finally, as has been stated above, the *Working Memory* should release an action command. Most likely, the various elements of the architecture will have generated more than one action tendency. The following sources of action tendencies have been described:

- the *Drives*, and the *Basic Emotions* module of the *Pre-Decision/Id* unit, producing affective action tendencies based on a quick and rough evaluation of input stimuli,
- the *Desires*, the *Complex Emotions*, and the *Acting-As-If* module of the *Decision/Ego* unit, producing refined affective action tendencies based on appraisals that take into account emotionally rated previous experiences, and cognitively derived action tendencies based on logical inferences, and
- the *Super-Ego*, containing behavioral rules that represent social ideals or standards.

Usually, according to psychoanalysis, the action tendencies produced by the Id, the Super-Ego, and the Ego will be in conflict with each other such that a "decision" amongst them has to be made. This is a dynamic battle between the different action tendencies. Generally, actions are motivated by the goals that can be reached with them (be it the fulfillment of a need or the accomplishment of a task). A current goal may require the execution of several steps of actions. In the architecture, the currently pursued goal – and thus sequence of actions – depends on the height of the tension being potentially discharged by the actions, and on the number of action steps it takes until this happens. The applied selection mechanism, how it may look in detail, should strongly favor actions that are associated with the highest tensions, and actions which do not need much steps to discharge a tension. Moreover, there should be a parameter that varies the relative strength of the *Pre-Decision/Id, Super-Ego, and Decision/Ego* unit on the action selection process (Fig. 3.1.16). Whether an impulsive action from the *Pre-Decision* unit – which can occur any time during the execution of a longer action sequence – can be inhibited by processes of the *Decision* unit could differ between different implementations, leaving room for optimizations. The same applies to the influence of the *Super-Ego* which delivers antagonistic action tendencies for some of the desired or planned actions (Fig. 3.1.17).

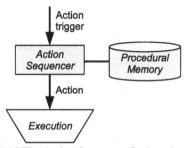

Fig. 3.1.16 The Action Sequencer feeds and reads the
Procedural Memory for dealing with routine actions.

The switch from an old goal to a new goal, and thus behavior, is also influenced by emotions. Especially the Complex Emotions module not only evokes direct action tendencies, but also outputs that strengthen or weaken some of the other action tendencies. Generally, positive emotions support the prolongation or repetition of ongoing interactions, negative emotions lead to the abortion or change of current system/environment interactions. Actions with positive associated emotions are looked for, actions with negative associated emotions avoided. A feared for negative outcome of an event, or a situation where none of the tensions can be released, may activate a "defense mechanism".

H. Actions and Behavior

Actions have influences on external circumstances and on internal states. They are built up of action primitives. The *Action Sequencer* module handles sequences of actions. On the one hand, it loads them from the *Procedural Memory* where they are stored. On the other hand, when occurring repeatedly, it discovers new action patterns and transfers them to the *Procedural Memory* from which, in future, they can be activated as a whole, thus forming new kinds of routine behavior.

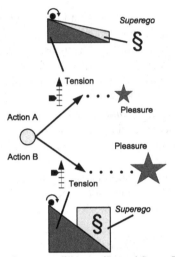

Fig. 3.1.17 If multiple goals are possible, tension and Super-Ego are competing. The higher a tension (plea-sure, to expect after action B), the more attractive, but the Super-Ego can inhibit this.

3.1.6   Implementation and First Experiments

Since it is not yet clear how large, how parallel, and how data-flow oriented the final model will be, it is not possible to make a final, sustainable decision on which technology mix is the best for implementing the ARS-PA architecture. Because of this, and because of the great flexibility and number of surrounding tools, a general purpose simulation environment seems to be the best choice for the time being. In the case of ARS-PA, JAVA-programs, embedded in the simulation tool "Anylogic" [42] were used. Such a simulation tool surely cannot offer the performance for the expected complexity of the future versions of the model, but helps to develop and test parts and details quickly and effectively. Later, a migration to more suitable technologies will surely be necessary. For now, the tools of the test implementation are block diagrams, software interfaces, data structures, and inter-process communication.

The first prototype of the ARS-PA architecture (called bubble family game, BFG)) is not supposed to fulfill a particular need (like serving as an alarming system, or household system). It is rather an object of study. In the beginning, everything is simulated. This is sufficient, as, at the moment, it is not the goal to enhance a real robot or another machine with an "ARS-PA program". So, all of the following three elements are simulated:

- the body of the artificial creature,
- the environment, and
- the mind of the artificial creature.

---

[42] www.xjtek.com

The body of the artificial creatures is a rather simple one. As there are no worth-to-talk-about physical dimensions and structures considered, the creatures are called "bubbles". Despite its simpleness, the body of the creatures (as explained earlier) is extremely important for the occurring drives and emotions, for the grounding of the symbols, and for developing intelligent behavior in general.

The bubbles are equipped with the following bodily features:

- a proximity sensor and eyes,
- feet (or some other means of movement),
- means to "eat" from energy sources, and
- an "on-board" energy storage.

The process of symbolizing the sensor values into meaningful chunks is considered as already done. The symbols the simulated bubble gets are like <energy source in front> or <other bubble on the left>. The bubble also has a homeostatic level, which is again relatively primitive. A kind of "battery" must be kept within a certain level. Living, walking etc. drains the energy level. To refill it, there are energy sources that can be used.

Similar to give the artificial creatures a "body", it is also necessary to put them into a setting that stimulates their senses and that allows for interactions with the surroundings, with other simulated bubbles, or even with other autonomous agents like humans (Fig. 3.1.18). In the prototypical implementation, the environment is a

- 2-dimensional, finite playground, with
- a number of different energy sources and objects (currently serving as obstacles), and
- a number of other bubbles.

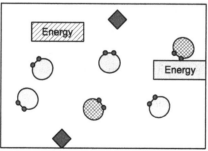

Fig. 3.1.18 The Bubble simulation environment: Artificial autonomous creatures (Bubbles), energy sources and obstacles are placed into a 2D world.

The setting might look trivial, resembling existing artificial life games, but it is necessary to offer stimulation of a certain analyzable and feasible complexity to the bubbles. Having one bubble in an intelligent private house would mean to hire a family just for living there in order to create enough input data for bubble research. The setting of multiple bubbles in a certain playground is – for the first step – much more efficient and practical.

The third and actually most interesting part of the simulations is the cognitive system of the bubble – its "mind". It implements – in its first version – the simpler parts of the ARS-PA architecture:

- *Drives*: hunger, seek, and play
- *Basic emotions*: pleasure; anger, and fear
- *Complex emotions*: hope, joy, and disappointment; shame, pride, reproach, and admiration

Most drives are directly connected with internal properties of the body (e.g. hunger is raised when the energy level drops below a certain threshold), others are raised when a (momentarily pre-defined) mixture of internal and external conditions applies. See [53] for details on the implementation. Concerning the desires which have been implemented so far, some of them only deal with individual bubbles and objects (for example, a bubble can get a desire for a very specific energy source), most of them, however, deal with the bubble as being part of a social group. These desires are supported by some rules which make the "social life" in the playground more interesting, examples of which are:

- There are energy sources that can only be "cracked" when two or more bubbles work together.
- Bubbles can form gangs, trying to dominate the playground.
- Some Bubbles desire company to play or to promenade together.

The purpose of the rules is to lead to situations which require cooperative behavior (and maybe also to such which require competitive behavior). We are especially interested in how to "configure" bubbles such that they can solve a problem or a task as a team. To do so, bubbles are generally designed such that being socially accepted, or achieving a task, are pleasurable states. Moreover, for solving cooperative tasks, bubbles have to remember relevant previously experienced scenarios such as:

1) I am hungry, but the energy source is too big to be "hunted down" alone.
2) The bubble over there is one that has been helped by me last time.
3) With high probability, it will help me now – so let's ask it.

Such action planning procedures are based on an architecture, described in [54]. Its main building blocks are:

- *Images*: snapshots of a set of recognized symbols[43],
- *Scenarios*: sequences of images,
- *Desires*: (activated) wishes to re-experience once pleasurable scenarios the satisfaction of which discharges a tension of a certain amount.

A bubble that recognizes the current happenings as a known scenario can activate an associate desire (e.g. when the stored scenario leads to a pleasurable, tension-discharging state). So, desires can be activated by recognized scenarios, and special sequences of scenarios, in turn, serve as a blueprint for the bubble how to satisfy – directly or in several steps – the desire's tension. The desires are stored

---

[43] In the current prototype, symbols and images are not yet learned, they are pre-given by the programmer and currently called template images.

with a "chance" of success which has influence on the pursuit of the desire. If the desire is finally satisfied, this chance is set to 100%, and the tension is discharged.

The ARS-PA prototype bubble stores and manages both scenarios and desires within an episodic (or autobiographic) type of memory. If they are not needed for a certain time, stored entries can be gradually forgotten. In turn, entries are more likely remembered, if they have had a big (emotional) impact on the bubble (e.g. changed its energy level, were a threat, etc.). Each bubble has its own, personal *Episodic Memory* module that stores its experiences and previous adventures. Entries are lists of images, stored as XML sets of perceived symbols. The *Episodic Memory* is implemented as directed graph of these XML sets. It has two parts, both of which consist of images as states, but differ in the edges:

- *Scenarios*: edges are *images* that match.
- *Desires*: edges are either *images*, or finished *actions*, that match (see Fig. 3.1.14).

Both graphs can be "activated", when a certain image, the INIT state, is recognized. After traversing all states, the sequence is fully recognized and the graph is deactivated.

Apart from potentially triggering a desire, a scenario graph can also increase the *tension* of an existing, active, one. It is not unusual that several (similar) scenarios are simultaneously active, since they are activated when the image recognition exceeds a certain confidence level indicating the similarity of a stored template image with the currently perceived input (e. g. 70%). The states of an active desire graph, in turn, can cause complex emotions. Both types of graphs have timeouts, which deactivate the particular graphs if they remain in one state for too long. A timed-out desire graph leads to frustration.

The *Episodic Memory* is implemented as an associative memory, which means that an entry X can be searched by its meaning and not by its address. The search is flexible and fuzzy, so searching for a "cup falling from the table" scenario will, for example, retrieve those memorized entries where it fell down and broke, where it did not break, and where it was caught by someone. It currently has no explicit sense of time, but it can tell the timely order of nodes (i.e. images).

Using the above implementation of an episodic memory, the bubble can select its action to take. See [54] for description on how to represent *actions* in a graph so that they do not come into obvious physical conflicts (like simultaneous eating, drinking, and talking).

The entire simulation loop has one concurrent block of functionality that consists of eight parts:

(1) *Perception*: *Symbols* are composed to *images*.

(2) *Image matching*: The sensed image is compared with stored ones (*template images*) and an ordered list from 0% to 100% is formed. If the system is waiting for a certain image (has its focus of attention on something), the respective recognition level is increased by further 5%. All *matching images* (e.g. recognition level above 70%) can *eventually* issue a *reactive ac-*

tion tendency, sorted by matching level, and importance. Images directly lead to *drives* and *basic emotions* and can even cause contradicting ones[44].

(3) *Scenario Processing*: If an edge of an active *scenario* graph is fulfilled, the state is *changed*. If a special "INIT image" is recognized, the corresponding INIT state is invoked (i.e. the *scenario is activated*). Fully recognized scenarios can activate a desire.

(4) *Desire Processing*: Similar to scenarios, but now open edges can *create desire action tendencies*. They are sorted, depending on the current intensity of the associated *tension*, and their *likelihood of success*. Complex emotions are updated.

(5) *Merge of action* tendencies: The sorted lists of all existing *reactive and desire action tendencies are merged*.

(6) *Super-Ego*: Takes *influence* on the *action tendencies*. Some are simply not allowed, others gradually inhibited. New action tendencies can also be generated. At present being very rudimentary, it will later deduct social rules out of the current scenario.

(7) *Conflict detection and resolution*: Finalization of the ordering of action tendencies from all sources, different sources can be differently weighted. If after this there is still a conflict – this could be possible because of the fuzzy/integer representation of action tendencies – a decision is made by random selection.

(8) *Execution*: At the moment, the BFG can only *execute one action tendency* per simulation step and bubble [45].

Although the bubble society may appear as "agents" on the first sight (which they actually are!), the intrinsically conflicting agent system of our interest is inside each individual agent (bubble). So the real (or better the more important) multi-agent-system is inside the bubbles.

All processes and functions inside the bubbles run – per definition – in parallel. Even if the underlying hardware only allows quasi-parallel execution, the software itself is designed in a parallel manner. So all processes of the particular bubble or machine happen simultaneously, if desired. Think of a walking robot that has to move its feet, keep the balance, talk with some other robots, avoid obstacles, think about its goals, plan its way to go, and maybe watch out for energy sources and dangerous enemies. All this happens in parallel, and parallelism is – for computer engineers – not a problem at all. This first prototype, however, is based on a simulation environment, where every simulation step is considered as an atomic and in-

---

[44] Think of the two stored images <<food near, body weak>, <hunger, lust>> and <<enemy near, body weak>, <hunger, fear>> and the three perceived symbols <enemy near>, <food near>, and <body weak>.

[45] The action "move from A to B" is executed via a "god mode", where the Bubble just moves there without thinking about every single step in between. It wants to go somewhere, decides to do so and it just happens magically. The decision is the main focus of this work, and not the robotics.

trinsically parallel action: Within one simulation step, interdependencies are not possible, all functionalities are considered orthogonal.

3.1.7   Conclusion and Outlook

We have seen how even in Freud's own lifetime and over the course of his career at least three wholly distinct frames of reference and structural frameworks for the psychic apparatus succeeded one another. After Freud's death, psychoanalysts coalesced around certain aspects of these three models, emphasizing the features they considered most important, in accordance with their ideological position within the discipline and giving rise to autonomous and quite distinct schools of psychoanalysis. At the moment, psychoanalysis is a highly diversified field divided up into schools of thought characterized by a profound gulf in communication between them and with the other humanistic and scientific disciplines.

On the strength of what has been said so far, how could the Golem of Artificial Intelligence be built on the basis of the *Torah* of (official and unofficial) psychoanalytical theory? We have seen that both Freud and post-Freudian psychoanalytical authors have strained to refine ever more the brilliant notions of psychoanalysis. It is also true that psychoanalysts have been less concerned to resolve the contradictions inherent in the various Freudian theories or the discipline's inconsistencies with data gathered in clinical practice with patients. Various psychoanalytical researchers have turned their attention to important topics, such as assessing psychoanalytical techniques in the study of children [55], the validation of clinical concepts in psychoanalysis [56], the evaluation of the effectiveness of psychoanalytic psychotherapy [57], and the empiric comparison between the neuroscience, neuropsychology and psychoanalysis [38]. We think it is open to question whether a psychoanalytic model is compatible or incompatible with any putative application in terms of creating artificial intelligence. It is important to point out the limitations in this approach in order not to create illusory (or even sciencefictional) applicative expectations. Indeed, in [13], misgivings in this regard are expressed as follows:

The mind as such is unconscious, but we can perceive it consciously by introspection. This is the capacity to "look inwards" (through introspection or selfawareness) which represents the core property of the mind. The "I" which we all perceive through introspection can also be perceived by means of our external senses as a physical body (made up of organs or physiological processes). The body is not the mind. Bodily processes are not intrinsically mental; a body's functions can even be performed by artificial devices. This is the reason we have stated that the mind as such is unconscious. The quality of consciousness can only be attributed to the perception of these processes while they are at work. Moreover, it should be added that these perceptions reach us in two forms, of which only one can properly be called "mental". The mind is therefore intimately connected to the point of observation in the first person. This is the only perspective giving a complete experience of all the things we observe, rooting it in the basic sensation that generates the self, which, in the final analysis, is based on our inner aware-

ness of living in a distinct physical body. A computer could acquire the property of consciousness only if it were possible to incorporate within it this capacity for self-awareness, which is grounded in a fleshly body.

Below they add:

The self this system is supposed to monitor should be a body (preferably one with a long developmental history), in order truly to be able to generate sensations.

And finally:

[...] unless a way is found to design a computer which has feelings (these being, moreover [...], strictly linked to consciousness), it will probably never be possible to build an artificial [mind] which is really efficient.

Apart from expressing provisos and reserves concerning machine consciousness, the above quotation expresses the view that bodily self-awareness and affectivity ("having feelings") play a key role for consciousness. On the other hand, being conscious (at least on the human level as one knows by introspection) is always strongly connected with thinking, that is, the linguistic processing of symbols representing chunks of information. It is this second aspect of consciousness that has predominated the efforts of Artificial Intelligence ever since its beginnings. We think it is now time, to concentrate also on the first aspect, and especially to *try to bring both of them together*. Thereby, we consider the presented neuropsychoanalytical approach as a very promising one, providing lots of valuable hints on how to do the combination.

Actually, the goal of this project is *not* to find consciousness in our bubbles, but rather having them behave in a humanlike intelligent way (i.e. deliberative, smart, adaptive, fuzzy, flexible, learning, etc.). So, luckily, the highly debated question of machine consciousness is out of the direct scope of this work. If consciousness ever turns out to be a must-have for intelligent behavior, it will of course have to be taken into consideration. See [58], for example, for a comprehensive article on consciousness in imitative machinery, and [59] for a path from designing a modular mind to pondering about consciousness.

For the time being, however, there is in our opinion lot of work to do on the way of creating intelligent machines without worrying too much about machine consciousness. Especially, the organizational structure and the processes used in intelligent artifacts have to be improved significantly – and they can. A key for this, in our opinion, is the mentioned combination of affective concepts, like emotions and desires, and symbol manipulating capacities (independent of whether this brings about consciousness or not).

The attempt in this project is to build machines that are supposed to "really understand" what is happening, where they are, if the situation is bad or good, what were the different possibilities they could do now, and which would be the hypothesized consequences in each case. The crucial point seems to be to equip machines with evaluative mechanisms such that they can autonomously acquire information and adaptively turn it into *meaningful* pieces of knowledge. These pieces ought to be, on the one hand, grounded in the ongoings of the real world, and, on the other hand, should form a (logical) structure of associations that allows the making of inferences without having to directly refer to the grounding of the symbols each time one uses them. This problem cannot be solved without learning. The importance of the question of how mechanisms leading to intelligence have come to exist has also been stressed as central for the psychoanalytic model of Sandler. We think that the presented architecture, and especially the affective concepts included, allow for a lot of different learning mechanisms to be "accommodated", some of them have already been briefly addressed. The right mixture between pre-defined knowledge, system adaptation guided by a supervisor during a training phase, and run-time learning has to be found. This topic, and how it can be supported by psychoanalytic insights, certainly has to be explored in more depth in future.

The ARS-PA model given in this paper is a first attempt to structure a neuro-psychoanalytical model for implementing it in a machine. What can we expect from such an implementation? The engineering community expects "smarter" systems. As discussed in the introduction, automation systems are about to reach a level of complexity which makes it hard to be solved with traditional methods. Autonomous systems like auto-pilots or digital assistants need context-aware decisions, memories, emotions, experiences, and other features commonly reserved for humans.

Psychoanalysts, neurologists, and psychologists can, in turn, hope for a new tool of research. Similar to animal experiments, they can use a machine, entirely described in software, to try things out. Comparing experiences from real human patients with experiments on a machine could help proving a diagnosis or an understanding of certain interrelationships. It is easy in a machine to "turn off" parts, or to temporarily destroy memories, or some inter-module communication. Like the tragic victims with head injuries of World War I lead to the first break-through results in neurology, we are convinced that a machine hosting an artificial mind according to a neuro-psychoanalytical model could help to refine and further develop the model and the understanding of the human mind in general.

Up to now, the implementation results show little difference to existing "artificial life" tools on the first glance. Only a small part of the model is implemented and moving bubbles have a strikingly similar appearance to a group of reactive, mobile agents. The difference lies within the organizational structure. Our bubbles are equipped with drives and emotions, imitating the internals of humans. Unfortunately the beauty of this concept will only be accessible when more parts of the model are implemented. Memories, associated with emotions, have proved to be a powerful tool for action planning, and so hopefully will the rest of the model. It is, however, still a long way to go, until for instance a rule of the Super-Ego inhibits a hunger reaction for some good reason. It is a non-trivial task to select the right order of implementation to get meaningful mid-term results during the project.

An aspect that might become relevant in (not so?) far future, but still worth considering, is system management. System management of contemporary systems typically includes network management, configuration, system extension, parameter upload and tuning, maintenance, debugging, and so forth. A technical system usually needs such treatments to work properly. We should expect the same for our to-come ARS-PA machines. The question is simply, what will this look like? The parameters of such a machine will be far too complex as that a human would comprehend their real meaning. Other parameterization interfaces must be found. Not unlikely these interfaces might be called "School" or "Therapy", just to fall into the trap of Problem No. 1. The language, used for these interfaces might even be natural language, not because it is easier to say *"Please tell me yesterday's events!"* than typing *"ReadMem (EV, T-86400, T)"*, but rather because it might be more suitable for expressing fuzzy requests.

So, obviously the project does not only attempt to build a system for complex problems, the system and the project themselves are complex as well. But however difficult it might be, it is definitely worth trying.

References

[1]   D. Norman, A. Ortony, and D. Russell, "Affect and machine design: Lessons for the development of autonomous machines," IBM Systems Journal, vol. 42 (1), pp. 38–44, 2003.

[2]   S. Turkle, Computers, ethics, & society. New York, NY, USA: Oxford University Press, Inc., 1990, ch. Artificial intelligence and psychoanalysis: a new alliance, pp. 128–145.

[3]   W. D. Hillis, The connection machine. Cambridge, MA, USA: MIT Press, 1986.

[4]   M. Minsky, Society of mind. Simon and Schuster New York, 1986.

[5]   S. Franklin and A. Graesser, "Is it an agent, or just a program? a taxonomy for autonomous agents," Lecture Notes Artificial Intelligence, vol. 1193, 1997.

[6]   P. Palensky, "Reactive energy management," Ph.D. dissertation, Vienna University of Technology, 2001.

[7]   W. S., "Pythia: A knowledge-based system to select scientific algorithms," ACM Trans. Math. Software, vol. 22 (4), pp. 447–468, 1996.

[8]   A. Buller, "Volitron: On a psychodynamic robot and its four realities," in Proceedings Second International Workshop on Epigenetic Robotics: Modeling Cognitive Development in Robotic Systems, C. G. e. a. Prince, Ed. Edinburgh, Scotland, 2002, pp. 17–20.

[9]   A. Buller, "Building brains for robots: A psychodynamic approach," in Proceedings Pattern recognition and machine intelligence PReMI 2005, K. P. et al., Ed. Springer-Verlag Berlin Heidelberg, 2005, pp. 70–79.

[10] C. Breazeal, Designing sociable robots. Cambridge, MA, USA: MIT Press, 2002.

[11] A. Lurija and F. Yudovich, Speech and the development of mental processes in the child. Penguin Books, 1971.

[12] D. Hebb, The organization of behavior. New York: Wiley, 1949.

[13] M. Solms and O. Turnbull, The brain and the inner world. Karnac/Other Press, Cathy Miller Foreign Rights Agency, London, England, 2002.

[14] H. Graf and L. Jackel, "Analog electronic neural network circuits," IEEE Circuits and Devices Magazine, vol. 5 (4), pp. 44–49, 1989.

[15] R. Sridhar and Y.-C. Shin, "VLSI neural network architectures," in Sixth Annual IEEE International ASIC Conference and Exhibit, 1993.

[16] N. Barnes, P. Healey, P. McKee, A. Oapos, M. Rejman-Greene, E. Scott, Webb, and D. Wood, "The potential of high speed opto-electronic neural networks," in Proceedings of IEE Colloquium on Neural Networks: Design Techniques and Tools, 1991.

[17] H. Caulfield, J. Kinser, and S. Rogers, "Optical neural networks," Proceedings of the IEEE, vol. 77 (10), pp. 1573–1583, 1989.

[18] E. Kandel, "A new intellectual framework for psychiatry," American Journal of Psychiatry, vol. 155, pp. 457–469, 1998.

[19] E. Kandel, "Biology and the future of psychoanalysis: A new intellectual framework for psychiatry revisited," American Journal of Psychiatry, vol. 156, pp. 505–524, 1999.

[20] S. Freud, The origins of psycho-analysis. New York: Basic Books, 1950.

[21] S. Freud, "Project for a scientific psychology," The Standard Edition of the Complete Psychological Works of Sigmund Freud, vol. 1, pp. 295– 391, 1950.

[22] J. Sandler, "The background of safety," International Journal of Psychoanalysis, vol. 41, pp. 352–356, 1960.

[23] J. Sandler and S. A.M., Internal objects revisited. London: Karnac books, 1998.

[24] J. Sandler and S. A.M., Memory of child abuse; truth or fantasy. London: Karnac books, 1998.

[25] A. Freud, "The ego and the mechanisms of defence," The Writings of Anna Freud, vol. 2, 1936.

[26] S. Freud, "Inhibitions, symptoms and anxiety," The Standard Edition of the Complete Psychological Works of Sigmund Freud, vol. 20, pp. 87–172, 1925.

[27] S. Freud, "The interpretation of dreams," The Standard Edition of the Complete Psychological Works of Sigmund Freud, vol. 4 & 5, 1900.

[28] J. Sandler, "On the concept of superego," Psychoanalitical Study of the Child, vol. 15, pp. 128–162, 1960.

[29] S. Freud, "The ego and the id," The Standard Edition of the Complete Psychological Works of Sigmund Freud, vol. 19, pp. 12–66, 1923.

[30] S. Freud, "The unconscious," The Standard Edition of the Complete Psychological Works of Sigmund Freud, vol. 14, pp. 166–204, 1915.

[31] S. Freud, "Constructions in analysis," The Standard Edition of the Complete Psychological Works of Sigmund Freud, vol. 23, pp. 257–269, 1937.

[32] S. Freud, "Remembering, repeating and working-through," The Standard Edition of the Complete Psychological Works of Sigmund Freud, vol. 14, 1914.

[33] J. Sandler and S. A.M., ""second censorship; three box model": technical implications," International Journal of Psychoanalysis, vol. 64, pp. 413– 426, 1983.

[34] S. J. and W. Joffe, "Tendency to persistence in psychological function and development," Bulletin of Menninger Clinic, vol. 31, pp. 257–271, 1967.

[35] J. Sandler and S. A.M., "Past unconscious, present unconscious, and vicissitudes of guilt," International Journal of Psychoanalysis, vol. 68, 331–342, 1987.

[36] G. Pratl and D. Lorenz, B. and Dietrich, "The artificial recognition system (ars): New concepts for building automation," in Proceedings of 6th IFAC Int. Conf. on Fieldbus Systems and their Applications, Puebla, Mexico, Nov 14–15, 2005.

[37] C. Rösener, B. Lorenz, G. Fodor, and K. Vock, "Emotional behavior arbitration for automation and robotic systems," in Proceedings. 4th IEEE International Conference on Industrial Informatics, 2006.

[38] M. Solms and K. Kaplan-Solms, Clinical studies in neuro-psycho- analysis. International Universities Press, Inc., Madison, CT, 2000.

[39] G. Roth, "Das Verhältnis von bewusster und unbewusster Verhaltenssteuerung," Psychotherapie Forum, vol. 12, pp. 59–70, 2004.

[40] A. Sloman, "Beyond shallow models of emotion?" International Quarterly of Cognitive Science, vol. 2(1), pp. 177–198, 2001.

[41] G. M. Edelman, "Naturalizing consciousness: A theoretical framework," PNAS, vol. 100, pp. 5520–5524, 2003.

[42] A. Damasio, The feeling of what happens. Body and emotion in the making of consciousness. Harcourt Brace & Company, New York, 1999.

[43] A. Damasio, Looking for Spinoza: Joy, sorrow, and the feeling brain. Harvest Books, 2003.

[44] M. Minsky, The emotion machine: Commonsense thinking, artificial intelligence, and the future of the human mind. Simon and Schuster New York, 2006.

[45] G. M. Edelman, Neural Darwinism: The theory of neuronal group selection. Basic Books, Inc., New York, 1987.

[46] S. Freud, "Triebe und Triebschicksale," Gesammelte Werke, vol. X, pp. 209–232, 1915.

[47] G. Pratl, "Symbolization and processing of ambient sensor data," Ph.D. dissertation, Institute of Computer Technology, Vienna University of Technology, 2006.

[48] M. Solms, "What is consciousness?" Journal of the American Psycho-analytic Association, vol. 45, pp. 681–778, 1997.

[49] J. Panksepp, Affective neuroscience: The foundations of human and animal emotions. New York: Oxford University Press, 1998.

[50] A. Sloman and J. Chappell, "The altricial-precocial spectrum for robots," in Proceedings IJCAI'05, Edinburgh, 2005, pp. 1187–1192.

[51] S. Freud, Ed., An outline of psycho-analysis. New York: W. W. Norton, 1940/1989.

[52] A. Baddeley, Human Memory: Theory and Practice. Psychology Press, 1997.

[53] W. Burgstaller, R. Lang, P. Pörscht, and R. Velik, "Technical model for basic and complex emotions," in Proceedings of 6th IEEE Conference on Industrial Informatics, 2007.

[54] C. Rösener, R. Lang, T. Deutsch, G. Gruber, and B. Palensky, "Action planning model for autonomous mobile robots," in Proceedings of 6th IEEE Conference on Industrial Informatics, 2007.

[55] A. Freud, "Normality and pathology in childhood: assessments of development," The Writings of Anna Freud, vol. 6, pp. 3–236, 1965.

[56] J. Sandler, "Hampstead index as instrument of psychoanalytic research," International Journal of Psychoanalysis, vol. 43, pp. 287–291, 1962.

[57] P. Fonagy, "Moments of change in psychoanalytic theory: discussion of a new theory of psychic change," Infant Mental Health Journal, vol. 19, 163–171, 1998.

[58] S. Blackmore, "Consciousness in meme machines," Journal of Consciousness Studies, vol. 10, pp. 19–30, 2003.

[59] P. Haikonen, Visions of mind. USA: Idea Group Publishing, 2005, ch. Aftificial minds and conscious machines.

3.2 Issues at the Interface of Artificial Intelligence and Psychoanalysis: Emotion, Consciousness, Transference

David Olds

To comment on the emerging interests shared by psychoanalysts and the current generation of robotics engineers from the point of view of a psychoanalyst, [46] *I will take up three topics where I think the two disciplines may have something to offer each other. (1) The importance of emotions in the cognitive apparatus, and in the potential design of AI machines, (2) The possibility of consciousness in advanced robots, and (3) The possibility of the psychoanalytic notion of transference in our relationships with robots.*

3.2.1 Introduction

The romance between the creator and intelligent creations goes back to the beginning of culture. Most civilizations have constructed a creation fantasy in which a highly intelligent deity brings into being creatures, slightly less powerful and intelligent than itself, with whom it is imagined to be in dialogue to the end of time. Adam and Eve are essentially robots in comparison to God. More homey myths involve humans and their creations such as the story of Pygmalion and Galatea, or of Pinocchio. The human relationship to robots has a history and a future. In the past, robots have been dramatic cultural icons. The most famous is probably Dr. Frankenstein's creation, who develops certain emotions and becomes as fatally destructive towards his adoptive family [16, 17]. He is depicted as a large humanoid with Halloween skeletal features. More, as robotics has advanced, so has the imagination of science-fiction writers. In Star Trek, we had Spock, and in the second version, Commander Data, who is a very human looking silicon creation with superhuman cognitive capacities. Even more recently in Steven Spielberg's movie "A.I.", we have a new Pinocchio, a nearly real boy, David, who is capable of attachment and love. The cultural meme of these stories expresses our collective transference toward the robot. The robot has the cognitive power of a super computer, but usually lacks real emotionality. Cdr. Data resembles many fictional and cinematic protagonists who have Asberger's syndrome, an autistic spectrum disorder, characterized by lack of emotion, but with high intelligence.

One thing that is clear is that the discipline of Artificial Intelligence has taken on a tremendous intellectual task. In order to create artificial intelligence one has to understand *intelligence* (of the natural kind), and that means not only psychoa-

---

[46] This paper was prepared for the ENF conference in Vienna, in August, 2007, as a response to a paper by Palensky, Lorenz and Clarici [1], which discusses the attempt to use neuropsychoanalytic theory to provide models and questions for artificial-intelligence scientists as they increase the effectiveness of machines that provide functions similar to those of the human brain. These authors give a succinct and sophisticated summary of psychoanalytic theory and its major contemporary variants, a summary compatible with the thinking of most analysts identified as ego psychologists.

nalytic theory, but all of cognitive psychology, neuroscience, and evolutionary biology and psychology as well. In the essay to follow, I would like to make three points: 1. We are beginning to contemplate the need for emotions in artificial minds. 2. Many now feel that consciousness is something in the distant future for robots; I would like to suggest that we may want to include consciousness-like phenomena earlier. Even if my argument is completely wrong from an evolutionary point of view, it may still provide for interesting discussion. 3. Psychoanalytic concepts may be important, not only in relation to how we construct advanced robots, but also with respect to how humans will transferentially relate to them, especially as they become more humanoid.

The ideas offered here are possibly naïve and very speculative, but they may be useful in an endeavor of brainstorming.

3.2.2   The Use of Emotions

Many robotics designers give important weight to emotions as part of the cognitive apparatus, since emotions may be needed for the more complicated tasks of sophisticated sentient beings. Solms and Turnbull [2] have it right that a biological brain system must have an *embodied* system of affects. The embodied systems with which we are familiar reside in animals of one kind or another, and they involve physiological signals to the organism, display signals to other creatures, as well as motor tendencies such as to fight or flight. These systems had to develop, evolving in the Darwinian way of natural selection. This differs from the AI way, which allows for programmers to implant complex systems ready made into their robots. Biological species have not had that luxury; they have to have an affect system – one that is adequate for survival – in place at all times if they are to survive. Then, with evolution, the systems can be modified for more effective function in the given environment.

It may be that in a competitive environment that our imaginary robots would have to live in, they would need such an affect system. But there is no reason this would have to evolve in the biological way, it could be implanted. But in order to do this we would have to know more about the nature and function of affects.

Let's say we are designing a robot. How would we implant or download an affect such as fear? Imagine that our robot is approaching a fire. It needs to know not to simply march into the fire. So it will need heat sensors, as well as some registry of properties of fire that make it dangerous. That will involve cognitive capacities that are already available to robotics engineers. The affect system is a value system, at its heart an approach-avoidance system. Gerald Edelman [3] in the early generations of his Darwin robot, had an approach value to light, the robot "seeks" light. For Darwin III, light is "good." Our robot will have certain things that are obvious to approach or flee. But it has to distinguish "friendly" light from "dangerous" light such as fire.

Now does our robot have to have all the affects that humans have? One could argue "No." It would need seeking and avoidance systems, the bases of our pleasure and pain systems, in order to survive, or to fulfill the purposes of its designers. In some circumstances it might even need an aggressive drive when faced with competition. But why would it need to have separation anxiety? Since it will not have an infancy, it will roll full-grown off the assembly line. However, with a fully mobile robot we might need adaptive devices such as a separation preventer, like a dog's invisible fence, to keep it from wandering off and getting lost. And, then would it need to have sadness resulting from the loss of a valued person or other robot? I am not sure that it would. This will probably be an issue leading to debate among future robotics engineers.

Each of the basic embodied affect systems present in mammals has at least four major aspects [4]. They are: the *physiological, autonomic pattern* within the body; the *display* of the emotion via facial and bodily expression; the *subjective awareness* of the feeling; and the *behavior* relevant to the affect. With *fear*, for instance, one will have autonomic responses, including shortness of breath, rapid heart beat, sweating, and muscle tension; one's face and body will display signs associated with fear (cringing body, wide open eyes, facial grimace, etc.); one will be consciously aware of danger or threat; and one will tend to withdraw or flee. Will a robot need to have this set of features in order to have an affect system? This would seem to be complicated. For one thing, as LeDeux [5] has pointed out, humans, and probably other mammals, have a two-tiered system, one for emergency reaction, and another for more deliberate response. In his well-known *fear* model, if one is walking in the woods, and suddenly comes upon an object that looks like a snake, one will jump back in order get away from the creature, before one is fully aware of what one has seen. A few seconds later one may look and see that it was only a stick, and therefore there is no danger. In the human model, two systems were activated. In one the percept, set to detect dangerous objects roughly and quickly, on a hair trigger for danger, travels to the thalamus and then directly to the amygdala, leading to the defensive reflex reaction. Milliseconds later the percept has traveled to the visual cortex, where it can be examined at more leisure, and it is seen to be a false alarm.

Now would a robot need such a system? In a mammal, the physiological response acts as the sign to the organism registering the danger - or the insult, or the loss, or the pleasure. Evolution has developed a sign system with includes the heart rate, skin changes, muscle tensions, breathing changes which make up the *embodied* code. We could imagine that a robot would not need such a complex system; it could do with a numerical symbolic code, which would label the danger and instruct what to do about it. It could register "fire" and this would code for a rapid escape response. This implies the robot would not have a central representation of the danger; it would be like the mammalian "reflex." This of course is the model that has been provided for us by behavioral theory, and it would seem that a robot could be designed in this manner.

This would be a system without a visual cortex, like the system that probably exists in a reptile. A reptile lives in a very simple social system, where just about everything operates on reflexes. Higher mammals who live in more complex social and technologically based systems seem to need more nuanced guidance systems, and this means central brain areas where "trial actions" or "hypothetical situations," can be entertained before any action takes place. This brings us to consciousness.

3.2.3 The Use of Consciousness

There has been considerable discussion in AI circles about the possibility of consciousness in robots. Recently a whole issue of the *Journal of Consciousness Studies* [6] was devoted to papers on *machine consciousness*. One idea I want to try out, in contribution to this discussion, is that consciousness is an *affect system*. It is unlike the other affect systems in that it apparently is more about the self and its attention mechanisms than about an evaluation of experience. It is the whole of subjectivity, not just one of the colors of subjectivity as are the other affects. But it is a kind of state phenomenon, which involves the whole body, like other affects. The affects each have a kind of valence or value orientation; the value may be about seeking/pleasure, about danger and need to avoid it, or sadness about loss. The valence of the consciousness affect is *salience*, i.e. what is worth attending to right now. There are guides to what is attended to, various indicators of saliency in the perceptual realm. These would include strength of signal (loud or bright or painful), novelty of signal, surprise value of signal, importance to some larger scheme, or relevance to some search procedure. All of these criteria are imposed consciously as in a deliberate search, or unconsciously as in free association, day dreaming, or dreaming.

Each of the usually conceived affect systems relates to a particular facet of possible experience, usually real-world situations in which the organism finds itself. Examples are danger, loss, pleasure, separation panic and others. Each one is instantiated in neurons, perception organs and output programs – display, physiology and behavioral output as well as its place in consciousness. If we think of consciousness as the affect related to salience or importance at the moment, we may have a concept that could be useful in AI research.

There are differences between consciousness and the other affects. One is that the other affects are variable in intensity, i.e. ranging from minimally noticeable to overwhelming. Consciousness we think of as *on* or *off* – *conscious or unconscious*. But is that true? We do speak of being intensely "in the moment," or "hyperaware," but also "dimly aware" "semiconscious" of a perception or idea, "disavowing" of a fact or obvious conclusion. At any of these variable levels, the issue is about: will I attend to this, based on my estimation of its salience.

Does this sound circular? Maybe so but is it different for any other affect? For instance, I am sad because of a loss; why should that be? Because loss is sad. I am angry because of an injustice done me; why? Because injustice is angry making. Evolution has given us affect systems that allow us to respond adaptively to the basic situations of danger, separation, territoriality, and sexual stimulation, among others.

With consciousness do we have a similar set of relevant phenomena? If we refer to the four-part model of affect mentioned earlier we see a similar schema.

When something is worth attending to, a person will be (1) high on the arousal/reticular activation scale, and awake; (2) will show signs of being attentive; (3) will be subjectively aware of it; and (4) behaviorally will focus visual and auditory attention, and take the necessary action. As Posner [7] has described, in discussing attention mechanisms, there is a set series of events leading up to attention and thence to consciousness. There is arousal; there is search, and orienting towards a stimulus implicitly judged to be salient.

The trigger to consciousness often is one of the other affects, and that affect will be part of some perception. That something is very good, such as sensual pleasure, or very bad, such as a rattlesnake, is often the stimulus to awareness. So affect may teach us something about consciousness and vice versa. And what we learn from comparing and contrasting these two mental phenomena may be useful for AI.

Then there is the issue of "where is consciousness?" Its neurology seems to be distributed widely as are the other affects. The system on which consciousness is based is rooted in the brainstem in the deepest and oldest brain structures. Consciousness must have evolved coincidentally with the early affects, which were manifest simply by approach and avoidance. Those two go back to unicellular organisms. Were they conscious? Maybe not in any way we would recognize. But somewhere up the evolutionary tree, the great decider must have been born, its job being to say what is important enough to attend to. We might then call it the third affect. With subsequent evolution, that affect allowed for more complex affects which have subjectivity in them. With Panksepp [8, 9], we assume mice have sadness, fear, separation anxiety, anger, and positive feelings of lust, mirth and pleasure seeking. Once consciousness is online then the other affects may be more useful; and again they didn't all come on-line at the same time. One could imagine anger coming along later in evolution than did fear, and even than sadness. And we are pretty sure that the social emotions such as pride and shame emerged even later.

We could at least imagine that the designers of advance robots could model the primary affects by the simple behaviors of approach and avoidance. Then we might be able to model consciousness. And, rather than saving consciousness until last, maybe we should tackle it early, along with the primary affects, as Mother Nature did. Maybe we should build in an internal reflector to register salience, which will operate in conjunction with negativity and positivity.

One thing about consciousness is that it has long been described in two ways. One way is as part of the activation/arousal system for just about any complex organism. We go from sleep to stupor to arousal, and up to awareness and self-awareness. This involves the massive apparatus described by Pfaff, [10] of energizing systems, which are the activating source of the drives, which he sees as a way of understanding Freud's concept of libido. The other aspect of consciousness refers to its content, to that of which we are aware. Mark Solms [11] pointed out that this content includes both the perceptual sphere, and also the inner-feeling sphere, so that consciousness brings together the perceived world and the self or inner world. And here we have the divisions inherent in the cognitive consciousness system, which includes conscious awareness, and the non-conscious, implicit, procedural mental content. We also have the dynamic unconscious, which is the potentially conscious content that is excluded by defense mechanisms. In our salience model we could say that the defenses function to produce "dis-salience," in other words to reduce salience. They also could produce "hyper-salience," either to keep a trauma alive, or to distract one from something painful.

So where does psychoanalysis come in? Once we have a somewhat independent, motile robot with consciousness, we will have a robot with conflict. And that's where we come in. In the simplest system there will be ambiguous signals, and consciousness will be part of sorting them out. It may do so on the basis of its own prerogative, deciding the salience of each possibility. Two not-yet conscious stimuli may compete for attention. Or two conscious choices may compete for realization. For instance with the robot approaching a fire, the light would be attracting while the heat, when it came sufficiently close, would lead to withdrawal. In our programming we would give the danger signal more salience in the consciousness register, and the robot would then avoid the fire.

This is an over simplified version of how salience would contribute to motivation and decision. In a robot, the programming would give heat more salience than light, but you see how complicated that becomes. Warmth might be better than extremely bright light, which might damage the photoreceptors. Each sense operates on a spectrum of expectable intensity. At the extreme ends of that intensity we get increased salience, often leading the organism to become aware, and to do something about it.

What are the purposes of consciousness in organisms? One is the attention to salience, to stimuli most likely to be important to an organism's survival and welfare. Another is to predict the future. Pally [12] has written about this function. In her view we operate most of the time with unconscious procedures, confident we know what is coming next. Whenever reality confounds us, and the expected does not happen, or the unexpected does happen, consciousness comes on line and corrects the error. Another way of saying this is that unexpectedness or surprise is one of the factors that produces salience.

There is another aspect of consciousness in which it tells us about the immediate future, what we ourselves are *about to do*. In Benjamin Libet's [13] experiments he showed that human subjects, in the process of making a decision, such as which of two buttons to push, an EEG readout shows that there is a *readiness potential* in the brain a few hundred milliseconds before one is *consciously aware* of the choice and the action one is about to make. Thus consciousness *follows* the readiness potential that signals the beginning of the action. Some have lamented that this means we are not the masters in our own brains, that consciousness is not the originator of actions; it is simply a read-out, like the inscription on the screen of a computer. But from an adaptational point of view, consciousness may be useful in that it tells me what I decided to do before I actually push the button, so I know what I am in the process of doing. If we take this to the level of survival value, we could imagine that, in a fistfight with a zombie, who was a competent procedural fighter, but who did not have such consciousness, this would give me as a conscious person an advantage. I would know where I am aiming my punch, and therefore where it is about to land, *before* it lands. The zombie would have to wait until his punch landed, see where it landed, and then figure out what to do next.

It would seem to be difficult to endow a robot with such split-second decision-making abilities. But maybe it would not be impossible. Somewhere in its circuitry, when a robot makes a choice between two actions, the choice is instantaneous, but the chosen action itself takes some time. It would seem possible to have an electronic signal of the *decision* be registered in its central processor before the action is finished, or even begun, so that the robot knows what it is about to do and can begin to predict the consequences.

In summary I want to suggest that consciousness is one of the basic *affect* systems, at the level of the usually recognized motivational systems based on approach and avoidance. It is largely housed in Luria's Unit I, which Solms quotes as the unit devoted to "modulating cortical tone and arousal" (2, p. 28; 14). This includes the brain stem, much of the limbic system, and part of the frontal cortex; it is the system that ascertains what is worth attending to. Consciousness has multiple levels as a neurological arousal phenomenon. Near to the top of the ladder are attention, awareness and self-awareness. The factor that determines attention or awareness is what we might call salience, a factor by which the perceptual/attention system selects figure-objects from background. Living motile organisms need to be able to select objects, and separate them from background, in order to deal with them in an adaptive way. Salience is differentiable so that some things are more salient than others, and this may determine perception, awareness and behavior. In building complex, epigenetic robots, we will need to instill modes of handling conflict. Such a robot may not need the full range of human affects, unless we are trying to build a Turing-eligible robot that would simulate a human being. Foreseeable robots will act as assembly line workers, plant managers and human helpers or mechanical servants. These will at times have conflicts, for instance in receiving two conflicting commands with two desirable goals. A consciousness-type system based on salience could help make choices. For instance, a more salient bad choice would supercede a less salient, good choice, and lead to avoidance, which would end up being self-preservative. One mode of learning in so-called epigenetic robots may eventually be via imitation. Again we may need a saliency detector to determine what to imitate.

Another point that may be relevant to the attempt to create machine consciousness is the "sensori-motor theory of consciousness." Julian Kiverstein [15] describes the work of Bach y Rita, who back in the nineteen fifties developed a vision prosthesis for blind persons. This apparatus included a camera worn by the person, which was connected to a flat console on the patient's abdomen. The camera delivered a pixilated grid-like set of electrical impulses to the grid, giving a kind of retinotopic "image." Each pixel, when activated, would deliver a vibration to the patient's abdominal skin. With some time and practice, the patient could walk around with the apparatus in place, and he would direct the camera as if it were a Cyclopean eye. At first the patient would experience a meaningless set of vibrations over his abdomen. But with time these patterns would register in his brain and he would actually see images. One interesting thing was, however, if he was not in control of the camera, he would never make the step to *seeing the images visually*, they would remain meaningless vibrations. Another relevant experiment, which has been repeatedly done, is to give spectacles to a sighted person with prisms in them that turn the visual field upside down. This is at first quite disorienting to the subject, who staggers around completely at a loss. However, with practice the person begins to see the world as right-side up and can negotiate in space perfectly. Then, when the glasses are removed, the subject has to go through the same process of accommodating to the reversed world, and righting it. Here again, the person must be free to move around independently under his own control. If you put the person in a wheel-chair, restrict head movement, and you control the wheelchair, he will not be able to make the spatial accommodation. This all has encouraged a "sensori-motor theory" of consciousness development (15). One sees the world and develops a skill at moving one's head and seeing that this changes the viewpoint in controllable ways. It seems likely that in the brain, the use of a connectionist type of architecture allows for the control of the visual image to eventually make that image predictable in a way that makes sense. Thus consciousness arises from this mastering of the visual world. How this works with other senses, has not been explored.

It is postulated that with a robot with a video vision system, with image storage capacity, that the robot's control of its visual field could generate a kind of visual consciousness – in other words a visual world that is predictable and controllable. Could that be the basis of consciousness? What about a congenitally blind person? Apparently such a person has to make do with sound and touch, and it could be that the same rules apply. The person controls his perceptions by moving around and changing what he feels and hears in a controllable way, and thereby develops a more tactile consciousness.

This is an intriguing speculation. It suggests that when one gains sensori-motor control of experience, that becomes the basis of visual consciousness, and each sense probably has its own version of this. The brain is conscious of its sensori-motor world via defined levels of activation; within that activation space, there is an active pursuit of salience, or of salient percepts. In mammals a saccade system keeps the visual field refreshed, and things that change between saccades may become salient.

These speculations about the evolution of consciousness may be completely wrong as explanatory models of evolutionary reality. In other words it may be that consciousness and emotions are much more different than I am suggesting. But we might use the idea anyway and provide robots with a kind of consciousness, based on the criterion of salience. It could be an avenue to machine consciousness.

3.2.4 *Transference*

Humans have a great and unsurpassed capacity for transference. Transference is one of the key foundation memes of psychoanalysis. Analysts consider transference to be a "transfer" of feelings and expectations from one person to another. This usually means that habitual feelings or expectations that a child experiences with early objects – parents or other caretakers – will influence the later experience with teachers, bosses, or lovers. Transference can actually alter perception and cause distortion in the experience with other people later in life. What does this have to do with the human relationship with robots? There are several major aspects that are important.

Most obvious is the fact that until recently the robot has existed only in people's imaginations. Since the robot is created by my imagination I can create it from scratch and make it like anything I please. In such a situation I have freedom from reality, and therefore the creation will be derived from my inner understanding and even my neurotic distortions of reality. So what inner models do I use? I will use some of the basic archetypes that structure my personality, including those derived from my psychic development from childhood, and those arising from my cultural milieu. For example, the robot of our fiction is usually a kind of omnipotent servant. And who do we know who is our omnipotent servant? Of course, our parents. Our first caretaker, our mother, is our first servant, and she is also omnipotent in our eyes. That the robot also is emotionally deficient is interesting. For fiction writers there is no obvious reason for this. Why not a robot with superhuman emotionality and capacity for empathy? Possibly, our idea of the mythical robot is subject to a kind of splitting, between archetypal good and bad robots. The negative side of the split is represented in fiction by villainous robot adversaries, and such creatures are not depicted as having the usually considered positive, endearing trait of empathy.

Now we already have increasingly sophisticated robots used as human companions. Sherry Turkle [18] gave a talk, at the American Psychoanalytic Association, about robotic companions and the human feelings they elicit. She has described some of the early game-boy type computers, who very soon become "he" or "she" and no longer "it." They also were deemed to "think" and to "want." Then there is the Furby, a small doll-like companion that was built to provide playmates for children, and also for nursing home inhabitants, to provide company for the aged and demented.

We have long appreciated the difficulty, maybe the impossibility, of providing robots with real emotions of the embodied kind that exist in animals. But maybe there is no need to. People do a very good job of projecting their own emotions, i.e. transference, onto their stuffed animals. Simulacra of animals, such as the Furby, seem to be almost as good as the real thing. And some people prefer Furbies to humans. As robot companions become more humanoid, they too may be preferred. And what will life be like for the humans who prefer them?

From a psychoanalytic point of view our fantasies about robots have led the way in our creation of robots, and these fantasies seem related to our experience of our parents, and our Gods. Another fantasy of the robot designer may be the idea that he or she is a god, and that this is Creation.

A conclusion we may draw from this is that the fantasies of roboticists are important to understand, because those fantasies will influence the robots created. As those robots grow more and more human-like, their architectures will carry the mark of the minds, the neuroses, and the character traits of their designers - the people who program their brains. At this point this kind of scenario is the stuff of fiction, but eventually it may be a hugely important aspect of the culture we will live in. And, it will take a psychoanalyst to figure it out.

References

[1] Palensky, P., Lorenz, B., & Clarici, A. (2007) Cognitive and affective automation: machines using the psychoanalytic model of the human mind. Paper presented at ENF 2007.

[2] M. Solms and O. Turnbull, (2002) The Brain and the Inner World. Karnac/Other Press, Cathy Miller Foreign Rights Agency, London. England.

[3] Almassy, N.; Edelman, G. M.; and Sporns, O. (1998). Behavioral Constraints in the Development of Neural Properties: A Cortical Model Embedded in a Real-World Device. *Cerebral. Cortex* 8(4): 346-361.

[4] Olds, D.D. (2003). Affect as a Sign System. *Neuro-Psychoanalysis*, 5:81-95.

[5] LeDeux, J. (1996) *The Emotional Brain*. London: Weidenfeld & Nicolson.

[6] *Journal of Consciousness Studies,* 14(7) (2007)

[7] Fossella, J & Posner, M.I. (2004). Genes and the development of neural networks underlying cognitive processes in M.S.Gazzaniga ed. *The Cognitive Neurosciences* 3rd edition Cambridge: MIT Press.

[8] Panksepp, J. (1998). *Affective neuroscience: The foundations of human and animal emotions.* New York: Oxford University Press.

[9] Panksepp, J. (2003). At the interface of affective, behavioral and cognitive neurosciences: Decoding the emotional feelings of the brain. *Brain and Cognition.* 52, 4-14.

[10] Pfaff, D., Martin, E., & Kow L-M. (4/7/07) Generalized brain arousal mechanisms contributing to libido. Paper given at the Arnold Pfeffer Lecture, New York, NY

[11] M. Solms (1997) "What is consciousness?" *Journal of the American Psychoanalytic Association,* vol. 45, pp. 681–778.

[12] Pally R. (2005) Non-conscious prediction and a role for consciousness in correcting prediction errors. *Cortex* Special issue on: Consciousness, mind and the brain (5): 617-738.

[13] Libet, B., Gleason, C. A., Wright, E. W. & Pearl, D. K. 1983 Time of conscious intention to act in relation to onset of cerebral activity (readiness potential): the unconscious initiation of a freely voluntary act. *Brain* 106:623-642.

[14] Luria, A. (1973) *The Working Brain*. Harmondsworth: Penguin.

[15] Kiverstein, J. (2007) Could a robot have a subjective point of view? *Journal of Consciousness Studies* 14(7): 127-139.

[16] Shelley, M. (1818), *Frankenstein, or the Modern Prometheus*. London: Penguin Classics, 1985.

[17] Litowitz, B.E. (1997). Learning from Dr. Frankenstein What Makes Us Human. *Annual of Psychoanalysis*, 25:111-123

[18] Turkle, S. (2004) Whither Psychoanalysis in Computer Culture? *Psychoanalytic Psychology: Journal of the Division of Psychoanalysis*, American Psychological Association, Winter 2004.

3.3   Discussion Chaired by Authors

In the following, the most important questions and comments of the discussion after the above paper presentations are summarized. In this context, "Q" is used for any type of question or comment, usually from the auditorium, while "A" stands for "answer", usually from the chairs.

Q1:   The reason for not attributing emotions to robots can be found in the very emotion these robots would have if they were able to feel what they really would feel, which would be rage. They would be enraged at humans for making them do all these things and not giving them any consideration for their own needs.

A1:   Robots may not actually have the whole range of emotions. The question is whether it would make any sense for a robot to feel rage, or sadness, for instance.

Q2:   The neural definition of basic emotions may be a crucial point in integrating hardware and software, as Prof. Dietrich said, and might be a micro program of how we function, where affects emerge from neocortical cognition. Where in Palensky's diagram would this micro program be?

A2:   The diagram is a metaphor for the transition from hardware to software. However, our model is platform-independent. If we had platform-dependent models, we would have to build the platform as well, but we do not understand what this platform should be.

Q3:   The biological platforms are of course infinitely complex, and no one is ever going to be able to simulate them. What we can do is implement different kinds of architectures, attractor envelopes, they are very large and embodied; representations of the organism. All the information processing that occurs is centered around an organismic concept. So my question is, is anyone building that kind of an organismic concept as the first foundational principle?

A3:   There are promising new technologies, however, what you are talking about goes much further. The way we work is that we always have plain software. If it is considered necessary to have this hardware, as it influences the functionality as well, we would just go one step down and simulate this hardware, as well. We certainly can never achieve to simulate real biological matter, but what we hope to achieve is to simulate as much as is necessary to obtain intelligent behavior.

Q4:   In the presentation, emotion was described as a slow response, while cognition was said to be fast. However, I should think that emotion tends to be a much quicker response than cognition tends to be. Additionally, I have realized that robots are usually conceptualized as either male or neuter, and, as for instance a Japanese toy, often require caretaking. I think that we should think about maternal robots, as in psychoanalysis the maternal function of recognizing what goes on in another person's experience and then attending to it. In my opinion, it would be important to be able to model this particular function in a machine.

A4: Just as far as science fiction movies or literature is concerned, female robots on the one hand and caretaking robots on the other, as in "Metropolis", "Bladerunner" and "I, Robot", have been anticipated.

4 Session 4

4.1 The Prometheus Phantasy – Functions of the Human Psyche for Technical Systems

Elisabeth Brainin and Gerhard Zucker (né Pratl)

Technical systems support us in our daily life and have been improved ever since. Still the available systems today are not able to understand their environment in a way that is similar to human understanding. Enabling technical systems to interact with the real world and with human users in a more human-like way is a challenging research topic. To do so we need models of human abilities. We need to understand how humans perceive their environment, how knowledge about the environment is created, stored and accessed, how motility works, how decisions are taken and actions are executed. And we need technical translations of these models that allow us to implement them in a technical, not a biological way. This paper identifies principles and functions of the human psychic apparatus that make up a set of basic requirements for developing technical systems that can perceive and interact with the environment – including human beings – in a more human-like way.

4.1.1 Introduction

Culture and in the long run technical skills and theory should serve mankind for the purpose of submitting nature under the control, rule and reign of mankind. It makes nature useful and should as a final aim to provide humanity with immortality. Being immortal, banning death is an ever existing aim of mankind.

Having that in our mind, knowing about the grandiosity and megalomania of this task, we want to restrict ourselves to only a few functions of the mental apparatus, which we shall define in the course of this paper. Nevertheless, we call our paper *Prometheus phantasy*, in mythology he was the first human (actually a titan) who dared the "hybris" (or sacrilege) towards the gods of antiquity, to create and form humans and to rule over nature. For this purpose he needed the fire and he robbed the divine fire for mankind! Thinking about the divine punishment Prometheus had to endure we will restrict ourselves to certain functions of the psychic apparatus, based on neuro-psychoanalytical principles.

In order to achieve human-like abilities in perceiving and interacting with the environment we need to be focused on a discipline that can provide us with the answers to our questions: we need a theory of the human mental apparatus. This theory has to abstract the functions of the human mental apparatus so that they become accessible to engineers.

Although many partial problems have been solved in the past by applying knowledge from different scientific domains, it is until now not possible to supply a complete functional description of the human abilities that are involved in perception and interaction. The research results from neuro-psychoanalysis that have been gained in the last years, however, are very promising and are a candidate to create the first complete and scientifically sound model of the human psyche. This process is by far not finished, but connecting neurology with psychoanalysis fills a gap that makes it now possible to access this knowledge from the engineering domain. Research is ongoing in a project at the Vienna University of Technology [1].

In terms of technology we are focused on computer technology. A computer has the basic abilities of information processing and storage and the mechanisms used are well researched. We are able to program such information processing machines arbitrarily. Although the current von Neumann machine might not be the most feasible one for the tasks at hand, it is still generic enough to be programmed in a way that fits our needs.

Neuro-Psychoanalysis tries to find a link between the abstraction of the human mental apparatus and the biological basis for it. A good example is "memory": different kinds of memory linked to different parts of the brain. Memory is one basis for all functions of the psyche. To fulfill all tasks of a human being, to build human relations, to love and attach to other human beings, to differentiate between inner and outer world, to perceive and to integrate stimuli from the inner and the outer world, we need perception, which is not dependent only on current sensory information (see, for example, [2]) and memory, which is the basis of "intelligence".

This paper starts with an explanation why we think that Neuro-Psychoanalysis and engineering should cooperate; then we describe two prototypic applications taken from the engineering domain to define the area in which we want the technical system that we described to be located. Next we start to blend engineering and Neuro-Psychoanalysis by tackling the issue of a biological or virtual body, respectively. After a brief description of the ever important issue of memory we compile a list of principles that we see as necessary to create a technical system with human-like abilities.

We are aware that the language that is used in engineering (that is, information technology, control engineering, and automation) and Neuro-Psychoanalysis is very different. We try to consider this fact by using only limited vocabulary from both sides and explaining terms that might be ambiguous or unknown on the other side.

4.1.2   Why Neuro-Psychoanalysis?

This paper is based on the following hypothesis: We use neuro-psychoanalytic models to create a technical system with human-like abilities. Neuro-Psychoanalysis gives us the theoretical framework for an abstraction of the human mental apparatus, which we use for our technical problems. The link between the biological basis for this abstraction to a theory of the human mind is Neuro-Psychoanalysis.

The models may be augmented by models, theories or knowledge from other disciplines (i.e. if Neuro-Psychoanalysis does not provide enough information for a technical implementation), but whatever is added must not contradict the results of Neuro-Psychoanalysis. Following this hypothesis we conclude the following: If we want to use the knowledge of Neuro-Psychoanalysis we have to create a system that obeys this knowledge; this means that it has to implement functions of the human psychic apparatus, which we shall define later and which are based on neuro-psychoanalytical assumptions.

If we want to apply the research results of Neuro- Psychoanalysis to a technical system, we need to build a system that follows the functional description of a human being. To be more specific, we cannot build a technical system if we don't implement the mechanisms of the human psyche. Therefore we compile a list of functions and principles that we think are the minimum requirements for a successful technical implementation. But we do not only define these functions and principles, we also try to provide a reasonable translation into the domain of engineering. We are aware that the list we create is by far not complete and will need extensions. But we want to define a *sine qua non*, a set of conditions without which it is *not* possible to create technical systems with human-like capabilities.

A biological being cannot exist without its body. We could therefore argue that we will only succeed, if we build a virtual body. Depending on our eagerness for precision, we could try to reproduce the biological body down to its most elementary mechanisms, which would end up – in its final stage – in a molecular copy of the body. This is not the path we have chosen, since it does not look too promising. We also do not attempt to model the human psyche on, for example, neural level (by simulating behavior of neurons). Our approach is to identify a functional description of the human psyche on a much higher level and then use this abstract description for creating a technical implementation.

One of the main challenges we face is the considerable difference between biological individuals and technical systems. There are maybe more things that separate them than they have in common. On a basic level biological individuals have to survive and to reproduce. In the long run this implies that they have to *attach* to each other and develop relations. While we might be able to argue that a self-sufficient robot also has a need to "survive" in terms of ensuring that its batteries get recharged on time, we will not be able to state this for a system that is installed in a building with a fixed power supply and no means for changing it. Even worse, we will not be able to argue that technical systems need to reproduce[47].

It appears futile to equip a machine with a copy of mechanisms that we find in biological individuals. Too many properties create conflicts when attempting to translate them. Therefore we abandon copying and look at abstracting functions. We want to define principles of the human mental apparatus, which we could use in the further course of our work. They are abstract concepts – as is the concept of a mental or psychic apparatus. We nevertheless stick to the concept of a body, but in an abstract way: There has to be something that represents the current state of

---

[47] We would appreciate it, if they could, but as we said earlier, we are looking at state-of-the-art technology where such abilities are out of reach.

the system and its current requirements. It is not a biological body, but is has functions that a biological body also has.

4.1.3 Applications

Why do we want to understand the human psyche, where is the benefit for engineers? The answer we see is in the current state of the art in technology: today machines lack the ability to "understand" the environment in which they operate. They are able to fulfill tasks they have been programmed to, but with very little flexibility. They can process a limited amount of information that gives them clues about the environment. Many sensors are restricted by design (a motion detector can only provide information about some kind of movement within its range, but is has no information whether this movement was triggered by a person or not). But even highly sophisticated sensors (e.g. cameras), which provide vast amounts of information do not solve the problem, because the available information cannot be processed. Today's surveillance systems are able to detect persons and separate them from the background. They can also distinguish between a person and the shadow that a person casts and they can detect partly overlapping persons or – to a certain degree – only partly visible persons. But these abilities are far from human-like abilities, especially when facial expressions, movements or gestures are relevant.

We briefly describe two possible prototype applications for a technical system with human-like abilities. Such applications are available today with all limitations of today's technology and therefore provide a good platform for future applications.

Application 1: Supervision System

The system supervises persons that have special needs, e.g. in hospitals or retirement homes. It knows where the persons are and if everything is all right. It creates a comfortable environment for its inhabitants. In case of an emergency or if a person needs assistance, it alarms the human operators.

The system is exposed to a real-world environment where it has
(1) to perceive objects and events in the environments.
(2) to interact with objects in the environment.
(3) to anticipate behaviors of humans[48] in the environment.

The system is mainly passive, meaning that the main focus in this application is perception, and "understanding" intentions of the supervised persons, but not only to be understood, but to be able to anticipate their actions; that means an interaction would be performed (see also [3]).

---

[48] In psychoanalysis humans are sometimes referred to as "objects": psychoanalytic objects can be persons (sometimes also inanimate objects) in reality or phantasy, which provide satisfaction (or not). Engineering uses "object" to refer to inanimate things. We explicitly add "psychoanalytic", if we refer specifically to "objects" in psychoanalytic sense.

Application 2: Self-sufficient Vehicles

Self-sufficient and autonomous vehicles are able to support persons in real-world environments. This support consists of tasks as they occur either in a household or in an office building – ensuring safety of inhabitants, leading visitors around the house, ensuring correct operation of infrastructure, to name a few examples. Additionally to the requirements of the first application, such a vehicle has a much stronger need to interact with the environment and the objects in it: it has to move around freely (which is considerable technical effort), take care about its power supply, and operate devices in the environment.

It is worth noting that both applications have already been implemented in many different ways and are therefore not a novelty. However, the level of sophistication in existing solutions is today not sufficient to make them broadly accepted. Therefore we return to the first hypothesis: we think that the best way to achieve humanlike abilities in a technical system is to follow a model that has the potential to describe the functions of the human psyche. Only then will it be possible to push forward the abilities of technical system to a level that can be called human-like. This is our Prometheus phantasy: we create entities that are similar to humans in the way they operate – just like Prometheus created humans[49].

4.1.4   A First Translation

The technical requirements for the systems are on the sensory side concerned with information processing, pattern extraction and a first level of perception of the environment, and the object and the events in it. On the motor and actuator side the system in application 1 has to control actuators like door openers, locks, air-conditioning systems, sunblinds, or user interfaces. For the vehicle in application 2 considerable efforts have to be put into the control of the mechanical parts like motor drives or other mechanical actuators.

Anticipation

Aside of the technical challenges we have to take a close look at the third point, which is crucial for achieving good results with the system: it has to anticipate behavior of persons. This implies two things: for once the system has to be able to perceive behavior (which is a considerable technical challenge); then has to create hypotheses about the results of behavior and anticipate future behavior. In other words, the system must hypothesize about motivations and intentions of the persons. It can do so only if it has by some means experienced similar situations and is able to abstract and generalize from this experience. While we could claim this for a biological being, it is unlikely for a machine to experience situations like a human being does. We therefore have to shortcut the learning phase and teach the system these situations without letting it actually experience them. We have to

---

[49] The name Prometheus means "the one who anticipates". We will come back later to the fundamental human function of anticipation.

teach it *experience* that it cannot gather by itself. Still the method to gain experience remains the same and is explained in the following two sections.

Virtual Body

We have stated earlier that a biological being cannot exist without a body. This is a priori not true for a technical system – depending on how the "body" is defined. Although a system does always need a physical representation on which it can operate (the hardware), it is not tightly linked to it – a program can run on many different hardware platforms and operate identically. The hardware merely allows one instance of the program to run, but it is not the only hardware that enables the program to run.

In case of self-sufficient vehicles (application 2) the hardware is more tightly integrated into the whole system, but the program can still be switched from one hardware platform to another without ceasing existence. Therefore the idea of a *virtual body* is a way to move on without design. We stated earlier that creating an exact copy of a biological body is not a reasonable approach for our goals. Instead we have to identify the functions that we need from the body and implement those as the abstraction of the biological body.

The interface to the environment is given by sensors that provide information about the environment and the objects and events in this environment. Additionally, the body has a set of sensors that provide information about the body itself. The state of the body is not known as such, but has to be reported by sensors – just like the environment has to be reported by sensors. Anything that cannot be detected by sensors remains undetected (e. g. human beings have no sensors for detecting electric fields). This is valid for both events that occur in the environment as well as events that occur in the body[50].

The body of a system does not necessarily have to be completely virtual. The characteristics of the sensors used are relevant for the information that can be gathered – the sensors define how the system perceives the environment[51]. And especially the motor part (which is more important in application 2) depends on the design of the hardware. Therefore the (psychoanalytic) body of the system is a combination of virtual body and real hardware.

Two Worlds

The human psychic apparatus has to mediate between the requirements of the body and the requirements of the environment. This can be modeled by creating two representations (two worlds, see also [4]): an inner world representing the

---

[50] Unlike the biological body there is not necessarily a clear and permanent border between body and environment. In principle we would be free to change the assignment of sensors and allocate some to the body that have be assigned to the environment before. This may also happen dynamically. We leave it to the reader to experiment with this idea of a virtual body.

[51] This is the idea of embodied intelligence; intelligent behavior depends on the design of the body.

body and an outer world representing the environment. The task of the human psyche is thus to integrate the inputs or stimuli of both worlds, extract relevant information and create decisions based on current information and information that has been gathered earlier. The psyche acts as a complex control system. The integration of sensory information from both sides is done in the Ego and it is a main part of what we call human intelligence.

The environment defines the outer world, or, more precisely, the inputs of the sensors that build the interface to the environment. The outer world is a model of the environment and will in general consist of more than just the current sensor stimuli (that is, all the information that has been gained earlier and is now stored in memory).

The same applies for the inner world, only that the inner world is more than merely a representation. It has to include more functions than the outer world; therefore we now take a look at the principles that make up the psyche and that are necessary for our purposes.

4.1.5   Memory

An important function of the human psyche (and thus also for the technical system) is *memory* which we will describe in the following part, using Mark Solm's and Oliver Turnbull's concepts [4]. Memory as we use it in this section is far from memory as it is currently used in computer technology: memory is not storage space, where information can be dumped and retrieved at any time[52] and in which information remains unchanged and unconnected. The term memory as we use it here covers several different mental functions, which should fulfill the following tasks: Encoding of newly acquired information; storage for retaining information; retrieval for bringing it back to the mind [4, p. 140]; consolidation of stored information – which makes the process even more complicated.

Working memory (or rather immediate memory, as it is also called) are memories of events or facts at this moment; long term memory (formerly called short term memory) begins a ready a few seconds ago. According to Hebb's law ("Cells that fire together, wire together") short term memory is transformed into long term memory [4, p. 146].

The semantic memory

As described by Solms and Turnbull semantic memory is the storage of our knowledge of the world, "objective" information such as general knowledge, abstract rules, dependencies between events, and the like. In connection with the principles of the psychic apparatus (which is covered below) semantic memory relates to the "conflict free Ego spheres" that are described by Hartmann [5, p. 10]. The bottom-up perception of babies is transformed in a top down perception in the course of development. Semantic rules are the domain of classic artificial intelli-

---

[52] This is what the term "Random Access Memory" – RAM – refers to: information is available at any time, no matter what information has been retrieved before.

gence, which give a good starting point for our technical system. But the representation of semantic memory needs to meet the requirements of the psyche that we envision; for example, it has to deal with incomplete and contradicting information.

The Procedural Memory

Procedural memory is the bodily memory [4, p. 156] for habitual motor skills. They are hard to learn and hard to forget, like riding a bicycle, walking, playing an instrument and the like. According to functional-imaging studies different parts of the brain are activated in procedural and semantic memory. If we adapt this memory in the engineering domain, we have to have knowledge about the "physical" entity that makes up the system and have knowledge how to operate our peripherals.

The Episodic Memory

This memory involves re-experiencing events of the past [4, p. 160], but "they do not exist as experiences until they are reactivated by the current Self" [4, p. 162]. This implies the complexity of the unconscious and repressed, which we will not cover in our paper[53]. Even if we have a Prometheus phantasy we are not able to imagine a machine with a feeling of Self. Nevertheless, our system needs to have memory about past events, in which it was involved. Retrieval mechanisms differ strongly from "traditional" computer memory. Information cannot be accessed, it has to be restored. It is also necessary to query episodic memory for events that are similar to some current event.

4.1.6   Principles of the Psychic Apparatus

If we want to build a machine, which is able to perceive and interact with objects in the environment and anticipate the motivations and intentions of persons, we have to know more about the function of "anticipation" and also which agency of the mental apparatus is responsible for it. In addition to the requirements that we have identified in the previous sections we now identify a list of psychoanalytic principles that are necessary.

(1)   *No activation without inhibition* is an elementary principle that is relevant for the psyche. Activation in the psyche is understood as a trigger between different parts of the human brain: a part of the brain that is active has the ability to activate another part. Inhibition refers to a mechanism that offers additional options by inhibiting functions of another system. The Ego, for example, has the ability to inhibit drive energies; this way a human being can have a "second round" of evaluating information, it does not have to follow the primary drive, but can modify processes that would otherwise run automatically (that is, determined by drives) [4, p. 99]. In a technical design we have to consider this principle by creating a ba-

---

[53] We also disregard the role of emotions and feelings for all kinds of memory.

sic control system with inflexible mechanisms, which can be overridden by another control system that operates on a different level, using more information (e.g. memory).

(2) *"Reverse afferentiation"* as the essential component of any organized motor action works with the principle of *reciprocal innervation* basic to all nervous action. This principle is closely related to the previous one; it requires a control loop in motor activities that feeds status information about the movement back to the controller.

(3) *Active anticipation:* as described by A.R. Lurija [6] requires frontal lobes[54] (for humans). They are responsible not only to the present but also to the future active behavior. In the EEG (electroencephalogram) we can observe expectancy waves, which increase with the expectancy of a stimulus. They appear in the frontal lobes of the brain [6, p. 93]. To the execution of motor activity the premotor cortex is necessary; it maintains the control of the execution, the general plan of movements. The frontal lobes participate in the regulation of activation processes for voluntary action. The increased attention is shown in the EEG by "depression of the alpha-rhythm" [6, p. 189] and the faster waves are increased in amplitude which is seen as evoked potential.

The technical system that adapts this principle has to perceive a person, extract information from posture, movement, mimic, or gestures and create a hypothesis about the current state of the person. On this base it has to extend the hypothesis into the future to achieve a prognosis about future behavior.

(4) *The pleasure principle:* reigns our psychic life, it means to avoid unpleasure. It the course of human development in extrauterine life[55] this task is becoming more and more difficult. "[...] pleasure and pain are the first mental qualities between which the individual distinguishes [...] one of the first mental achievements is the ability to imagine [...] to hallucinate wish fulfillment" [7, p. 58-59]. In extrauterine life need satisfaction becomes incomplete. "Wish fulfillment is delayed, and pleasure is rationed and curtailed. [...] these disappointments and frustrations become embedded in the Ego structure [...] it will be the task of the Ego to react in a similarly restricting, delaying, rationing function toward the strivings of the id. Probably these are the first beginnings of what will develop later into the reality principle" [Freud A, p. 60]. This leads to the forming of the Ego as a psychic agent.

(5) *The Ego as a psychic agent:* As the ego becomes "perceptive and knowledgeable for the environment, it makes wish fulfillment safer than it is for a wholly impulsive being [...] It is only that, in return for greater security, the individual has to put up with the restrictions of wish fulfillment which are imposed by the modes of functioning [...] of the ego" [Freud A, p. 60]. The ego is the agent which allows observing the other psychic agents like the Id and the Super Ego in self perception. In course of development we cannot see pure undistorted Id impulses.

---

[54] A structure in the human brain

[55] Life outside the womb, i.e. after birth

What is transmitted via the Ego's ability of observation are Id impulses modified by the defensive measure of the Ego. Only the psychoanalytic observer can distinguish between the separate institutions: the Id, the Ego and the Super Ego. The Ego as an observer and mediator notes the onset of instinctual impulse, the heightening of tension and the accompanying unpleasure, and the relief of tension through gratification.

The Ego functions as an integrative force between inner and outer world, includes perception and self perception (i.e. perception of the inner world), the executive motor functions – everything what we refer to as human intelligence. But in the psychic apparatus the Ego is described by Freud in a struggle against painful unendurable ideas or affects [8, p. 41-61]. Later it became the term repression, which serves the Ego as protection against instinctual demands.

In the engineering domain we need to have basic control mechanisms on top of which there are additional control mechanisms; the top mechanisms allow delaying fulfillment of lower requirements. The pleasure principle drives the technical system in an adapted way. It does not crave for food or love, but has to fulfill other synthetic needs, depending on its intended application. "Pleasure" and "unpleasure" are translated to the satisfaction of synthetic needs; if a need is satisfied, the system is successful, if not, it is "unsuccessful".

(6)   *The reality principle:* Freud mentions this principle, which replaces the pleasure principle under the Ego's instincts of self-preservation. The intention of obtaining pleasure is not abandoned, but achievement of postponement of pleasure becomes possible, which allows us the hypothesis of conflict free spheres in the mental apparatus. The technical system itself would represent reality and functions on the basis of the reality principle. Since technology is always embedded into the real world, the reality principle is fundamental and intrinsically contained in all designs.

(7)   *Postponement of satisfaction* becomes possible [9, p. 10]. In another basic work "The Formulation on the Two Principles of Mental Functioning" Freud says about Thinking: *"Thinking was endowed with characteristics which made it possible for the mental apparatus to tolerate an increased tension of stimulus while the process of discharge was postponed. It is essentially an experimental kind of acting [...]"* [9, p. 221]. This is not so far from our former mentioned inhibition-activation principle, but the Ego gives the human mind the possibility of postponement! In the engineering domain we need to create a mechanism that does not continually aim at maintaining balance, but also tolerates a threshold of tension to make space for alternatives (which leads us to another psychic function, the *stimulus barrier*, which we describe below). The system then can try out different alternatives and select one that fits best; trying out does not necessarily mean to apply it to the "real world" (the environment), but – depending on the sophistication of the outer world representation – can be done in a simulation; the results of the simulation have to be sufficiently close to what would have happened in the real world. Postponement serves another aim, which is another principle of the human mental apparatus:

(8)   *The economic principle of saving expenditure of energy:* Freud speaks about *phantasying*, day-dreaming and perhaps we could add theorizing: "A general tendency of our mental apparatus, which can be traced back to the economic principle of saving expenditure [of energy], seems to find expression in the tenacity with which we hold on to the sources of pleasure at our disposal, and in the difficulty with which we renounce them. With the introduction of the reality principle one species of thoughtactivity was split off; it was kept free from reality-testing and remained subordinated to the pleasure principle alone. This activity is phantasying, which begins already in children's play, and later, continued as daydreaming, abandons dependence on real objects" [10, p. 234].

In our concept of adapting certain Ego functions to a technical system we use the described mechanism of daydreaming (i.e. "saving expenditure of energy"); but instead of daydreaming we are trying to conceptualize a technical frame, working as an Ego.

Heinz Hartmann defines *conflict free spheres* in the Ego, which are responsible for the fulfillment of reality demands, without being in conflict with instinctual demands. Hartmann underlines the autonomous centers in the mental apparatus, conflict free spheres, which are independent from instinctual sources in their origin, but in a steady interrelation with the fate of the instincts. Aggressive as well as sexual energy can become neutralized via Ego functions [5, p. 133]. *Adaptation-Control-Integration* (which means the synthetic function of the Ego) and thought processes (i.e. Ego processes) work with desexualized energy (sublimated). So we rather focus on this concept, which includes the anticipatory function of the Ego [5, p. 52]. And so we are back at our starting point: active anticipation.

Now we have to define what should actually be anticipated: For anticipation of another person's motivations and intentions a memory is needed and also empathy for the feelings of the other person. We gain empathy from the body-movements, from the facial expression, from the sound of the voice etc. of another person. But there also exists a communication between humans, which is from the unconscious to the unconscious of the other person.

The anticipatory function of the Ego would be a part of Panksepp's Seeking System as described in: [11, p. 144 - 164]. As there is no unconscious of an apparatus beyond the human mental apparatus, we will restrict ourselves to the Ego functions described above.

(9)   *The Ego serving as a stimulus barrier:* This is an important function not only for our purposes. This shield against "traumatizing" stimuli serves against too much excitation or arousal, which could lead to feelings of unpleasure. Even Hartmann mentions here the dialectical relation between quantity and quality, the turn of quantity into quality, which is a leading idea of dialectic, formulated by Hegel. This idea and the function of stimulus barrier can serve us in constructing our technical system in the following way: even a computer has limited resources in terms of computing power and storage space. This implies that it may happen that the machine is confronted with too much information, which would delay execution of actions or creation of decisions. As we are not able to always find the

globally best solution, we have to add a mechanism that creates the technical counterpart of unpleasure in such overload situations, which then triggers a barrier for filtering information. We do not assume that we will be able to create a system that can actually be traumatized, but it would improve overall system performance.

4.1.7 Conclusion

This paper shows the attempt of combining two scientific disciplines by defining a common seam that will hopefully bring both sides closer together. We have chosen a "least common denominator" to define principles of an abstract psychic apparatus, which is feasible for a technical system. Both sides had to work carefully to give the other side a chance to take a peek at one's own methods and models. The result is a list of requirements originating in Neuro- Psychoanalysis and being adapted to the engineering domain. The list is certainly not complete and we have deliberately left out issues that we think cannot be solved yet; we have not touched the topic of consciousness at all and we deliberately stayed in conflict free spheres in the Ego. But although the functions are not yet accessible to a technical implementation, we are aware that they exist and we know that Neuro-Psychoanalysis has models that describe them. We can use this knowledge to verify that the functions we have defined are valid and consistent.

References

[1]   G. Pratl, B. Lorenz, and D. Dietrich, "The artificial recognition system (ARS): New concepts for building automation" in Proceedings of 6th IFAC Int. Conf. on Fieldbus Systems and their Applications, Puebla, Mexico, Nov 14-15, 2005.

[2]   O. Sacks, "The Man Who Mistook His Wife For a Hat", Summit Books/Simon & Schuster, Inc., New York, 1987

[3]   G. Pratl, "Symbolization and processing of ambient sensor data," Ph.D. dissertation, Institute of Computer Technology, Vienna University of Technology, 2006

[4]   M. Solms and O. Turnbull, "The Brain and the Inner World", Karnac/Other Press, Cathy Miller Foreign Rights Agency, London, England, 2002.

[5]   H. Hartmann, "Ich-Psychologie – Studien zur psychoanalytischen Theorie", Klett, Stuttgart 1972

[6]   Lurija, "The Working Brain", New York, Penguin, 1973.

[7]   Freud, "Problems of infantile Neurosis, A Discussion", The Psychoanalytic Study of the Child IX, 1954

[8]   S. Freud, "The Neuro-Psychoses of Defence", The Standard Edition of the Complete Psychological Works of Sigmund Freud, abbr. SE Volume III (1893-1899): Early Psycho-Analytic Publications, 1894

[9]   S. Freud , "Beyond the pleasure Principle", SE Vol.XVIII, 1920

[10] S. Freud, SE Vol XII, 1911

[11] J. Panksepp, "Affective Neuroscience: The Foundations of Human and Animal Emotions (Series in Affective Science)." Oxford University Press, New York, New York, 1998

4.2 Return of the Zombie – Neuropsychoanalysis, Consciousness, and the Engineering of Psychic Functions

Yoram Yovell

Neurobiologists and philosophers tend to equate the mind with consciousness. Accordingly, the mind-brain problem has been formulated as "finding the neural correlates of consciousness." In their efforts to emulate the mind, engineers are attempting to design machines that would possess some of the properties of the human mind. Since building technical systems that are conscious appears out of reach at this time, it might be useful to examine whether indeed being conscious is synonymous with having a mind. Neuropsychoanalysis suggests that consciousness is not synonymous with the mind. It views consciousness as an unstable property of one part of the mind. This definition has considerable explanatory power. There are situations in which high-level, complex, and distinctly human cognitive and emotional processing of information may occur without conscious awareness. Furthermore, clinical evidence suggests that consciousness is not a unitary phenomenon, and it may occur on several levels. It might therefore be possible to design and build technical systems that are not conscious but nevertheless emulate important aspects of the mind.

4.2.1 Introduction

Over the last two decades, significant advances in the neurosciences, in cognitive psychology and in neuropsychology have increased our understanding of the functioning of the human mind [1]. Together, these advances in the cognitive neurosciences have led philosophers, neuroscientists, psychologists and now engineers to take a fresh look at the mind-body problem [2].

Also known as the mind-brain, or psychophysical problem, it is the riddle of how the mind and the brain are connected, i.e., how the subjective experience of being a person and having a mind is connected to the physical organ of the brain. Despite numerous attempts to solve it – or to make it go away – it has remained an elusive, enduring problem of Western philosophy [3]. It is also as relevant as ever for anyone who attempts to diagnose or to treat mental problems [4]. The mind-body problem is usually defined in the current literature as *the problem of the physical basis for consciousness*. Indeed, several contemporary authors, both philosophers and neuroscientists, have approached it as such [5], [6], [7], [8].

In his book "The conscious mind: In search of a fundamental theory" [9], philosopher David Chalmers, who is familiar with recent neuroscientific advances, divided the mind-body problem into an "easy" problem and a "hard" problem. The easy problem, according to Chalmers, is *discovering the physiological processes and anatomical loci in the human brain that are responsible for consciousness.* Over the last two decades, the cognitive neurosciences have made preliminary but significant steps toward achieving this goal. The easy problem is of course anything but easy, and yet it appears likely that the cognitive neurosciences would in due course provide us with some satisfying answers to it. It is in this sense that it is regarded as "easy". In contrast, the hard problem is not about "what" or "where", but about "how". The hard problem is *finding out how the abovementioned physiological processes and anatomical loci in the human brain produce consciousness*.

The volume, diversity and complexity of the scientific and philosophical literature that deals with these issues are such that I would not attempt to review it here. Still, I will mention that it is currently a matter of uncertainty and debate whether or not the hard problem is a scientific problem at all, i.e., whether it is in principle approachable by science. Our profound and frustrating inability to account for the way in which subjective experience is connected to the material substance and physical processes of the brain has been labeled *the explanatory gap* [10].

4.2.2   The Prometheus Phantasy

It is therefore not surprising that in their original and thought-provoking paper, "The Prometheus phantasy – functions of the human psyche for technical systems" [11], Pratl and Brainin have chosen to explicitly avoid the subject of consciousness altogether. In their paper, they pioneered some of the initial conceptual groundwork that is required for developing technical systems with more humanlike properties. Specifically, they examined several principles that appear to underlie the organization and functioning of the human mind, and attempted to formulate them in terms that could be translated into technical language. This approach brings with it an exciting potential to generate technical systems that would operate according to the principles that govern the action of the mental apparatus. Nine such principles were delineated: *no activation without inhibition, reverse afferentation, active anticipation, the pleasure principle, the ego as a psychic agent, the reality principle, postponement of satisfaction, the economic principle of saving expenditure of energy,* and *the ego serving as a stimulus barrier*.

To my reading, the translation of some of these mental principles into technical terms is more straightforward than others. For example, the technical model for the principle of "no activation without inhibition" appears to be tightly coupled and convincingly correlated with its neuropsychoanalytic formulation. In contrast, the statement that the reality principle is intrinsically contained in all technical systems "since technology is always embedded into the real world" appears to lack the specificity and complexity of the psychoanalytic formulation of this principle.

Despite these concerns, Pratl and Brainin's efforts are impressive. Using a top-down approach, they seek to create technical systems that would mimic salient and overarching aspects of the organization and functioning of the human mind, as revealed and defined by neuropsychoanalysis. This is in contrast with neural network and other approaches that attempt to study brains, as well as humanlike technical systems, by modeling the individual elementary building blocks of the brain (i.e., neurons) and the rules that govern their cellular interconnections (i.e., synapses) [12].

4.2.3   Consciousness and the Design of Technical Systems

Pratl and Brainin state that "even if we have a Prometheus phantasy we are not able to imagine a machine with a feeling of SELF." Elsewhere, in the context of modeling the stimulus barrier function of the ego, they note: "We do not assume that we will be able to create a system that can actually be traumatized." In other words, the machine that they envision may operate according to principles that govern the activity of the human mental apparatus, but it is very unlikely that it would develop consciousness, or a subjective sense of agency. This reasonable assertion raises two sets of questions that underlie the subject matter of our discussion:

(1) Is it possible, in principle, to construct a conscious technical system?
   (a) How can this be done?
   (b) How can we know for sure that we have done it?
(2) Is it possible, in principle, to be human without being conscious?
   (a) How human-like can the functioning of a technical system be, if it lacks the human traits of consciousness and a sense of agency?
   (b) Are there different levels of consciousness, such that some of them may be relevant to projects that are currently technically feasible?

The first group of questions above is probably familiar to all lovers of science fiction, and perhaps to most moviegoers (consider Blade Runner; I, Robot; 2001 Space Odyssey; etc.). I suspect that at least some of our engineering colleagues were drawn to this field in the hope of finding an answer to them. In the spirit of selfdisclosure, I must confess that I myself am not free from fascination with this possibility. Some authors [13], [14] believe that if we could build a conscious machine (i.e., answer question 1.a.), then we have solved the hard problem. The Turing Test [15], as well as Searle's Chinese Room metaphor [16], address question 1.b. above, and are well-known formulations of the basic requirements and potential pitfalls that are inherent in any attempt to achieve such an amazing feat.

It is therefore understandable why Pratl and Brainin acknowledge that "we have deliberately left out issues that we think cannot be solved yet; we have not touched the topic of consciousness at all". However, it appears to me that the sheer magnitude, fame, and controversy that characterize the first group of questions above are such that they might have overshadowed and obscured the more modest and perhaps more accessible second group of questions.

In what will follow, I will try to address the second group of questions from a neuropsychoanalytic perspective – the perspective that holds that large parts of the mind are unconscious. I will then speculate about the relevance of the relationship between mind and consciousness for the "Prometheus phantasy" – the development of technical systems with more human-like properties.

4.2.4   Is It Possible To Have A Mind Without Being Conscious?

Perhaps the most important contribution that Freud had made to the scientific study of the mind was his realization that equating the mind with consciousness, as philosophers have done (and are still doing) [6], [9], [13], is probably incorrect. Freud had come to suspect that the traditional equation between the mental and the conscious was not a statement of fact, but rather a result of a biased point of view. As Solms [17] noted, Freud articulated this viewpoint early on in his career, and he remained faithful to it throughout his life. A year before his death, shortly after he was forced to flee from his beloved Vienna, Freud wrote a fragment that was never incorporated into a complete paper. In the fragment he outlined his views about the relationship between consciousness, mind and brain:

- "If ever human thought found itself in an impasse it was here. To find a way out, the philosophers at least were obliged to assume that there were organic processes parallel to the conscious psychical ones, related to them in a manner that was hard to explain, which acted as intermediaries in the reciprocal relations between 'body and mind', and which served to re-insert the psychical into the texture of life. But this solution remained unsatisfactory.

- "Psychoanalysis escaped such difficulties as these by energetically denying the equation between what is psychical and what is conscious. No; being conscious cannot be the essence of what is psychical. It is only a quality of what is psychical, and an inconstant quality at that – one that is far oftener absent than present. The psychical, whatever its nature may be, is in itself unconscious and probably similar in kind to all the other natural processes of which we have obtained knowledge" [18].

Freud based his assertion that a large and active part of the mind is unconscious on three groups of psychological phenomena: gaps in conscious thinking, slips of the tongue, and confabulations that accompany post-hypnotic suggestion. He then applied Occam's Razor to the counterargument, which is sometimes still invoked by philosophers [6, p. 161], that what is unconscious is by definition not mental.

According to Solms [17], [19], this fundamental psychoanalytic idea is a presupposition. From the point of view of our engineering colleagues, it is certainly good news: If consciousness is not synonymous with the mind, then an individual (or, for that matter, a technical system) may be mindful, at least to some extent, without being conscious. Going back to the second set of questions I have outlined above, they appear much more friendly in this light.

4.2.5   *"Oh, Dr. Sperry, You Have Some Machine!"*

Neuroscientists and neuropsychologists have known for some time that it may be possible to function like a person without being conscious, or, at least, without being in the kind of self-aware, reflective state that Damasio [20] calls extended consciousness [21], [22].

Solms and Turnbull [23] quote Galin [24], who described an experiment that is highly relevant to our discussion. It was carried out by Roger Sperry, who pioneered research in split-brain patients. These individuals have undergone surgical section of the pathways that connect their two cerebral hemispheres as a treatment for intractable epilepsy. With the use of a special projector, it was possible to present them with visual stimuli to each of their cerebral hemispheres separately. Within a sequence of neutral geometrical figures that were presented to the right (nonverbal) cerebral hemisphere of a female patient, Sperry inserted a nude pin-up. The patient blushed and giggled. When asked what she saw, the patient answered, "Nothing, just a flash of light." When asked why she was laughing, she laughed again and said, "Oh, Dr. Sperry, you have some machine!"

The important point, from our perspective, is that the patient's right hemisphere was capable of fully processing the visual information of the nude figure, together with its embarrassing meaning and the nuances of its inappropriateness on a social level, without her being reflexively aware of what she was seeing. This prompted Solms and Turnbull [23, p. 83] to conclude that "an entire hemisphere can, in a sense, function unconsciously."

Sleepwalking is another fascinating example of performing on a human level while being minimally conscious. A sleepwalking person may not only walk around obstacles, open doors and climb ladders, but also drive a car. He is not consciously aware of what is he is doing, does not respond to commands and requests, and has no memory of the episode later [21]. Still, he may manage not to hit other cars or people while on the road. In sum, there appear to be instances in which people perform rapid, complex, high-level, and distinctly human cognitive and emotional tasks without being conscious of what they are doing, either in real time or retrospectively.

This last point is important, since several lines of evidence suggest that much of what we ascribe to our conscious free will is actually the product of similar rapid, stereotyped and unconscious sensorimotor processes. Consciousness appears to set in only hundreds of milliseconds later. We often tell ourselves (and the world) that we have acted out of our own conscious volition, while in fact we are merely rationalizing an unconscious event that was not under our willful control when it occurred [25].

4.2.6   Why Aren't We Zombies, Then?

Directors of horror movies, philosophers and neuropsychologists have labeled individuals who function on a human level without being consciously aware of what they are doing as 'zombies'. This is a rather unflattering name, and it suggests how much we value our conscious awareness and willful control of our actions (or, if I may, the illusion that we have such awareness and control in real time). Zombies lack a true sense of agency, and one may argue that they lack subjectivity altogether [26].

Our discussion raises the question whether consciousness is actually needed for anything. Eliminative materialism, an approach to the mind-body problem that once dominated the natural sciences and is sometimes still invoked today, holds that consciousness is indeed superfluous. According to this viewpoint, consciousness is merely an epiphenomenon, a byproduct of brain activity, and it has no causal influence on anything real [27].

Neuropsychoanalysis is categorically opposed to the viewpoint of eliminative materialism. Specifically, it maintains that extended consciousness is essential for our survival [19]. By combining internal feeling states with external sensory perceptions, consciousness enables us to ascribe value to objects in our surroundings, and thereby guide our actions on the basis of how they make us feel inside. Furthermore, extended consciousness enables us to generate complex solutions to abstract problems. By invoking traces of the past and allowing us to consider the consequences of actions we have not yet performed, it enables us to plan for future actions without relying on the ancient, costly and painful method of trial-and-error.

In contrast, the kinds of tasks in which zombies excel do not necessitate deep reflection. Rather, they require skillful, stereotyped, immediate action. Such actions are not mere reflexes, however; they are interactive, in that they rely on real time feedback loops and on an ongoing, rapid processing of information from the environment. Milner and Goodale [22] have characterized the parts of the brain that are responsible for such rapid and unconscious information processing as 'on-line systems', by analogy with software that processes information in real time.

Returning to question 2.a. above in light of our current discussion, it appears that it is indeed possible to function like a human in the absence of reflexive awareness or an extended consciousness. While this conclusion is encouraging for the Prometheus phantasy, I must add that we can only consider unconscious mental processes within the contextual framework of a conscious mind. In other words, the fact that a computer functions without consciousness (as all computers currently do) does not mean that it is undergoing unconscious mental processes, or that it possesses a mind.

4.2.7 How Much Consciousness?

The current neurobiological study of consciousness, like its philosophical counterpart, to which it is intimately linked, is a hotly contested field. Perhaps more than in any other area of the neurosciences, it is not unusual for different observers to reach diametrically opposed conclusions, and to hold on to them. For example, neurobiologists and Nobel laureates Sir John Eccles and Eric Kandel differ sharply about the relationship between consciousness and brain physiology. Kandel believes that a complete understanding of brain physiology will one day fully account for consciousness (i.e., a reductionist, monistic approach) [1, pp. 398-402]. Eccles, on the other hand, believed that consciousness in and of itself causes small but significant changes in brain physiology (i.e., an interactionist, dualistic approach) [28]. While these differences are impressive, there appears to be a consensus about the basic definitions of consciousness in the medical and psychological literature. Broadly speaking, consciousness may be defined by its *content* (what we are conscious of) as well as by its *level* (how conscious we are) [23, pp. 79-104]. Of these two ways to approach it, the latter is the more relevant one for our discussion here.

Levels of consciousness may be defined clinically, as in the Glasgow Coma Scale or the Reaction Level Scale [29]. They range from deep coma through mild stupor to alert, reflexive awareness. They may also be defined from an evolutionary perspective [30, pp. 34-35], according to which the most basic form of consciousness, which we share with all mammals, is affective, and it does not depend on the functioning of the neocortex. Higher and more selfreflective levels of consciousness characterize primates, while the highest form, which includes a sense of time, the capacity for abstract thinking, and the ability to phantasize, is characteristic of humans only.

As Panksepp [30] suggested, all levels of consciousness appear to depend on a small and phylogenetically ancient part of the brainstem, the periaqueductal gray (PAG). The PAG generates representations of the body and its internal state. Full, introspective and reflexive consciousness depends on language, and hence on the function and connections of the language areas of the left cerebral hemisphere, as well as other cortical areas. However, it is likely that such higher-order consciousness is not necessary for elementary emotional experience. Returning to Pratl and Brainin's statement that "We do not assume that we will be able to create a system that can actually be traumatized," the fact that there are several different levels of consciousness may suggest that this goal is not as unreachable as it might seem. In other words, if a mouse can be traumatized, then it might not be unreasonable to expect that one day machines would be, too. But in order to have any kind of consciousness, it is likely that a technical system would have to contain, as the PAG does, some representation of a *body* that has *needs* [23, p. 94].

4.2.8   Conclusion

The goal of creating conscious technical systems appears far away at this time. However, the fact that humans may function on a human level without conscious awareness suggests that it may be possible for a machine to have a mind, or parts thereof, without being fully conscious. In addition, the fact that consciousness is not a unitary phenomenon but rather a property that exists on different levels means that reflexive consciousness is not the only way to be conscious. As a consequence, it might be possible to design technical systems that display humanlike behavior without anything that resembles human consciousness. Zombies should not be underestimated.

References

[1]   Kandel E.R., Schwartz J.H., Jessell T.M., Principles of neural science. 4th ed., New York, McGraw-Hill, 2000
[2]   Hameroff S.R., Chalmers D.J., Kaszniak A.W. (Eds.), Toward a science of consciousness. III: the third Tucson discussions and debates. Cambridge, MA, MIT Press, 1999
[3]   Heil J., Philosophy of mind: A contemporary introduction. London, Routledge, 1998
[4]   Kendler K.S., A psychiatric dialogue on the mind-body problem. Am. J. Psychiatry, 158(7): 989-1000, 2001
[5]   Chalmers D.J., Facing up to consciousness. J. Consciousness Studies, 2: 200-219, 1995
[6]   Searle J.R., The rediscovery of the mind. Cambridge, MA, MIT Press, 1992

[7] Crick F., The astonishing hypothesis: The scientific search for the soul. New York, Simon and Schuster, 1994

[8] Edelman G.M., Tononi G., A universe of consciousness. New York, Basic Books, 2000

[9] Chalmers D., The Conscious Mind: In Search of a Fundamental Theory. New York: Oxford University Press, 1995

[10] Levine J., Materialism and qualia: The explanatory gap. Pacific Philosophical Quarterly, 64: 354-361, 1983

[11] Pratl G., Brainin E., The Prometheus phantasy – functions of the human psyche for technical systems. These proceedings, 2007

[12] Knight R.T., Neural networks debunk phrenology. Science, 316 (5831): 1578-9, 2007

[13] Dennett D.C., Consciousness explained. Boston, Little Brown, 1991

[14] Fodor J.A., The elm and the expert. Cambridge, MA, MIT Press, 1994

[15] Turing A.M., Computing machinery and intelligence. Mind, 59: 433 -460, 1950

[16] Searle J.R., Minds, brains, and programs. Behavioral and Brain Sciences, 3: 417-457, 1980

[17] Solms M., What is consciousness? Journal of the American Psychoanalytical Association, 45: 681-778, 1997

[18] Freud S., Some elementary lessons in psychoanalysis. S.E., 23: 283, 1938

[19] Solms M., What is the "mind"? A neuro-psychoanalytical approach. These proceedings, 2007

[20] Damasio A., The Feeling of What Happens. New York, Harcourt Brace, 1999

[21] Koch C., Crick F., The zombie within. Nature, 411: 893, 2001

[22] Milner A.D., Goodale M.A., The visual brain in action. New York, Oxford University Press, 1995

[23] Solms M., Turnbull O., The brain and the inner world: An introduction to the neuroscience of subjective experience. London, Karnac, 2002

[24] Galin D., Implications for psychiatry of left and right cerebral specialisation: A neurophysiological context for unconscious processes. Archives of General Psychiatry, 31: 572-583, 1974

[25] Libet B., Do we have free will? In: Libet, B., Freeman, A., Sutherland, K. (eds.), The volitional brain: towards a neuroscience of free will, pp. 47-57, Imprint Academic, Thorverton, UK, 1999

[26] Skokowski P., I, zombie. Consciousness and Cognition, 11: 1-9, 2002

[27] Churchland P.M., Matter and consciousness: A contemporary introduction to the philosophy of mind. Rev. ed. Cambridge, MA, MIT Press, 1988

[28] Eccles J.C., How the self controls its brain. Berlin/New York, Springer, 1994

[29] Starmark J.E., Stalhammar D., Holmgren E., Rosander B., A comparison of the Glasgow Coma Scale and the Reaction Level Scale (RLS85). Journal of Neurosurgery, 69(5): 699-706, 1988

[30] Panksepp J., Affective Neuroscience: The Foundations of Human and Animal Emotions. New York, Oxford University Press, 1998

4.3 Discussion Chaired by Authors

The following is a short summary of the most important questions of the auditorium after the above paper presentations. Again, "Q" is used for any type of question or comment, while "A" stands for the answers from the chairs.

Q1: In our everyday language, we have a rich vocabulary to cover a wide variety of consciousness-phenomena. However, in science, consciousness suddenly becomes "it", as if it were one thing, either present or absent. But we need to understand the great variety of mental functions and the requirements for implementing this variety, and once this has been done, it will not be necessary any longer to implement this one consciousness, as we will have the various functions.

A1: It is the point of view of psychoanalysis that consciousness is a lot wider than that, that looking at consciousness from this point of view is missing the point. It is a dimension, it is something that evolved over time and has many facets. However, what I was trying to convey was that do not need to get to the higher levels in order to achieve machines that would act in human-like ways.

Q2: I am sure there was not enough time for Dr. Yovell to make all the qualifications, but of course you believe that the right hemisphere is affectively conscious? There is great confusion in the literature about the use of the term consciousness. Some for instance believe that you need awareness to have an experience as opposed to just phenomenal experience. What I thought very important in your presentations was Dr. Zucker's statement that everything is connected in the brain, because this is a fact and I think we have to keep this in mind with whatever we try to simulate.

A2: Of course the right hemisphere is conscious. However, it is interesting that there are widely conflicting views on this, ranging from the assumption that the right hemisphere certainly is conscious, while maybe a little bit less than the left hemisphere, to the notion that the right hemisphere has a consciousness far inferior to that of a chimpanzee. Which is, as we see in clinical studies of split-brain patients, untrue. But it is interesting how even neuroscientists are in wide conflict about this issue.

Q3: I have a slightly different question to the expert on zombies. Is it possible that a "zombie" can speak and talk and socialize?

A3: There is the distinction between so-called "Hollywood Zombies" and "Philosophical Zombies". Hollywood Zombies usually don't talk and are a little slow in whatever they do. Philosophical Zombies on the other hand are people who look and function exactly like you and me, without the subjective experience. Externally, we have no test to prove that these people are zombies. Philosophical zombies is the philosopher's way of saying that we might never be able to tell whether anyone has subjective experience or not. We as neuro-psychoanalysts do not agree with that. We know how seminal it is to have consciousness, and if you did not have it, you could not do most of the things that make us human.

5   Discussion Sessions

Anna Tmej

In the following, the discussion sessions at the ENF 2007 are very shortly summarized. Within every discussion session, several questions to the session participants had been formulated beforehand to make the discussion easier. All comments are presented in alphabetical order, supplemented by the respective number of the question being discussed. If one person makes more than one statement, this is indicated by additional numbers (e.g. A1.1, A1.2, etc.).

5.1   Psychoanalysis and Computer Engineering

Session 1a; Chairs: M. Solms and R. Dillmann

5.1.1   If an Artificial Intelligence is not Conscious, Can it Perform Truly Mental Functions? What Makes a Function Mental?

A1:   It depends on the observational perspective taken. Observed from an external point of view, the mental apparatus will be seen as a "thing", a physical organ. However, perceived from the point of view of being that "thing", it will be perceived as a mind, this is what "the mental" is: the state of being the mental apparatus as opposed to looking at it from outside. A function is "mental" if it is described from the point of view of being that thing, which, however, does not necessarily imply that these mental processes are also conscious, they might as well be unconscious.

B1:   For engineers, the distinction between a conscious state and a mental state does not seem crucial at the moment. Rather, the question is how to make a machine recognize its environment and the patterns it should operate in. And to achieve consciousness or a mental state, it will probably be necessary to find out how the brain processes information from its environment.

C1:   I do not agree. When humans interact with machines, and they share the same reality, it is very difficult to understand why the machine reacts the way it does. So more information will be necessary, maybe background information, or about getting to conclusions, about decision-making. Maybe the machine should be able to describe its experience so that humans will be able to evaluate what the machine does and compare the machine's experience to their own. So, it is crucial to design machines that will be able to represent the wide variety of different views of reality and to share this with humans.

5.1.2   What is the Specific Contribution that Psychoanalysis Can Make to Computer Engineering as Opposed to Psychology in General?

A2.1:   If you only study the brain, its mechanisms, connections etc., without taking into account the subjective experience, there will be things about how it works that you will miss and therefore fail to implement. Those things have to do with intentionality, motivation, drives, which are very important aspects of how visual processing works. And this is where psychoanalysis comes in.

D2:   I think I understood that not all mental phenomena are necessarily conscious and that we as engineers need not worry about it. But it will remain an unsolved problem if we just describe mental phenomena that can be implemented in computational representations that need not be conscious, as this does not really pinpoint the specific features of consciousness and the properties of being aware, which is actually, as a research question, the interesting part of the discussion.

A2.2:   Maybe there has been a misunderstanding. I have certainly not been trying to say that it does not make a difference whether we are conscious or not, quite the contrary. And here, it is important to take the psychoanalytic, the introspective point of view. For if you ask the brain why it does the things it does, and it tells you, it will be possible to know about functions it would have not been possible to know if you only had looked at the brain from an external point of view. Not taking into account the conscious state it experiences would completely mislead you about, for instance, the causal sequence that led to a certain behavior. And from this psychoanalytic study of the brain, it will be possible to draw conclusions about the use and the functionality of consciousness. And the more we understand this functionality, the more we will be able to create something that functions more closely to the way a human mind does.

E2:   As time goes by, there will be more and more interaction between humans and robots. However, human beings are not psychically "normal", we all have our symptoms, our problems, and we enjoy this in our day-to-day communication. So I have been trying to develop a mathematical model for what I call "psychopathic robots", as maybe it would be useful to have kitchen robots, for instance, with their own symptoms and problems, to improve interaction quality. So is it possible to incorporate psychical deviations, repression etc. into robots?

C2:   In terms of effectiveness, this will not be useful. It may be useful, however, to provide a robot with the ability to perceive and adapt to preferences in humans, which can of course vary inter- and intra-individually.

5.1.3   What Contribution Can Computer Engineers Make to Psychoanalysis?

A3:   As a representative of psychoanalysis as a discipline, I can say that I have found the dialogue with the engineers very valuable: on the one hand we have had to clarify our concepts in a way we had never attempted to before, and on the other hand this cooperation enables and forces us to test the worth of our models.

5.2   The Mammal in the Machine

Session 1b; Chairs: J. Panksepp, R. Trappl

5.2.1   *Humans Learn Very Much After Birth While Animals Don't. How Can We Locate a Technical System in This Respect?*

A1.1:   I disagree: as all mammals have a neo-cortex, they, too, learn enormous amounts after birth. As far as we know, the neo-cortex is tabula rasa to begin with and is completely programmed by experience. Even the visual cortex and the auditory cortex specialize only after birth, they are not genetically determined.

B1:   The relationship between the genome and adult competence varies enormously. Some animals have rich competences right from the start, and others, both mammals and birds, are apparently very incompetent after birth. However, they still have some competences that are pre-programmed, such as the birds hatched on a cliff: they have to be able to fly the first time they try. So evolution discovered that some things can be pre-programmed in a lot of detail while others cannot, and discovered how to deal with the latter other than tabula rasa. I am speaking about for instance programming a system to find out what is in the environment by experimenting on it, however with some constraints to what can be learned.

A1.2:   What I am trying to say is that the instinctual part, in mammals, is not in the neo-cortex. And with this I am emphasizing that we have a lot of general purpose computational space, which of course can do nothing without the instinctual tools, such as emotions etc.

C1:   In what way do you think that emotions play a primary role in learning?

A1.3:   I am speaking about Pavlov's and Skinner's concepts of reward and punishment, and these are the main ways animals learn. Rewards and punishments in that sense are affective changes in the brain. And these affective changes tell the animal what in the possibility of neutral actions and neutral stimuli should be highlighted as being important.

5.2.2   *How Can We Modularize Mental Functions For Engineering Purposes?*

D2.1:   A good example would be the famous BDI model which psychoanalysts usually are not aware of. BDI stands for beliefs, desires and intentions, which are the three systems without which sensible actions would not be possible. It is, naturally, a very simple model, which however can be developed to quite some richness. What is missing, though, are the emotions. Therefore, we are working on implementing emotional appraisal systems into BDI models. Emotions serve various purposes: they add specific value to information of any kind with respect to the needs of a specific being, probably they also serve to reduce the complexity of search spaces and also help convey information about the personal status over long distances.

E2:   When human beings need to make a decision facing very complex problems, and they do not have enough information to make a decision or even have too much information, interestingly there are mechanisms that make decisions possible, on a gut level, as one would say. Somewhere, a feeling is created on which the decision is then based. So it would be interesting to think about how we could model this kind of emotion in decision making in a machine. How does this kind of feeling develop in a human being's brain?

A2:   This would require building real feeling into the machine. I do not see any immediate solution to this problem.

D2.2:   Emotions often help to react faster than would be possible if we did deliberately. So if we want to have a robot react very quickly we should give it the possibility to react quickly under uncertainty, for instance in risky or dangerous situations, and then later on decide whether this reaction was appropriate or whether it actually was a normal situation that only looked risky. This would be one example where we should put in something like an alarm which triggers a different reaction and which would bypass long inferential processes which could delay an appropriate reaction.

5.3   The Remembering Body

Session 1c; Chairs: O. Turnbull, L. Goldfarb

5.3.1   How is "Memory" Organized Functionally?

A: There are multiple, parallel and independent memory systems: short-term memory or immediate memory system, at least one of them for phonological memory and one for visual-spatial memory, episodic memory which is usually associated with conscious awareness, procedural memory for the motor-learning of complex acts, emotion-related learning which is sometimes important for decision-making and is sometimes experienced as some form of intuition, and semantic memory.

B: Another fundamental question might be what it is that has to be stored in our memory systems. We need to understand what the primary evolutionary processes are that drive everything, including storage. And my hypothesis is that the basic subconscious processes are classification and accompanying processes. Emotions can be thought of as colorings of classes of events we have experienced, situations and the like, which allow us to very quickly classify more complex situations.

5.3.2   Coding of Memories: What is Known on the Level of Neurons, on the Level of Semantic Memory?

A: Semantic memory concerns encyclopedic memories that people have, which can be split up into relatively independent components separating objects according to their structure, their functionality, animation etc.

C: I think that the term "semantic memory" is used as a label for a rather disparate class of things, the differences between which might be important relating to their implementation. For example, you know what kinds of objects there are, which defines an ontology. However, in addition to this ontology you also need to know about the laws which relate those objects, and their constraints. So these two classes are very different from each other, and again are different from episodic memory, which makes use of the ontology, but this use is very different from the learning about generalizations. I suspect there is even another subdivision which has to do with the differences between the concepts that relate to what you can perceive and act on and the concepts that provide deep explanatory power such as "gene" and "electron" or more generally those kinds of things that scientists develop, which you cannot perceive or act on. So perhaps they have to be implemented in a different way, as they don't have the same connection to sensory and motor capabilities.

A: I think what might be useful here is the multiple memory systems concept, which describes that objects are not only stored in one memory system but that a given object will evoke memories from a disparate range of experiences. For the question of generalizations, however, and knowing about possibilities which are not there in the original raw material, I think there is a role for the concept of "executive functions".

B: I believe we develop formalisms in which the distinctions between semantics and syntax disappear. And if everything is in classes, as I proposed before, then the way we study scientific concepts is no different than the way we study anything other. We will just create artificial classes and study them. What we need to understand, of course, is the right formalism to study classes, as they have structure and structure representations. Conventional formalisms such as neural networks do not answer the question of what a class is.

5.3.3   Who Governs the Updates (Add/Delete) of Memories?

A: The key-determinants would include processes such as time itself, probably also the question of overwriting, as any material is of course stored in the same block of neurons, and the way in which short-term electrical storage is, in the brain, transformed over time into the more physical substantiation that happens as dendrites form to connect together the different nerve cells.

5.3.4   Do We Need Different Technologies to Implement Different Types of Memory?

A: Many classes of memory are laid down in brain systems or, in many cases, in cortical systems, that use more or less the same chemistries as the memory systems in neighboring cortices that have different functions. So, the core technologies or pharmacologies that underpin memory are not the central determining factor of what designs memory systems. Probably a more central factor is the proximity to the classes of input in determining what sorts of memories are recalled: phonological short-term memory, for instance, is located very closely to the primary auditory cortex.

5.3.5   How Much Embodiment Does Memory Need?

D: There is one class of memories that is different from all other classes of memory, in that it needs to be more embodied than they do, which is emotional memories. There is a fundamental difference between memories of emotions and emotional memories. The first is an episodic memory, but for the latter you need to feel something. So you cannot have an emotional memory unless you have some kind of a body that experiences that feeling.

5.4   Emotions, Drives and Desire (Silicone in Love)

Session 1d; Chairs: Y. Yovell, M. Ulieru

A1: Being able to engineer control systems for machines that emulate human behavior triggered by emotions, drives and desires would provide us with a fundamental paradigm shift. The advantage would be that the machine would fight against entropy. And the question follows, whether it would be useful to have such a machine. In homecare, for example, would this kind of robot be able to help an elderly person, would it be useful to cure this person or help them deal with loneliness? Would it be able to love that person? And, the other way round, would the person love the machine?

B1: Neuro-psychoanalysis can contribute to this question by clarifying the way emotions are wired in the brain, and clarifying this might help the engineers conceptualize and emulate them. First of all, it is important to make the distinction between channel functions and state functions in information processing in the human brain, and to understand that emotions have a very unique wiring.

C1: I think, for these discussions to be productive, we need to talk about biological information processing.

D1: All the things we call emotions are examples of affective states and processes of which there is a great diversity. They are all concerned with control functions, however, we must distinguish between current control and dispositional control. When thinking about emotions, most people think about current control only; that is about what is happening now: I am angry, for instance. But we do not think about the other side, which however is very important in human life, that is thinking about grief, infatuation, jealousy, obsession etc., states that last a longer time. I think these states are usually neglected, but by doing so we leave out affective states and processes that are most important for human beings.

A2: I really think we need to learn how to engineer affective memory. This is an essential mechanism we are currently missing.

E1: Aren't most concepts discussed here, such as love, too vague to work with for us engineers? And also, I wonder why should a machine love a person?

A3: I believe that for machines that are meant to work with elderly people, it is very important to make them as likeable as possible. And regarding your other question, the main thing is that for the past years of our work, we have ignored emotions, drives and desires completely. I believe that the reason for this is that we as engineers are trained to highly respect logic, and to believe that emotions stand against logic. But I have learned that emotions somehow represent the logic of life, and for me, this is the big paradigm shift.

B2: For mammals, love is absolutely essential. If you want to understand a mammalian brain, you will have to understand something about love. Through millions of generations, we have been selected to be capable of loving and of being loved. In the mammalian world, if there was nothing to attach a mother to an infant, the genes would not be passed on to the next generation.

F1: I understand mind as evolving over time, and as having a history. I was wondering whether this aspect of a changing mind, changing by interacting with the environment, is of any concern to this project, is there any component of a change in time?

C2: In the former model we developed, the concept of history was central to the representation, and therefore for memory and everything else. In this model, to represent something is to represent a formative process, the similarities of objects are perceived as similarities of these formative processes. And this is how objects are classified, by judging whether they have similar formative histories.

G1: Returning to the importance of love for life and artificial life, one of the reasons why we have so much drug addiction in our world is that people do not have love and companionship. Most of opiate addiction is because of loneliness, alienation and the inability to find emotional satisfaction. If these people cannot find another person, I do not see anything wrong with them finding companionship in a robot who has an interest in their life and has been programmed with a certain amount of emotional resilience. Right now, people fulfill this companionship with a pet, and the reason for this is that pets can show the same emotions.

5.5   Getting a Grasp

Session 2a; Chairs: E. Brainin, K.-F. Kamm

5.5.1   Can We Model the Brain Without Considering its Physiological Basis?

A1:   Referring to Solm's statement that the mind alone knows what it feels to be itself and can provide a verbal report about its subjective state, one important principle of the nervous system would be that there is no activation without inhibition.

B1:   Activation and inhibition might be too simple to describe the nervous system; it would be more feasible to try to describe it in terms of oscillations, interferences and so on.

C1:   It is important to distinguish between the model of the system and its functions. In computer technology terms, a program that is based on a model may result, for instance, in an activation, which is the function of the model, its application. Oscillation is a result of the functions, while what should be of interest are the functions themselves.

D1:   There is the question of what kind of information is processed inside the brain, and then whether there should be the possibility of, e.g., evolution within this simulated mechanism, which will make the modeling process much more complex.

E1:   In principle, as it is possible to model all kinds of systems and structures, it should also be possible to model the brain. On the other hand, the structure of the brain is very complex, considering all its functions, dynamical behaviors, and connections, but considering all the possibilities there are in electrical engineering nowadays, there certainly is potential.

5.5.2   Does Intelligence Require Embodiment?

E2:   First of all, there is agreement that "embodiment" may also include technical systems. Certain aspects of intelligence do not need embodiment to function correctly, such as logical judgment or algebra. However, other aspects, such as intuition and emotional intelligence, or image analysis certainly do require embodiment.

A2:   This could be explained considering the Ego, which is the integrative agent for outer and inner stimuli, and which is foremost a bodily one. This notion stands behind all psychoanalytic reflections about perception and about how perception can be processed and integrated: The Ego is like an interface between outer and inner reality.

B2:   It is important to see intelligence also in its social function. When constructing intelligent machines, it must be taken into consideration that these must work in a human environment.

F2:   The interpersonal aspect of intelligence should not be neglected, which will also need embodiment to function correctly.

C2: Taking psychoanalysis as the basis for our model, it is clear that a body is certainly needed to have drives etc., which is very important.

5.5.3 Which Properties and Needs of a Body Are Crucial to Achieve Intelligence in an Organism?

A3: A body is needed to achieve human intelligence, a feeling of being oneself and of being in relation to the environment. Without stimuli from the outside, there would not be any development whatsoever, and here, again, the Ego takes up an important role as a stimulus-barrier.

G3: Our categories of thought, as, for instance, seeing relations between different categories, depend upon the actual physicality of our body, such as inner forces, stressed by Freud: somatic needs and drive pressures, but also the features of our physical experience, having an inside and an outside. A body of some sort would certainly be needed as an underpinning for intelligence, whether technical or physical.

H3: Also, the Unconscious, which differentiates human from artificial and animal intelligence, will have to be considered.

E3: With regard to the crucial needs of the body, certainly the relationships to another person, the first relationship to the parents, are needed to develop intelligence. It is a critical question how it will be possible to achieve this aspect in technical systems.

5.5.4 How do Drives, Emotions, and Desire Contribute to the Development of Human Intelligence?

E4.1: It seems clear that drives, emotions and desires are the driving forces for the development of human intelligence, as are the relationships to others, and the wish of every biological system to grow.

I4: It is clear that, for humans, drives, emotions and desires are necessary, but why should a machine have emotions? It does not seem sensible to have a machine that gets angry at the user.

C4: Human intelligence certainly is the highest in our realms, so it seems only natural to try to implement human capabilities, which root in human consciousness, in artificial intelligence. Therefore, it is also necessary to bestow emotions upon machines.

A4: Another important aspect for the different purposes of these machines is the anticipation of another (human)'s behavior, which again requires empathy, being able to anticipate another human being emotionally.

G4: Drives, emotions and desires could also allow the machine to imagine beyond into areas of innovation and creativity.

E4.2: Empathy with the user would also be a prerequisite for enhanced user-friendliness of computers.

J4: It seems that there is both a desire to put emotions in a machine and a fear of doing so. Doing it selectively seems to be a suggestion, but then there seems to be the apprehension that we might not be able to and might end up with an angry machine.

5.6   Free Will

Session 2b; Chairs: W. Jantzen and M. Vincze

SHORT PRESENTATION:

5.6.1   The Libet Experiment – W. Jantzen

Interpreted in Luria's tradition, the Libet Experiment only examines decision-making, not free will. The prerequisites for voluntary action are goal-directedness, an obstacle and contradictory motives. Additionally, voluntary function depends on motives appropriated in former social interactions, resulting in a hierarchical structure of motives, so that one motive can be overcome by another.

5.6.2   An Engineer's Position – M. Vincze

The original subsumption model was purely behavior-oriented, purely reactive. From an engineering point of view, of course, no free will can be found in this model. Everything has been pre-programmed, and the robot will immediately react to whatever is presented to it. As there are clear limitations to this kind of model, it has been broadened to include a "deliberative layer" to permit a choice between certain behaviors. Observers might interpret consciousness or free will or emotions into the machine, but from the point of view of the programmer, neither actually exists.

5.6.3   Space-Time Continuum in Mental Functions – W. Jantzen

Functional systems always exist in the present, go back to the past and create a possible future. The present situation is compared to the remembered past, from which processes of perception, assessment, and decision arise; deciding entails a space of possible future. Needs are attributed to the space-time organization of the past, emotions to the present, and motives to the future. Emotions play a crucial role and entail a decision between expected processes within the body and expected processes in the outside world. On this base, the concept of will can be introduced, which is, according Leontyev, founded in physiological tone. Two kinds of wills can be distinguished: one to overcome the tone between conflicting motives before deciding, and the conscious or "free" will to master the problems that arise from conflicting situations in our anticipated future.

DISCUSSION

5.6.4   If There is Such a Thing as Free Will – What are the Mental Functions Behind it?

A1.1: I am searching for a link between Prof. Jantzen's theory and engineering. A machine with a feedback system evaluating the effect of what it has been doing might be the key; I am speaking about Heinz v. Foerster's so-called Eigen values.

Interactions in Eigen value processes are very complex, but I think that we need this kind of mathematics to find the link.

B1: Finding a mathematical approach of this kind has been the goal of artificial intelligence for the last 50 years. However, we are trying a different approach. We certainly need Foerster's results for the lower abstract levels of our model, but for the higher abstract levels we exclusively use Freud's model. Here, no mathematics are needed, we can directly integrate the model into the computer.

C1: We must think of the whole system as producing Eigen behavior. The starting point for Eigen behavior of the higher functions must be present in the lower functions from the beginning, corresponding to Trevarthens assumption of an Intrinsic Motive Formation (IMF) which consolidates itself on brainstem level between the 5th and 8th embryonic weeks and is directed to a "friendly companion". For this reasons "free will" has to be understood only within the system of the emotional and cognitive processes as a whole.

D1: Thinking about free will, observations of neurological patients suggest that humans will always try to claim personal agency for their actions, whether electrically stimulated or tics, as in Tourette syndrome.

A1.2: And of course, actions caused by tics are not subject to free will. But via feedback systems, it is possible to reach higher levels of free will. So what you just spoke about are very good examples of Eigen behavior.

5.6.5   *Would a Machine With its Own Drives, Emotions and Feelings be Willing to Work for Man?*

E2: We know in our society that men work for other men; this is probably regulated partly by fear and partly by need. So whatever it is that operates in men would probably also work with machines. However, in the last meetings, it has become clear that there is great fear of granting free will to machines or robots, and especially to grant them the free will to feel emotions maybe contrary to what we humans might want the machine to feel at the moment, for instance anger.

C2: Robots will be willing to work for us if we build robots with feelings of attachment and companionship. They have to be constructed to be in resonance with human beings and thus develop a common space of understanding and re-regulation, for example.

F2: Many of us gain hope from the idea of free will in robots, thinking that it will enhance the intelligence of our machines, others are afraid of it. The answer may lie in applying the three Asimov laws, however, here another question arises: will the Asimov laws restrict the robot's intelligence in any way? I do not think so, as we adhere to rules, as well, without damage to either our feeling of free will or our intelligence.

G2: Considering the fact that there will be obsolescence relatively rapidly with regard to the development of robots, what do you do if you have a robot who is attached to and in resonance with a human and knows that it is going to be surpassed but wants to survive? Also, what about the human who is attached to the robot?

5.6.6 Would a Human Want a Robot to Show Free Will?

E3: Regarding the Asimov laws, this sounds like humans want a relationship with robots comparable to the one with human slaves or domestic animals. Their free will must not surpass our free will, and we have the power to kill them if we want them or when they're obsolete.

C3.1: I agree that there are many problems when we think about the future of robotics and society. What we can do is think about humans and their relationships. From this we can learn how to construct machines to support humans. Of course, we must keep in mind the dangers that may arise from this. We still have to learn a lot more about psychic processes, mass psychology, for instance. What do we expect from men, what do we expect from robots?

F3: If we want to make robots more intelligent and make them develop we will certainly have to consider giving them some rights.

B3: I wish to clarify something: we do not want to copy brains or anything like that. We only wish to apply the model of the mental apparatus, which is something completely different. We are not interested in building human beings.

H3: I wonder whether additionally to the Asimov law there should also be a law about not making robots too human. Is there a threshold over which they become human, while before that they were machines?

F3: This has been discussed many times, and so far, no answer has been found.

I3: I have the feeling about this entire day that somehow we, engineers and psychoanalysts, have been talking at cross-purposes. I still think we just need mathematics.

C3.2: What I have been trying to show you is an architecture of a functional system developed from physiological functional system theory, which is also used in modern cybernetics. What I did was transfer this system onto a higher level taking into consideration all the research I know. These systems only work in an ecological niche. And for humans, this niche is social. If you do not take this into consideration, you will try to construct a mind within a machine with all Cartesian errors, and will never succeed.

Part III Responses to the ENF 2007

List of Authors in Alphabetic Order

Ariane	Bazan	Université Libre de Bruxelles, Belgium
Andrzej	Buller	Cama-Soft A.I. Lab, Gdynia, Poland
Dietmar	Bruckner	Vienna University of Technology, Austria
Tobias	Deutsch	Vienna University of Technology, Austria
Dietmar	Dietrich	Vienna University of Technology, Austria
Joseph	Dodds	Institute of the Czech Psychoanalytical Society, Czech Republic; Charles University; University of New York in Prague, Czech Republic
Georg	Fodor	Vienna Psychoanalytic Society, Austria; University of Cape Town, South Africa
Robert	Galatzer-Levy	Institute for Psychoanalysis, Chicago Illinois, United States; University of Chicago, United States
Wolfgang	Jantzen	Universität Bremen, Germany
Roland	Lang	Vienna University of Technology, Austria
Brit	Müller	Vienna Circle of Psychoanalysis, Austria; Vienna University of Technology, Austria
Charlotte	Rösener	ACS Technologies, Austria
Matthias J.	Schlemmer	Vienna University of Technology, Austria
Ron	Spielman	Australian Psychoanalytical Society, Australia
Anna	Tmej	Vienna Psychoanalytic Society, Austria; Vienna University of Technology, Austria
Mika	Turkia	University of Helsinki, Finland
Mihaela	Ulieru	The University Of New Brunswick, Canada
Markus	Vincze	Vienna University of Technology, Austria
Takahiro	Yakoh	Keio University, Japan

1 Introductory Words

Brit Müller

As an introduction to the papers, their relation to the concepts developed by the authors of this book shall be outlined, the background of which is described by Dietrich et al. in their article (see Chapter 2.10). Searching for a bionic model of the mental apparatus, the hierarchical model of the brain as developed by Luria conforms well to the model concept of computer engineers. Hardware, micro program and software constitute an analogy to Luria's three levels of projection, association and tertiary cortex. In both models, the higher the localization of actions, the more difficult it is to univocally allocate functional units to physical units. The neuroscientists Luria and Freud agree on this issue, which however does not mean that Luria developed his neuropsychological concepts from his knowledge of Freud's early, pre-analytic writings, as Jantzen points out in his paper (see Chapter 2.8).

The articles of Turkia (see Chapter 2.1) and Spielman (see Chapter 2.2) address drive representations central to Freud's theory: affects on the one hand, and ideas represented by memories on the other. The authors of this book agree with Turkia's notion that feelings as a result of consciousness are an integral part of any computer engineering model of the mental apparatus, which Dietrich et al. point out in their article (see Chapter 2.10). In contrast to Turkia, however, the ARS project does not aim at modelling these feelings independently from a complete model of this mental apparatus. In a model concept based on psychoanalytic ideas, affects and their interplay with other functional units have a well-defined position and can only be usefully described by considering this relationship to other functional units (see Chapter 2.10).

Similarly, Spielman's functionality of memory (see Chapter 2.2) fundamentally corresponds to the basic theoretic scientific principles of the ARS project, in that concepts from different sciences (in this case psychoanalysis and computer engineering) reciprocally put each other to the test and thus use the ensuing effect of *"Verfremdung"* (alienation) (Wallner 1991) to more concretely identify and possibly correct or broaden the structures, strengths and weaknesses of the respective theory construction. Spielman employs such a theoretic scientific method when he compares the three memory systems, the encoding, storing and the retrieval of information, with the physical concept of the generation of electro-magnetic waves and again connects this to neurophysiological notions to arrive at his hypothesis.

However, Turkia's and Spielman's respective paths to the bionic model are fundamentally different from the one chosen by the ARS project. The project postulates a consistent model with a continuous architecture (Dietrich et al. 2008). To achieve such a model, computer engineers and psychoanalysts work together in an interdisciplinary approach whereby psychoanalysts and neuropsychoanalysts describe a differentiated, complete model of the psychic apparatus and then model it with the computer engineers. Prerequisites and reasons for this way of interdisciplinary work are extensively described in Dietrich et al.'s article (see Chapter 2.10).

Turkia and Spielman, on the other hand, choose and try to describe singular functions of the apparatus, always employing the knowledge of the respective other sciences. Turkia resorts to psychoanalytic affect theories, Spielman to the physical concept of electro-magnetic waves. Schlemmer and Vincze (see Chapter 2.3) choose a similar approach to analyze the fundamental functional principles of the mental apparatus described by Solms and Turnbull (2002) according to their relevance and applicability in the area of computer engineering.

The psychoanalytic contribution by Galatzer-Levy is in line with the approach of Dietrich et al. to construct a total model of the psychic apparatus. Galatzer-Levy (see Chapter 2.7) describes basic psychoanalytic concepts that essentially characterize the psychic apparatus, independently from the respective psychoanalytic schools and their different focuses and interpretations of psychic experience. For the technical implementation (Galatzer-Levy specifically refers to the Turing Test), these concepts will have to be considered as components of the psychic apparatus to function according to psychoanalytic principles.

Bazan combines neuroscientific and psychoanalytic findings in her critical response to the topic of an inner and an outer world of the body, their respective contents and distinctive features, as it was presented at the ENF conference (see Chapter 2.9).

An interesting approach which, however, contrary to the ARS project does not aim at modelling a complete mental apparatus oriented towards a consistent psychological model, is argued by Buller (see Chapter 2.4). He employs Freud's concept of the pleasure principle, the interplay of pleasure and unpleasure and the striving of the psychic apparatus for homeostasis to develop four laws of machine psychodynamics. This way, without a complete model, one aspect of psychic dynamics, that is ambivalence and conflict, can be captured.

The ARS project aims at modelling the mental apparatus of a single individual. It may be a perspective for the future to broaden the application of psychoanalytic concepts by group analytic concepts to find even more satisfying results for certain tasks. The question will have to be tackled under which circumstances, seen from a group analytic perspective, a multi-agent-system might be effective or might achieve something different than a single agent. The article of Rösener et al. (see Chapter 2.5) traces this question along the lines of Freud's ideas of a mass psychology, which again forms the basis of all group analytic concepts developed later on by Bion or Jaques, extensively illustrated by Dodds in his article (see Chapter 2.6).

References

Dietrich, D., Zucker, G., Bruckner, D., Müller, B. (2008). Neue Wege der kognitiven Wissenschaft im Bereich der Automation [New Paths of Cognitive Science in the Area of Automation]. *Elektrotechnik und Informationstechnik (e&i)*, 5, 180-190.

Solms, M. & Turnbull, O. (2002). *The Brain and the Inner World*. London: Karnac.

Wallner, F. (1991). *Acht Vorlesungen über den Konstruktiven Realismus* [Eight lectures on Constructive Realism]. Wien: WUV.

2   Collected Papers

2.1   A Computational Model of Affects

Mika Turkia[56]

Emotions and feelings (i.e. affects) are a central feature of human behaviour. Due to complexity and interdisciplinarity of affective phenomena, attempts to define them have often been unsatisfactory. This article provides a simple logical structure, in which affective concepts can be defined. The set of affects defined is similar to the set of emotions covered in the Ortony-Collins-Clore model, but the model presented in this article is fully computationally defined, whereas the OCC model depends on undefined concepts.

Affects are seen as unconscious, emotions as preconscious and feelings as conscious. Affects are thus a superclass of emotions and feelings with regards to consciousness. A set of affective states and related affect-specific behaviours and strategies can be defined with unconscious affects only.

In addition, affects are defined as processes of change in the body state, that have specific triggers. For example, an affect of hope is defined as a specific body state that is triggered when the agent is becomes informed about a future event, that is positive with regards to the agent's needs.

Affects are differentiated from each other by types of causing events. Affects caused by unexpected positive, neutral and negative events are delight, surprise *and* fright, *respectively. Affects caused by expected positive and negative future events are* hope *and* fear.

Affects caused by expected past events are as follows: satisfaction *results from a positive expectation being fulfilled,* disappointment *results from a positive expectation not being fulfilled,* fears confirmed *results from a negative expectation being fulfilled, and* relief *results from a negative expectation not being fulfilled. Pride is targeted towards a self-originated positive event, and shame towards a self-originated negative event. Remorse is targeted towards a self-originated action causing a negative event. Pity is targeted towards a liked agent experiencing a negative event, and happy-for towards a liked agent experiencing a positive event. Resentment is targeted towards a disliked agent experiencing a positive event, and gloating towards a disliked agent experiencing a negative event. An agent is liked/loved if it has produced a net utility greater than zero, and disliked/hated if the net utility is lower than zero. An agent is desired if it is expected*

---

[56] The author would like to thank Krista Lagus (Helsinki University of Technology), Matti Nykänen (University of Kuopio), Ari Rantanen (University of Helsinki) and Timo Honkela (Helsinki University of Technology) for support.

278

to produce a positive net utility in the future, and disliked if the expected net utility is negative.

The above model for unconscious affects is easily computationally implementable, and may be used as a starting point in building believable simulation models of human behaviour. The models can be used as a starting point in the development of computational psychological, psychiatric, sociological and criminological theories, or in e.g. computer games.

2.1.1 Introduction

In this article, computationally trivial differentiation criteria for the most common affects for simple agents are presented (for introduction to the agent-based approach see e.g. [3]). The focus is in providing a simple logical or computational structure, in which affective concepts can be defined.

The set of affects defined is similar to the set of emotions presented in the OCC model [1], which has been a popular emotion model in computer science. However, as e.g. Petta points out, it is only partially computationally defined [4]. For example, definitions of many emotions are based on concepts of standards and norms, but these concepts are undefined. These limitations have often not been taken into account. The OCC model may be closer to a requirements specification than to a directly implementable model. Ortony himself has later described the model as too complicated and proposed a simpler model [5], which may however be somewhat limited.

The missing concepts are however definable. In this article, the necessary definitions and a restructured model similar to the OCC model are presented. A simple implementation of the structural classification model is also presented.

The primary concept is the concept of a computational agent, that represents the affective subject. An agent is defined as possessing a predefined set of goals, e.g. self-survival and reproduction. These goals form the basis of subjectively experienced utility. An event fulfilling a goal has a positive utility; correspondingly, an event reducing the fulfilment of a goal has a negative utility. All other goals may be seen as subgoals derived from these primary, evolutionarily formed goals. Utility is thus seen as a measure of evolutionary fitness.

An agent is defined as logically consisting of a controlling part (nervous system) and a controlled part (the rest of the body). To be able to control its environment (through controlling its own body, that performs actions, which affect the environment) the agent forms a model of the environment. This object model consists of representations of previously perceived objects associated with the utilities they have produced. All future predictions are thus based solely on past experiences.

An affect is defined as a process, in which the controlling part, on perceiving a utility-changing event in the context of its current object representations (object model), produces a corresponding evolutionarily determined bodily change, i.e. transfers the agent to another body state.

Specific behaviours and strategies can be associated with specific affective states. The set of possible affective states and associated actions may be predefined (i.e. innate) or learned. Innate associations may include e.g. aggression towards the causing object of a frustrating event (i.e. aggression as an action associated with frustrated state). Learned actions are acquired by associating previously experienced states with the results of experimented actions in these states.

Emotions and feelings are defined as subclasses of affects [2]. Emotions are defined as preconscious affects and feelings as conscious affects. Being conscious of some object is preliminarily defined as the object being a target of attention (see e.g. [6]). Correspondingly, being preconscious is being capable of attending to the object when needed. In contrast, unconscious affects are processes that cannot be perceived at all due to lack of sensory mechanisms, or otherwise cannot be attended to, due to e.g. limitations or mechanisms of the controlling part.

Thus, emotions and feelings are conceptualized as requiring the agent to be capable of being conscious of changes in its body states [7]. As an affect was defined as a physiological state change triggered by a perception of a predefined event or an object constellation, we can also define a system where agents are not conscious of their affects, but still have them. These unconscious affects suffice to produce a set of states, to which affect-specific behaviours and strategies may be bound. In effect, such agents are affective but not emotional.

Relations between the concepts of affect, emotion, feeling and consciousness were defined above. Another question is the differentiation of affects from each other. This is achieved by classifying the triggering object constellations. The constellations include the state of the agent, which in turn includes the complete history of the agent. In other words the idea is the following: affects are differentiated from each other by both the event type and the contents of the current object model, i.e. by the structure of the subjective social situation. This subjective social situation is formed by the life history of the agent, i.e. the series of all perceived events.

To preliminarily bind this conceptual framework to psychoanalytic object relations theory (e.g. [8], [9]), we note that objects and their utilities in relation to self form a network of *object relations*.

2.1.2 Simulation Environment

Ontological Definitions

Let us assume a *world* that produces a series of *events*. The world contains *objects*, some of which are alive. Living objects that are able to act on the world are called *agents*. Agents' *actions* are a subset of events. An event consists of a type indicator and references to causing object(s) and a target object(s).

An Agent as a Control System

An agent is seen as a *control system*, which consists of two parts: a controlled system and a controlling system (in computer science, this idea has been presented by at least [10]). This division can be done on a functional or logical level only. Let us thus define, that an agent's controlling system is the brain and the associated neural system, and the controlled system is the body. Physically the controlling system is part of the controlled system, but on a functional level they are separate, although there may be feedback loops, so that the actions of the controlling system change its own physical basis, which in turn results in modifications in the rules of control.

An agent usually experiences only a part of the series of events in the world; that part is the *environment* of the agent. Experiencing happens through *perceptions* that contain events. These events are *evaluated* with regards to target agent's needs. The value of an event for an agent's needs can be called its *utility* (the utility concept used here is similar to the utility concept used in reinforcement learning [11]). The utility of an event or action is associated with the causing object.

In a simplified model, fixed utilities can be assigned to event types. In this case the evaluation phase is omitted.

The basis of utilities represented in the controlling system (i.e. brain) are the needs of the controlled system (i.e. body). Utilities direct actions to attempt the fulfilment of the needs of the body. Utility is thus to be understood as a measure of change in evolutionary fitness caused by an event. An agent attempts to experience as much value as possible (maximize its utility) during the rest of its lifetime. Maximizing utility maximizes evolutionary fitness, i.e. self-survival and reproduction, according to a utility function preset by evolution of the species in question.

In order to attempt this utility maximization, the agent has to be able to affect the environment (to act), so that it can pursue highly valued events and try to avoid less valued events. It also has to be able to predict which actions would lead it to experience highly valued events and avoid low-valued (meaningless or harmful) experiences. To be able to predict, the agent has to have *a model of the environment*, which contains models of perceived objects associated with their utilities.

As the agent can never know if it has seen all possible objects and event types of the world, all information is necessarily uncertain, i.e. probability-based in its nature. Therefore, when an agent performs an action, it expects a certain utility. The actual resulting utility may differ from expected, due to the necessarily limited predictive capability of the internal model of the environment.

At any moment, an agent selects and performs the action with the highest expected utility. Thus, every action maximizes subjective utility. Also, any goal is derived from the primary goals (needs), i.e. self-survival and reproduction.

Lifetime utility means the sum of all value inputs that the agent experiences during its lifetime. *Past utility* means the sum of already experienced value inputs.

Future utility means the sum of value inputs to be experienced during the rest of the lifetime. Since this is unknown, it can only be estimated based on past utility. This estimate is called *expected future utility*. Then *expected lifetime utility* is the sum of past utility and expected future utility. An agent maximizes expected future utility. If it would maximize expected lifetime utility, it would die when the expected future utility falls below zero.

Temperament and Personality

Personality is the consequences of learned differences expressed in behaviour. Thus, personality is determined by the learned contents of the controlling system. *Temperament* is the consequences of physiological differences expressed in behaviour.

Norms and Motivation

Norms are defined as learned utilities of actions, i.e. expected utilities of action. Fundamentally, norms are based on physiological needs, as this is the only way to bootstrap (get starting values for) the values of actions. Utilities can be learned from feedback from agent's own body only. However, the utilities determined by internal rewards may be modified by social interaction: an action with a high internal reward may cause harm to other agents, who then threat or harm the agent, lowering the utility of the action to take into account the needs of the other agents. Thus, norms of an individual usually partly express utilities of other agents. In a simplified model there is no need to represent the two components separately. They may however be separated when modeling of internal motivational conflicts is required.

A *standard* is defined as a synonym for norm, though as a term it has a more personal connotation, i.e. internal rewards may dominate over the external rewards. *Motivation* equals the expected utility of an action. Motivation and norm are thus synonymous.

The Processing Loop

The processing loop of the agent is the following: perceive new events, determine their utilities, update object model, perform the action maximizing utility in the current situation. As a new event is perceived, the representation of the causing object is updated to include the utility of the current event.

The object representation currently being retrieved and updated is defined as being the target of attention. After evaluating all new objects, the object with the highest absolute utility (of all objects in the model) is taken as a target of attention.

Object Contexts

If an agent's expected future utility, which it attempts to maximize, is calculated as a sum of utilities of all known objects, it can change only when new events are perceived. However, if it is calculated from conscious objects only, or taking the conscious objects as a starting node and expanding the context from there, keeping low-valued objects unconscious becomes motivated. Now e.g. the idea of *repression* becomes definable.

Thus, introduction of an *internal object context* enables internal dynamics of the expected future utility. Two kinds of dynamics emerge: first related to new objects, and second related to context switches, which happen during the processing of new events.

It can also be defined, that agents can *expand* the context. This expansion is conceptualized as an action, which is selected according to the same principle as other actions, i.e. when it is expected to maximize future utility. This may happen e.g. during idle times, i.e. when there are no new events and all pending actions have been performed, or when several actions cannot be prioritized over each other. Expansion or contraction of the context causes context switches and thus potentially changes in expected future utility.

An especially interesting consequence is that the idea of context expansion during idle times leads to the amount of stress being related to the size of the context (an "emergent" feature). When the agent is overloaded, it context expansion may not take first priority. It "does not have time" to expand the context, i.e. think things thoroughly. Therefore, consciousness of objects' features diminishes; consciousness "becomes shallow". This shallowness includes all object representations, also the self-representation.

Overloading has another consequence. New percepts must be evaluated and appropriate actions selected, but there may be no time to perform these actions, which are then queued. The priorities of the queued actions may change when new events are evaluated. Therefore, at each time point a different action has first priority. Actions taking more time than one time unit are started but not finished, since at the next time point some other action is more important. Therefore, the agent perceives that it is "too busy to do anything", a common feature of *burnout*.

In practice, expansion is done by traversing the object network from the currently prioritized object towards higher utility. For example, an agent has perceived a threatening object and thus expects negative event in the near future. It targets an affect of fear towards the object. As a result its body state changes to "fear" state.

One way of conceptualizing action selection would be to think that a list of actions is browsed to see if there is an action that would cancel the threat. Another way is to think of the action as a node in the object network. Taking the feared object as a starting point, the network is traversed to find a suitable action represented by a node linked with the feared object, the link representing the expected utility of the node. If the node is has the highest utility of all the nodes starting from this object, it is traversed to. If the node is an action, it is performed. If it is another object, the expansion continues.

As the expected future utility is calculated from the objects in the context, the threat is cancelled when an action with a high enough utility is found, although it may not yet be performed (the utility should be weighted by the probability of succeeding in performing the action). This in effect corresponds to a discounting of expected utility. Another, probably better, option would be to take the affective state as a starting node. If the agent has previously experienced a state of fear, it has a representation of this state (an object), and actions associated with the state. Personality was previously defined as the learned contents of the controlling part of the control system.

Personality is therefore formed by adding new objects and their associated utilities to the object network. In the psychoanalytic tradition this is called *internalization* [9]. The continuing process of internalizing new, more satisfying functions of the self may be called *progression*. In progression, an agent's focus shifts on the new objects, since the old objects turn out less satisfying in comparison. Correspondingly, if the new functions later turn out to be useless and better ones cannot be found, the agent turns back to the old objects; this may be called *regression*.

2.1.3   Affects, Emotions and Feelings

Affect as a Bodily Process

Affects are defined here as predefined bodily processes that have certain triggers. When a specific trigger is perceived, a corresponding change in body state follows. This change may then be perceived or not. If it is perceived, the content of perception is the process of change.

In other words, an affect is perceived when the content of the perception is a representation of *the body state in transition*, associated with the perception of the trigger. This is essentially the idea of Damasio [7].

The triggers are not simple objects, but specific *constellations of object relations*. A certain constellation triggers a certain emotion. For example, fear is triggered when a negative event is expected to happen in the future. There is thus an object relation between the agent and the feared object, in which the object has a negative expected utility. This relation may be seen as an object constellation. In principle the current affect is determined by the whole history of interactions between the agent and the objects, not just the current event, since if e.g. the expected utility was very high in the beginning, a small negative change would not suffice to change the object relation from hope to fear. Alternatively, if an agent knows how to avoid the threat (has an appropriate action), then fear is removed when a representation of the suitable action is retrieved from memory. In such case the agent was expecting to be able to cancel the effects of the expected negative event, and expected utility rises back to a level corresponding to the sum of utilities of the event and the reparative action.

These differences are however related to triggers only. What makes an *experience* of fear different from an experience of e.g. hope are the perceived differences in bodily reactions associated with these emotions, i.e. a representation of body state associated with one emotion is different from the representation of a representation of another emotion. This is essentially the 'qualia' problem, which in this context would be equal to asking why e.g. fear *feels* like fear, or what gives fear *the quality of fearness*. The solution is that the 'quality' of feeling of e.g. fear is just the specific, unique representation of the body state. There cannot be any additional aspects in the experience; what is experienced (i.e. the target of attention) is simply the representation.

Differentiating by Levels of Consciousness

Relations between affects, emotions and feelings are defined according to Matthis, who defines affects as a superclass of emotions and feelings [2]. Differentiation is made with respect to levels of consciousness. Emotions are preconscious affects, whereas feelings are conscious affects. There may also be affects that cannot be preconscious or conscious (i.e. cannot be perceived); these are labelled unconscious.

Now we seem to face the problem of defining consciousness. However, the agent only has to be *conscious of some objects*. E.g. Baars has suggested, that being conscious of an object corresponds to that object being the *target of attention* [6]. Let us thus define that conscious contents are the contents that are the target of attention at a given moment. Correspondingly, preconscious are the contents that can be taken at the target of attention, if they become the most important at a given moment.

When an object is perceived (as a part of an event), an agent searches its internal object model to see if the object is known or unknown. It then attempts to estimate the utility of the event (good or useful, meaningless, bad or harmful) by using the known utility of the object. The internal model of this object is then updated with the utility of the current event. If there is no need to search and no unchecked objects are present, attention is targeted towards the object or action which has the highest absolute value of expected utility. The idea behind this is that utility is maximized by pursuing the highest positive opportunity or dodging the worst threat. If only one goal is present, a higher positive event cancels out a lesser negative event. Multiple goals create more complicated situations, which are not discussed in this article.

Multilayered Controlling Systems

For body states in transition and in association with a perceived triggering object constellation to be taken as targets of attention, the controlling system needs an ability to inspect its own structural configurations and their changes in time. Therefore another layer is needed, that records the states of a lower layer of the controlling system. These records of state change sequences can then be handled as objects and attention can be targeted at them, thus making them preconscious or conscious.

Unconscious affects are then first-layer affects that cannot be perceived by the second layer. This may be due to e.g. fixed structural limitations in the introspection mechanism. Defined this way we can also say that there may be affective agents that are not emotional. In particular, all agents with one-layer controlling system would be affective only. An affective agent can thus be fully unconscious. However, an emotional or a feeling agent needs consciousness.

Differentiating by Object Constellations

Classification presented here contains mostly the same affects as the OCC model [1], but the classification criteria differ. The classification is presented in Fig. 2.1.1, which may be compared with the classification proposed in the OCC model [1, p. 19].

The differentiation criteria are: *nature of the target*: whether the target of affect is an event, or an object or agent; *time*: whether the event has happened in the past or is expected in the future; *expectedness*: whether the object was known or unknown, or whether a past event was expected or unexpected; *goal correspondence*: whether the event contributed positively or negatively to agent's goals; *self-inflictedness*: whether the event was self-inflicted or caused by others; *relation to the target*: whether the target object or agent of the event was liked or disliked.

A simplified implementation of these criteria can be constructed as follows: agents do not form memories of events as a whole, but only record utilities of causing objects. Future expectations are thus implicit and consist of object utilities only. In other words, agents do not expect specific events, but expect a specific object to have a utility that is the average of the previous events created by it. An object is expected, if a model of it exists, i.e. it has been perceived before as a causing object. Goal correspondence is implicit in the utilities, as agents only have one goal: maximization of the utility. Goal structure and goal derivatives are thus abstracted away in this simplification.

Affects Related to Events

The first differentiation criteria for event-related affects are: whether the event was targeted towards self or towards other; and whether the originator of the event was self or other.

(1) Events targeted towards self:

(a) Unexpected past events: Fright is an affect caused by a negative unexpected event. Correspondingly, delight is an affect caused by a positive unexpected event. Surprise is caused by a neutral unexpected event. Whether or not it is an affect is often disputed. If it is associated with e.g. memory-related physiological changes, it would be an affect. Another criteria is, that it is associated with a typical facial expression; in this sense it should be classified as an affect.

(b) Expected future events: An expected positive future event causes hope. Correspondingly, an expected negative future event causes fear.

(c) Expected past events: Relief is an affect caused by an expected negative event not being realized. Disappointment is an affect caused by an expected positive event not being realized. Satisfaction is an affect caused by an expected posi-

tive event being realized as expected. Fears-confirmed is an affect caused by an expected negative event being realized as expected.

(2) Events targeted towards others:

(a) Disliked objects: Envy is targeted towards a disliked agent that experienced a positive event. Gloating is targeted towards a disliked agent that experienced a negative event.

(b) Liked objects: Pity is targeted towards a liked agent that experienced a negative event. Happy-for is targeted towards a liked agent that experienced a positive event.

(3) Self-caused events: Remorse is targeted towards a self-originated *action* that caused a negative event to self or someone liked; events positive for disliked objects are considered negative for self. Pride is targeted towards a self-originated *action* that caused a positive event to self or a liked object; events negative for disliked objects are considered positive for self. Shame is targeted towards *self* when a self-originated action caused a negative event.

(4) Events caused by others: Gratitude is targeted towards an agent that caused a positive event towards self or someone who self depends on (i.e. likes). Correspondingly, anger is targeted towards an agent that caused a negative event.

Affects Related to Agents and Objects

In addition to event-related affects, also the originators and targets of events are targets of affects.

(1) Past consequences related affects: Consequences of events cause the originators of the events to be liked or disliked. Like and dislike can be thought of as aggregate terms, taking into account all events caused by an agent. Dislike or hate is targeted towards an agent, who has on average produced more harm than good. Accordingly, like or love is targeted towards an agent, who has produced more good than harm. The difference between e.g. like and love is that of magnitude, not of quality; i.e. love is "stronger" liking. A possibly more appropriate interpretation of love as altruism, i.e. as prioritizing needs of others instead of own needs, is currently out of scope of this model.

(2) Future prospects related affects: Future prospects are estimated on the basis of past experiences; therefore they are determined by the past. However, if we set the point of view on the future only, we can differentiate disgust from dislike and like from desire. Desire is an affect caused by a positive future expectation associated to an object. Accordingly, disgust is an affect caused by a negative future expectation.

(3) Identification-related affects: Identification-related affects are currently out of the scope of the computational implementation, as the concept of identification has not been implemented. Agent wants to identify with an object, that has capabilities that would fulfill its needs; in other words, if the object can perform actions that the agent would like to learn. Admiration is defined as an affect targeted towards an agent or object that the agent wants to identify with. Accordingly, reproach is targeted towards an object that the agent does not want to identify with.

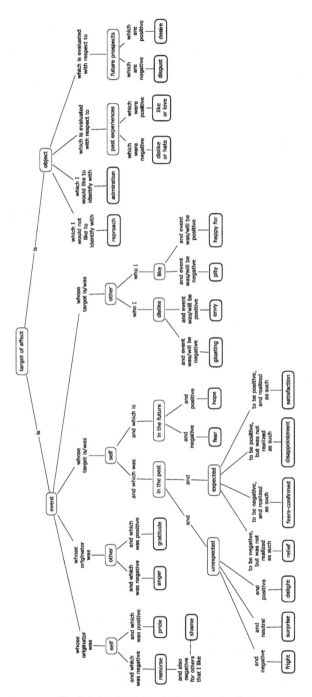

Fig. 2.1.1 Affects in relation to each other

(4) Self-referencing concepts: Above concepts referred to external objects or relations between objects. Affective concepts can also refer to the object itself: e.g. *mood* refers to the state of the agent itself. Examples of mood are happiness, sadness, depression and mania. A simple definition of happiness could be that the average utility of all events (or events in the context) is above zero (below zero for sadness, respectively). Depression could be defined as a condition where no known objects have a positive utility.

Affects and Time

Often mood is thought of as being somehow qualitatively different from emotions. In this paper, the longer duration of mood is thought to be simply a consequence of the stability of the contents of the object model, which in turn depends on the environment. If the environment does not affect the relevant needs, the affective state does not change.

2.1.4   Demonstration

A simple browser-based implementation is available at http://www.cs.helsinki. fi/u/turkia/emotion/emotioneditor. In this simulation the user provides events that change the affective states of three agents.

An example run is as follows. Agent 1 gives agent 2 a utility of 1. Since in the beginning the agents don't have models of each other, this positive event is unexpected, and agent 2 is thus delighted. Also, it now beings to expect a positive utility from agent 1, is begins to like agent 1. In turn, agent 2 gives agent 1 a utility of 1; agent 1 is similarly delighted. Now, agent 3 gives agent 1 a utility of zero. Agent 1 is surprised, and its attitude towards agent 3 is set to neutral. Agent 3 then gives a utility of -1 to agent 2, who is frightened, and begins to dislike agent 3. Although agent 1 is an outsider in this event it reacts to it, since it likes agent 2. Thus, agent 1 targets an affect of pity/compassion towards agent 2 and anger towards agent 3.

Agent 2 now gives a utility of -2 to agent 3, who is frightened, and begins to dislike agent 2. Agent 2 gloats over the misfortune of agent 3 and feels pride of its own action. Agent 1 feels pity towards 3 and anger at 2 (due to neutral attitude being defined equal to liking).

Finally, agent 1 gives a utility of 2 to agent 3, who is delighted. Agent 1 feels happy for agent 3 and pride for its own action. Agent 2 feels envy towards the disliked agent 3 and anger towards agent 1. At this point, all agents have expectations of each other. Agent 2 now accidentally gives a utility 2 to the disliked agent 3, after which it feels remorse and anger towards self and envy towards 3. Agent 1 feels happy for 3 and gratitude towards 2.

At this point, agents don't have utilities for themselves. To demonstrate affects related to expected past events, agent 2 gives itself a utility of 2. It is now delighted, likes itself, and expects a similar result in the future. When performing the same event again, agent 2 feels satisfaction and joy. However, now giving itself a utility of 1, it is disappointed and feels remorse.

As a result of the previous event history, agent 3 expects a utility of 2 and is in a good mood. It does not have expectations of itself. When giving itself a utility of -4, its average expectations change to -2, its expectations towards itself to -4 and its mood to bad. When giving itself a utility of -4 again, its fears are confirmed. When giving itself a utility of -2, it feels relief. The event sequence was thus (1,2,1), (2,1,1), (3,1,0), (3,2,-1), (2,3,-2), (1,3,2), (2,3,2), (2,2,2), (2,2,2), (2,2,1), (3,3,-4), (3,3,-4), (3,3,-2), where the first argument of the triple is the causing agent, the second is the target agent, and the third is the utility of the event. As mentioned before, the resulting intersubjective utility expectations form a network of object relations.

2.1.5   Conclusion

This article presented definitions of affects, that remove the limitations of the OCC model and are easily computationally implementable. They can be used as a starting point in the development of computational psychological, psychiatric, sociological and criminological theories, or in e.g. computer games.

References

[1]   A. Ortony, A. Collins, and G. L. Clore, *The Cognitive Structure of Emotions*. Cambridge: Cambridge University Press, 1988.

[2]   I. Matthis, "Sketch for a metapsychology of affect," *The International Journal of Psychoanalysis*, vol. 81, pp. 215–227, 2000.

[3]   S. Russell and P. Norvig, *Artificial Intelligence: A Modern Approach*. London: Prentice Hall, 1995.

[4]   P. Petta, "Identifying theoritical problems," in *Proceedings of the first HUMAINE Workshop on Theories and Models of Emotion*, June 2004. [Online]. Available: http://emotion-research.net/ws/wp3/

[5]   A. Ortony, "On making believable emotional agents believable," in *Emotions in Humans and Artifacts*, R. Trappl, P. Petta, and S. Payr, Eds. Cambridge: MIT Press, 2003, pp. 189–212.

[6]   B. J. Baars, "Some essential differences between consciousness and attention, perception, and working memory," *Consciousness and Cognition*, vol. 6, pp. 363–371, 1997.

[7]   A. R. Damasio, *Looking for Spinoza: Joy, Sorrow and the Feeling Brain*. Orlando: Harcourt, 2003.

[8]   P. Fonagy and M. Target, Eds., *Psychoanalytic Theories. Perspectives from developmental psychopatology*. London: Whurr, 2003.

[9]   V. Tähkä, *Mind and Its Treatment: A Psychoanalytic Approach*. International Universities Press, 1993.

[10] A. Sloman, "The mind as a control system," in *Philosophy and the Cognitive Sciences*, C. Hookway and D. Peterson, Eds. Cambridge University Press, 1993, pp. 69–110.

[11] R. S. Sutton and A. G. Barto, *Reinforcement Learning: an Introduction*. London: MIT Press, 1998.

2.2 The Physics of Thoughts

Ron Spielman

*From a "reverse engineering" perspective, it can be argued that memories –
and consequently thoughts – are stored in a hypothesised system of* electro-
magnetic wave forms *"generated" by neural networks. It is argued that the empirically experienced speed of memory encoding, storage and subsequent retrieval, as encountered in the psychoanalytical consulting room, as well as in general
life, is too complex and rapid to be served by known electro-physiological or biochemical brain mechanisms. There are adequate analogous examples of complex
data being stored and transmitted in electro-magnetic wave forms to confer plau-
sibility on the hypothesised memory system proposed. These wave forms would be
readily "accessible" in the brain and would be able to subserve all the empirically and clinically experienced functions described in the paper.*

*In the field of Philosophy of Mind, debates have been between "dualists" and
"monists". The monists have prevailed. A sub-set of monists favour an "emergent" position: the mind is an emergent epi-phenomenon of brain function.*

*This paper will adopt this "emergent phenomenon" position. However, this
still begs many questions. It will be the intent of this paper to explore some of
these questions, especially from the point of view of* "reverse engineering". *That
is, the paper will outline some empirically established "facts" about some mental
processes – especially those involving memory function - and ask what neurophysiological mechanisms would be required to support the existence of these empirically observed phenomena. These are the "demand characteristics" of the system
in question.*

*Further, some of the empirically observed phenomena addressed will be some
of those encountered clinically in the practice of psychoanalysis and psychiatry.*

Any attempt to emulate the mind *must begin with the "best possible" understanding of how the human mind does indeed function! Broadly speaking – in relation to memory - the main question will be "where does this aspect of the mind
reside?" – and what kind of known neuro-physiological and physico-chemical
processes could support this aspect of mental function.*

Sigmund Freud, in his Project for a Scientific Psychology [1], *hoped that
psychic processes would ultimately be shown to be resultant upon biological processes – but at the time of his writing, that "project" was beyond known science.*

2.2.1   The Current "Project"

Let me offer an analogy: a transistor radio, not "switched on", can be clearly
understood to be a physical entity, reducible to its many components. With a
charged battery in place to provide energy, and now switched on, the transistor radio can be "tuned in" to any available "broadcasts" of the appropriate frequency
electro-magnetic spectrum of radiation. Whatever music or voice transmission
from the relevant broadcasting "station" is "received" by the apparatus – and in

turn "played" by the radio - will be the counterpart of an "emergent" property of the brain etc. in the radio. That is, the music or voice "emerges" as a result of the physical arrangements and connections of the components of the radio. Importantly, however, the actual *content* of whatever music or voice is "produced" bears no relationship whatever to the physical components of the radio – it bears only a relationship to the frequency waves broadcast elsewhere and subsequently received by the physical components of the radio.

The radio will produce sound only as it "exists" in the form of radio frequencies which are in the neighbourhood of the radio *at that very instant in time.* The radio merely transforms the inaudible radio wave frequencies into audible sound[57].

Consider now, a transistorised compact disc player. Again, there are physical components capable of producing sound. This time, there is no external "broadcast" to be received. Once again, an energy source (a battery) is required. Also, a "CD" (compact disc) needs to be inserted into the apparatus and the apparatus activated. The CD player will accurately and reliably produce an audibly transformed replication of whatever sounds were inscribed ("recorded") by the relevant laser-based mechanisms on the compact disk. This disc can be played and replayed – or, another disc can be inserted and likewise reliably played to reproduce the original sounds.

This "analogy" has been laboriously outlined to establish a number of points necessary for the paper's arguments.

(1) The radio can respond to the relevant wave frequencies of the moment (the present) – but has no "memory" of the past. It has a present – but no past.

(2) The CD player has the capacity to play and replay "stored" information – it has the equivalent of a "memory". However, its "memory" is static and unalterable. It has a past – but no present.

(3) The nature of the sounds produced are entirely related to either the properties of the frequency waves received by the radio, or the laser-encoded information stored on the CD player – and in no way are related to the physical components of either apparatus.

This analogy is offered in order to distinguish the requirement for brains to encode and record "pasts" as well as respond to "presents". Using this analogy, each *psyche* will be "tuned in" to different present experiences (analogous to, but in no way equivalent to, the notion of "tuned in to the present 'broadcasts'") and have vastly different "pasts" (analogous to the recorded information encoded on any one of a myriad compact discs). Thus, individuals "respond to" and are "activated by" a combination of their present and past experiences – their *memories.*

The main question of this paper is "where does this past experience (memory) *reside* and *of what is it composed?*"

Let us establish some further empirically indisputable facts on which the paper will rely:

---

[57] As the human eye and brain together transform invisible electro-magnetic (light) frequencies into *experiences* of seeing objects!

(1) *Past experience is (largely) what differentiates human individuals and constitutes their personalities.*

As disparate a group as clinical psychoanalysts and Jesuit priests have no doubt that childhood experience is what ultimately shapes the adult personality. Psychoanalysts since Sigmund Freud have repeatedly confirmed that childhood experiences contribute massively to adult behaviour, while Jesuit priests aver: "give me a child until the age of seven, and I'll give you a Catholic for life". The alternative would be an organism analogous to the transistor radio: an organism that responds only to the present. What is being established here is the fact that the higher (and some not so high) organisms have *memory*.

(2) Memory is not *veridical.*

Many studies have established that memory itself is somewhat (if not considerably) "plastic". That is memory does not necessarily reliably record true events. Psychoanalysts since Freud have worked with the notion that memory can in fact be unconsciously "altered" during life in response to later experience (Freud's concept of *Nachträglichkeit* [1]).

(3) Memory is not highly localised in the brain.

Again, many studies have established that quite large quantities of the cerebral cortex can be removed or damaged, without significantly diminishing memory. Such studies have been conducted with lower animals, while comparable naturalistic studies in humans have resulted from brain damage incurred in accidents, in wartime, and as a by-product of brain surgery for neoplasm or epilepsy[58].

(4) Memory is *dynamic* – and not static.

Psychoanalytic observation in the consulting room has allowed the disciplined empirical observation of behaviour which occurs in day-to-day life. One of these empirical observations was made at the outset of the psychoanalytic research program by its earliest proponent, Sigmund Freud. This is the phenomenon called "the repetition compulsion". This involves the apparent compulsion to repeat certain experiences of the past with new figures in the present. Generally, these experiences are of an unsatisfactory or traumatic nature.

It is well known – both within psychoanalysis and within psychiatry and psychology – that many past traumatic experiences remain unresolved in some significant manner and exert an *active influence* on present experience and behaviour. The point here is that not only the *fact* of the trauma and its associated affects, but also *the content* and *significance* of the past trauma are represented in the present behavioural manifestations of it. Post-traumatic Stress Disorder (PTSD) is an exemplar of this phenomenon.

---

[58] This non-localisation has to do with *stored* memories. *Gateways* for the encoding of memory are probably much more localised, as (for example) the respective damage sustained in Wernicke's Encephalopathy and Korsakoff's Psychosis would suggest. Localization of a necessary memory *reader* is (as yet) a largely unknown phenomenon.

2.2.2　The Argument Proposed

A "reverse engineering" approach to the problem of "the mind". If we can agree that "no brain – no mind" and that the individuality of persons resides in their experience of themselves as a "self" (from the inside) or on our characterisation of them as a "personality" (from the outside), then our problem involves describing the workings of "the mind" as it is involved in the manifestation of personality and the experience of self – and, then, trying to conceptualise what kind of neurophysiological system would have had to be "engineered" (by evolutionary processes) in the first place to produce these workings:

(1) A system of rapidly *encoding* experience
(2) A system of rapidly *storing* experience
(3) A system which rapidly *accesses* stored experience
(4) A system where stored experience *actively finds expression in the present* (in some recognisable form)

As the perspective of this paper is a psychoanalytic one, it needs to be acknowledged that *both* conscious and unconscious mental processes are being taken into consideration. That is, that present behaviour is readily influenced by unconscious mental processes. These involve both unconscious perceptions (as in unconsciously determined interpretations of current stimuli) and unconscious motivations (as in unconsciously influenced decisions in choice-making behaviour).

Thus, point 4 above is a system requirement which takes into account the *empirically derived* fact that stored experience is *actively* involved in determining the nature of perceived experience and in contributing to the motivational influences on present behaviour. This empirical phenomenon – derived largely from psychoanalytic clinical experience – involving a *compulsion to repeat* past experience in some form requires that past experience have some mechanism of pressing forward its influence.

To summarise the main burden of this argument: on empirical grounds alone, there needs to be a memory "system" which functions very rapidly (especially in regard to *retrieval* of memories and experience) and which permits the stored encoded memory of experience to have recognisable influence on the present-time experience and behaviour of the individual.

To date, it would appear that the main focus of attention on memory systems has been on biological systems involving either bio-chemical mechanisms or electro-physiological mechanisms.

It is contended here that both bio-chemical and electro-physiological systems would conceivably fare well in delivering encoding and storage requirements – but would *not* be able to deliver on the requirement of *rapidity of response* as empirically experienced by any introspective person. Empirical experiences of memory retrieval are extremely rapid – especially if one considers those instances when long un-remembered (but not forgotten!) memories "pop" into mind with extreme rapidity in response to a relevant stimulus.

Further, when one tries to conceptualise (imagine) the number of likely "connections" between neurones and the associated number of synaptic "crossings" which would be required, in even a relatively simple "piece" of encoded memory, then a finite amount of time would be involved (each and every synaptic crossing would require a finite measurable number of milliseconds, or even *nano*seconds), which – it is asserted here – would far exceed the *actual* time as empirically experienced by any introspective individual.

What kind of system could – conceivably – meet the demand requirements of encoding, storage and rapidity of retrieval, together with the requirement of actual unconscious influence on the person's behaviour?

2.2.3 Encoding, Storage and Retrieval

Nowadays, there are many readily recognisable forms of encoding of data within a number of biological and physical modalities.

Biological Examples

The genetic code is comprised of sequences of nucleic acids, linked in large numbers to form the (now) familiar *gene*. Immuno-globulins encode past experience with identifiable antigens to form appropriate antibodies. Neither of these bio-chemical systems require rapid responses (in terms of milli-seconds), in either encoding or retrieval.

Examples of neuro-physiological systems which have evolved to encode and store "memory" are the cerebellar neuronal connections which store past experience of development of skilled movements. These are seen at their most highly developed in, say, elite athletes or virtuoso musicians, each of whom are able to repeat the most complex practiced movements with extreme rapidity. Retrieval may be extremely rapid, but encoding is probably painstakingly slow, usually involving many repeated practice sessions[59].

Non-Biological Examples

The field of recording of music and speech in the radio and TV arena would be readily recognisable instances of encoding and storage of experience. Audio-tapes and video-tapes are common examples of encoding sound and vision in electromagnetic wave forms on suitable magnetic tapes and retrieving them with the relevant equipment. High degrees of reliability (fidelity) can be achieved in storage and reproduction. In this instance, encoding and storage are almost instant, as is retrieval. However, plasticity and flexibility of "responses" is zero. In contrast, a skilled sportsman can alter a previously stored response where required.

---

[59] In the case of a batter facing a baseball pitcher or a batsman in cricket facing a fast bowler, the *actual* distance from pitcher or bowler to respective batter or batsman, bears a calculable relationship to the response times biologically possible for perception of speed and direction of the ball and the required response.

A familiar sight now in supermarkets is the almost ubiquitous "bar code". Here, considerable amounts of data can be encoded in the thickness and frequency of the vertical "bars" comprising the barcode. Here, encoding is via a man-made arbitrary system. This is analogous to any secret code ever devised – say in the course of espionage – where the encoder must make the nature of the artificially designed code known to the decoder. Else the message cannot be retrieved!

2.2.4   The Special Problem of Retrieval of Encoded Memory

As has already been stated, the rapidity of empirically experienced memory retrieval places a special burden on any hypothesised memory system.

The assertion here is that known mechanisms based on either biochemical (as in protein or other chemical substances) or neurophysiological (as in cerebral neurones and their interconnections) storage are likely too slow to deliver the ultrarapid retrieval rates as empirically experienced.

To mention only one common phenomenon: slips of the tongue and other parapraxes. These phenomena are experienced as almost instantaneous. For a memory system which depends on a form of "searching" (as in computers) for retrieval, ultra-high speeds would be required which would be beyond biological tissues. Otherwise, a degree of physical propinquity would be necessary to overcome speed-of-search considerations. For both biochemical and neuronal based storage systems, the similar stored elements would need to be physically somewhat close and localised. Current memory research suggests memory is diffusely located and not localised at all.

What kind of memory system could be independent of such localised physical biological phenomena, such as biochemical substances or neuronal connections?

A Hypothesis:

"Give me a fruitful error any time, full of seeds, bursting with its own corrections. You can keep your sterile truths for yourself."[60]

It is hypothesized, now, that experience is (somehow) encoded within the interconnections between cerebral neurones (of which there are known to be more than sufficient numbers present in the brain). *Neural networks* have the important property of being able to modify themselves when exposed to new experience. That is, they can learn.

However, this encoded stored information is not *statically stored* – but, by the nature of the operation of the neural network, exerts a *dynamic presence* within the brain. Put crudely, the "output" of these putative neural networks is always "switched on" and having the equivalent of an electro-magnetic wave form "presence" within the brain. In some way comparable to the barcode phenomenon, *unlimited* amounts of data (information) can be encoded within frequency wave forms. This is the everyday case with telephonic and other broadcast and transmitted communications which saturate the world's atmosphere and subserve many

---

[60] Vilfredo Pareto – quoted by Francis Crick [2]. The present hypothesis is offered in the same spirit.

complex requirements. Remote control of domestic appliances relies on this (infra-red frequency) technology. Consider, for example, the finely tuned and complex control of immensely far distant spacecraft by the medium of electro-magnetic transmissions from "mission control" on the planet earth.

The marvel is that these myriad electro-magnetic "transmissions" do not (apparently) interfere with one another and maintain their integrity over vast distances and considerable time. Perhaps, something analogous is possible within cerebral tissues.

A potential system which meets the empirically derived requirements suggested above could be provided if some biological mechanism of generating the equivalent of frequency wave forms could be conceived.

Many current works refer to "oscillations" of varying sorts in cerebral tissues (e.g. Crick [2]). Although the authors of these works almost certainly imply electro-physiological "oscillations", the speculative "shift" ventured here is toward "electro-magnetic" energy forms. Reasons for this, and supportive "evidence" from a variety of sources will be offered. Advances in understanding of cerebral neuro-physiology now encompass the phenomenology of neural networks and concepts involving connectionism. Indeed, work of Zeki and Bartels [3] hypothesises a neurocomputational model of consciousness involving "multiple micro-consciousnesses". The hypothesis of the present paper benefits highly from a pre-existing hypothesis which could translate equally well into the "demand requirements" for "multiple simultaneous un-conscious-es". These now almost more than putative phenomena seem now well recognised as playing a possibly very important role in cerebral mechanisms of a variety of types.

2.2.5 Empirical Evidence in Support of the Hypothesis

In a broad sense, the hypothesis calls for mental contents (thoughts, memories, ideas, etc.) to be borne (encoded and stored) in some form of "electro-magnetic wave" phenomenon – rather than in the more familiar "electro-physiological", membrane-potential based propagation of nerve impulses. Granted, there is, as yet, no direct evidence of this kind of "energy" being generated in brain tissue ... but perhaps it has not yet been looked for definitively enough.

Nevertheless, in popular language the phrase "brain waves" is often enough heard. We know that the *linga franca* often reveals "folk wisdom" which may in turn prove to have some scientific validity. What of the phrases "that rings a bell" and "that strikes a chord". These two phrases reveal an aspect of the individual's experience of the operation of their own mind which must have widespread resonance (sic.) among fellow human beings.

The point here is that popular language (at least in English) seems to point toward imagery related to properties of the electro-magnetic wave spectrum.

2.2.6   "Unconscious Influence"

One relatively simple example of what is meant by "unconscious influence" is manifest in what has become popularly known as the "*Freudian Slip*"[61]. This phenomenon occurs when a person intends to say something and finds himself saying something else – but closely related. It is held that these "slips" are far from random, and are instead motivated by some underlying influence.

As an example, a patient who suspects she may have been sexually abused in the night when she was very young says of the possible perpetrator "I could hear his foot faults coming". She had intended to say "foot steps" – but the (*unconscious*) underlying accusation is contained in the word "fault". Another patient, who had difficulty expressing her anger made the slip "livid memory" – for "vivid memory". Someone else spoke of going to a restaurant with their family. The restaurant was called "Nippers". However, she said "we all went to *Nipples* for dinner." The possible unconscious association of adults going for a meal with an infant being fed at the breast is not too far fetched (especially for a psychoanalyst!).

Each of these examples (the latter two most clearly) involve a substitution of a word or phrase which *sounds* similar to the intended one. The point here is that some ready substitution of similarly encoded stored data can be conceived of as having occurred.

In addition, the readiness in which the substituted word or phrase appears to be "available" implies that little "searching" for the substitute has occurred. It must have somehow been almost "there", available for use. If the electro-magnetic wave form of storage of data were the case, then similarly encoded data would be accessed through the closely similar "patterns" in the respective wave forms.

2.2.7   Pattern Recognition

It is well known that the physiology of the human brain in many of its perceptive modalities relies heavily on pattern recognition. This, no doubt, is of evolutionary significance in that there would be considerable advantage in readily recognising situations and circumstances which have been previously encountered.

All learning revolves around the comparing and contrasting of new experience with old and developing ways of dealing with novel experience.

From a psychoanalyst's perspective, this general phenomenon of pattern recognition is manifest in the well known specific clinical phenomenon of *the transference relationship*. Here, the patient "recognises" some aspects of the current relationship with the analyst as "belonging" to a pattern of past experience (say, to do

---

[61] It should be made clear here that in the current context "unconscious" is being used primarily in the sense of "un-aware", but that this includes the psychoanalytic sense of unconscious – as in "repressed" and associated usages.

with the relationship with the mother or the father) and experiences the current relationship "as if" it were the past relationship[62].

A major application of the proposed hypothesis here allows for even discrete parts of a complex encoded experience of the past to be "recognised" as a pattern in relation to a current experience, and for the whole of the past experience to become "activated" (e.g. Post-Traumatic Stress Disorder, where even a discrete aspect of the traumatic past experience can rapidly bring the whole of the past traumatic experience into consciousness). Lesser instances of this phenomenon are encountered in every-day life as in the "that strikes a chord" or "that rings a bell" experience.

Psychoanalysts since Freud have relied on this mechanism manifest as *free association*, whereby trains of thought are pursued, driven by the similarity of the new thought to the foregoing thought. The therapeutic benefit deriving ultimately from the unconscious links between (initially) apparently unconnected thoughts leading to uncovering of clinically relevant material. The "linking" will ultimately be found to have depended on some underlying "pattern" in the chain of associations. While in the earliest days of psychoanalytic technique, this pursuit of "free associations" was definitively undertaken by requesting such associations, nowadays, empirical clinical experience has found that in the (non-directive) psychoanalytic setting, all material is inevitably linked by unconscious associations and does not have to be artificially requested.

Patterns of relating, recognisable to psychoanalysts as deriving from early part-object experiences, constitute the object of psychoanalytic exploration in the consulting room. It is contended here that the encoded experiences which relate to these early part-object relationships (especially so in the case of the more primitive personality disorders) are constantly "active" in the mind/brain of all individuals and exert their influence in current day behaviour via a mechanism relying of pattern recognition – the patterns being greater or lesser portions of sequences of electro-magnetically encoded "brain waves".

2.2.8   Support for the Proposed Hypothesis from Clinical Psychiatry

Obsessive Compulsive Disorder

The disorder known as Obsessive Compulsive Disorder (OCD), is quite common, and may involve certain ideas, thoughts and behaviour patterns becoming involved in a reverberating loop of some kind.

Transcranial Magnetic Stimulation (TMS)

In recent times, there have been increasing numbers of reports in the psychiatric literature of utilising TMS in a manner similar to ECT. Instead of delivering

---

[62] This is necessarily a considerable simplification of a complex clinical experience for the purposes of this argument.

an electrical "jolt" to the brain (as in ECT), the brain is exposed to an intense magnetic field. The results are not universally acclaimed.

Nevertheless, as with ECT, TMS is at this point based on no particular rationale – other than empirical claims of efficacy. The present hypothesis would provide both TMS and ECT (more so TMS) with a sound rationale. If depression is contributed to by reverberating thoughts of, say, worthlessness and hopelessness, these may be (at least temporarily) disrupted by the application of either magnetic or electrical fields.

Clang Associations

Clang associations are a well known symptom usually associated with Schizophrenia. The patient classically speaks in a manner which includes rapid associations of ideas which seem based on the *sound* of key words involved in the associations.

2.2.9   Another Speculation

Central nervous tissue (which ultimately develops into the brain) derives embryologically from the *ectoderm* – that is from the external layers of the developing embryo, which are then invaginated and become deeply internal structures.

The speculation involves a notion that these human "ectodermal" tissues – as a result of evolution - derived *from* very early life forms which were "bathed" in electro-magnetic radiations (wave forms). It is hypothesized that specialized receptors for vision and hearing were "*induced*" by these very electro-magnetic wave forms. The retinas and the lining of the internal auditory canal each derive from *ectodermal* tissues – and are sensitive to electro-magnetic wave frequencies. So, might not the cerebral tissues have an analogous aspect of its function in an "electro-magnetic" kind of manner?

2.2.10   Some Potential Benefits Deriving from the Model

(1) As already alluded to above, the hypothesis would provide a rationale for the effectiveness of both TMS and ECT.

(2) Some empirically encountered phenomena, such as *dreaming* might find support in the hypothesised mechanisms: for example, the presence of patterns of relationships encountered in dreams reported in psychodynamically based treatments which bear a relationship to the therapeutic relationship might be explained. Similarly, the presence of so-called "day residues" in many (if not all) dreams could result from the "presence" of such memory traces in the sleeping mind, as the work of encoding experience is taking place during sleep (as hypothesised by some theories).

(3) In addition to the actual content of thoughts and memories being encoded in the hypothesised wave forms, so too could "qualifying considerations" be appended: for example, varying degrees of "permission" to access to consciousness could be attached to discrete memory wave forms. It is con-

ceivable in this model that some code for "not to be admitted to consciousness" could be at the base of the well recognised clinical phenomenon of
repression – and perhaps, explain the more extreme cases of Repressed
Memory Syndrome.

(4) The plasticity of memory – and especially that variant described by Freud
[1] as Nachträglichkeit (the retrospective revision of memory in the light of
later development) – could be accomplished by revised encoding in neural
networks and consequent alteration of the form of the resultant wave-
forms.

(5) For modern psychoanalysis, in particular the current theories reliant more
on object relationships than the earlier instinct-based theories, this hypothesis would offer a mechanism whereby the empirically observed phenomena – especially those involving unconscious influence of past experience in present day relationships – could be readily supported.

(6) Finally, "slips of the tongue" and other parapraxes would find ready explanation in a manner comparable to the "clang association" described in the
phenomenology of schizophrenia. The unconsciously substituted phrase in
the slip of the tongue would be as "present" as the consciously intended
one. Similarities in the encoded wave forms would make the choice between the consciously intended word or phrase and the unconsciously offered word or phrase an issue at the point of retrieval of the memory. The
choice favouring the unconsciously determined phrase would be influenced by "present" unconscious feelings, ideas or motives, which would be
vying for expression with all other thoughts. ideas, etc. through the atten-
tion/consciousness decoder.

2.2.11   Possible Research Directions

The entire hypothesis advanced rests on the possibility of the existence of electro-magnetic wave forms in cerebral tissues. Detection of energies in such wave
forms in deep space has been achieved by astro-physicists. It should be feasible
for bio-medical engineers to utilise comparable technology to detect the presence
of even minute quantities of such electro-magnetic energies during the course
(say) of open brain surgical operations.

2.2.12   Activity of Other Biological Tissues Known to Involve Electro-
magnetic Frequency Radiations

(1) Retinal tissue converts electro-magnetic wave frequencies into nerve impulses which are transmitted along the optic nerve to higher brain centers –
as do auditory receptors convert sound frequencies into impulses along the
auditory nerve.

(2) Bats and some mammals (e.g. dolphins and whales) use echo-location as a
means of navigation.

(3) Bees seem to rely on detection of magnetic frequencies.

(4) Although not yet definitively proven, there is suggestive evidence that users of mobile telephones are more than usually susceptible to tumours of the auditory nerve tissues as a result of exposure to levels of concentrated electro-magnetic radiation from their mobile telephones.

(5) Similarly, people living in proximity to high tension electricity installations are suspected of being more susceptible to blood disorders (leukaemias) and other cancers, as a result of the electro-magnetic fields generated by these installations.

2.2.13 Conclusions

Drawing upon empirical experience derived from psychoanalytic theory and practice, the "demand characteristics" of human memory encoding, storage and retrieval are outlined and the concept of "reverse engineering" used to hypothesise what is required to "design" a model of human memory.

It is hypothesized that existing known neuro-chemical physiological processes cannot support the empirically required functions of speed of encoding and retrieval of memory – and especially the dynamic (as opposed to static) phenomenology of past experience so strongly influencing present behaviour in humans. Argument by analogy and from empirical data is advanced to support the hypothesis.

This hypothesis rests on the concept that only a system whereby cerebral tissues generate electro-magnetic wave forms (presumably via the operation of neural networks) can the known high-speed phenomenology of human memory – and hence thoughts – be subserved.

References

[1] Freud, Sigmund (1950 [1895]) Project for a Scientific Psychology. Standard Edition of Complete Works, Vol. 1: 281-391.
[2] Crick, Francis (1994) The Astonishing Hypothesis: the scientific search for the soul. Charles Scrivener, New York.
[3] Zeki, S. and A. Bartels (1998) The asynchrony of consciousness. Proc. Royal Soc. of London 265: 1853-1585

2.3 A Functional View on "Cognitive" Perceptual Systems Based on Functions and Principles of the Human Mind

Matthias J. Schlemmer and Markus Vincze[63]

We outline functionalities of an artificial system that seem to be important for cognitive behaviour. We argue from the perspective of perception as perceptual capabilities are crucial for viable performance and situated behaviour. The first part of this paper is concerned with general anthropomorphic considerations and notional questions related to terms like "conscious", "intelligent" and "cognitive". Then, starting from the three basic properties and five broad functional principles that Mark Solms defines in his paper and described in his talk at ENF, we set out to link the requirements for a cognitive artificial system to those cardinal aspects of the mind. Although there are some robotic capabilities still missing in order to achieve an implementation of some functions of the human mind, the results indicate interesting synergies and cross-links between (neuro-) psychoanalysis and engineering.

2.3.1 Introduction

At the First International Engineering & Neuropsychoanalysis Forum (ENF), notions like mind and consciousness were central. Besides, cognition has become a buzzword – especially in engineering sciences. The aim of this paper is to respond to M. Solms' paper [4] and talk at ENF [3, #1, Ch. 4][64] in order to give a technical view on his depiction of the model of the mind, i.e., the three basic properties and the five broad functional principles, and what a possible synthetic implementation affords. The intention is to confront five elements of building such a cognitive system with those properties and functional principles defined by Solms. We will deliberately use the term "cognitive" system to underline the difference to "conscious" or "intelligent" – see Section 2.3.3.

The main viewpoint of this work is to strictly separate function from implementation. This seems to be necessary as this separation was a bit fuzzy throughout the discussions at ENF 2007. Consequently, the argumentations of the synthetically important aspects will be led from a functional angle. As Solms outlines in his paper, the functional perspective gives the basis for discussing mind-related aspects between (neuro-) psychoanalysts and engineers. The main idea is that reconsidering the functional goals for systems after reading the papers and listening

[63] This work is supported by the European Project XPERO under contract No. 029427 and by the Austrian Science Foundation under grant #S9101 ("Cognitive Vision").

[64] As this paper refers to the presentations at ENF, a lot of citations are taken from this book (i.e., [4]) as well as the ENF-DVDs [3], which provide the complete video footage of ENF 2007. In the references, # indicates the number of the DVD (as there are three), Ch then indicates the Chapter of the DVD, or, if applicable, the respective Session or Discussion Session is given.

to the talks of ENF brings up new ideas to tackle the problems currently faced by Artificial Intelligence.

This paper is structured as follows: First, in Section 2.3.2, general considerations concerning the aim of an interdisciplinary approach of (neuro-) psychoanalysts and engineers are given. The goal of "constructing a mind" is herein questioned and some of the constraints found by engineers are mentioned.

After a (strictly engineering-biased) excursus into the notions *consciousness*, *intelligence* and *cognition*, some aspects of cognitive robots that we judge to be important as prerequisites for a "Cognitive System" are delineated. Namely, *intentionality*, *symbol anchoring*, *abstraction*, *generalization* and *prediction* are tackled and linked with the three basic properties that Solms defines for the human mind.

Section 2.3.4 uses an example from robotic vision to guide through the five functional principles mentioned by Solms, namely the evolutionary biological principle, the principle of drives, the principle of feelings/emotions, the principle of motivations and the principle of cognition. A discussion on open issues (Section 2.3.5) and the summary (Section 2.3.6) conclude the paper. We are aware that such a discussion must be incomplete at this point of bridging engineering and neuropsychoanalysis; the intention is to get a more detailed analysis of the viewpoints of both disciplines.

A word of warning: Although Solms stressed at the beginning of his talk [3, #1, Ch. 4] that the psychoanalytical input to the interdisciplinary approach of ENF is that it is able to draw analogies between *being a human* and *being something else* (i.e., putting the subjective perspective in the foreground, which is then broken down to the functional level), in this paper we will argue that the subjective perspective is not what interests us most. We are rather trying to focus on the capabilities of humans and how such a model of the mind could help engineering to achieve similarly good behaviours. As the psychoanalytical approaches go further than just the subjective perspective, the specialized view on the phenomenon of the mind that allows us to look at its functions very differently than usually is very fruitful. This is particularly interesting because of the fact that some aspects can only be given by behaviouristic or by neurological studies – two views inaccessible for technical sciences. An interdisciplinary approach is thus highly relevant.

2.3.2   Constructing the Mind: The Ultimate Goal! – Or is It?

There is an old saying that only he who knows his goal will also find the way. Before investigating specific questions about the modelling of the mind, we surely have to be aware of the goals we are pursuing. E. Brainin said in a session at ENF that engineers should reflect on their wish to build the human brain, mind or psychic apparatus [3, #2, Discussion 2]. This request is very important to be answered as it will affect to a large extent how the investigation of the artificial construction of cognitive systems is led. Especially in conjunction with the question of the mind, it must be clear what to do and what to head for.

In Part II Section 1.3, point 1, Solms mentions that the ultimate goal of Dietrich et al. is to construct an artificial mind. To say it straight away: the authors of this work – looking at the question from the robotic side – do not go that far. Let us first decide the "grade" of the mind that we would like to build. At least from our point of view, the true goal of dealing with cognitive issues in artificial intelligence and robotics lies in the following: To build a machine that is able to serve purposefully on the human's side in a way that reflects deep understanding of things and scenes.

This has to be underlined right at the beginning of this paper as this affects the whole approach to the problem: The aim is the construction of a robot for successful operation and service to humans. Matching this system with humans in a way that it shall be *comparable* other than in the behaviouristic, but in the anthropomorphic sense – a question often tackled by philosophers – can therefore not even be posed. The only comparison that we might want to attempt is whether in a given situation the artificial system *performs* as a human would do in the same situation – either in a way of "understanding the scene" or really "acting". This can also be seen in conjunction with D. Dietrich's emphasis on a *unitary view*. His opinion is that we need such a unitary view to be able to construct the human mind. It is of course reasonable to believe that a patchwork – as it is usually done in artificial intelligence – might not work as well as a system designed top-down from a unitary view. However, looking at the *functioning* of the system at the human's side, the unitary view need not necessarily be the same nor even be oriented towards the human mind. What definitely is very doubtable is Dietrich's statement that the lower-level functions "[...] have nothing to do with the higher functions" [3, #1, Ch. 3]. Of course there are many abstract layers between the low levels and the highest level – but if the latter really had nothing to do with the former, what do we need a high-level model for? And even worse: Philosophically spoken this means in the last consequence that we actually exclude monistic views as the solution of the mind-body problem.

A solution would rather be to search for such a unitary model, followed by the definition of the necessary functions implied or implemented by this model, and then to design those "modules" with having the total model in mind (!). This naturally leads us back to the emphasis of the functions that should be focused on, but with the difference that the starting point is now a high-level model. Whatever the approach, the viability of the *human* model as blueprint for cognitive systems must be looked at very carefully in any case.

To relativise the possibly implied consequence that an interdisciplinary research between engineers and (neuro)psychoanalysts is useless if not looking at the full-fledged *mind* (including consciousness) but only at its *functions*, we have to clarify this: We do believe that there is a strong point in Solms' statement that it is not a matter of indifference whether there is consciousness or not. Studying the brain for answering problems without taking into account subjective experiences might lead us astray – according to him, it is not an accident that one can feel consciousness. The correct approach seems therefore to implement functions with the help of studying and understanding the mind as good as possible – and here we are in deep need of interdisciplinary approaches. However, this is also possible *with-*

out the target of fully *re-constructing* the mind. Therefore, the importance of consciousness can and should be assumed, which is very likely to be seen, e.g., if we look at mental causation: One point (in Solms' impressive example of a man feeling dreadful and then actually shooting himself) is that consciousness gives us the possibility of *knowing* – and even more important – *feeling* that there is a mismatch between inner and outer world and language gives us an additional opportunity to talk about that. If we implement such a primitive kind of inside-outside-balance (which is *one* feature of consciousness) we are exactly on the way of achieving reasonable prediction – a capability we define as being very important (see below).

Considerations on Anthropomorphism

It has already been indicated: A lot of the questions that are concerned with the interdisciplinary attempt of engineers and (neuro-) psychoanalysts tackle anthropomorphic considerations, not superficially in the sense of appearance but rather of "architecture". We should be cautious to rebuild humans as – what R. Dillmann stressed correctly in [3, #3, Session 1a] – robots should not replace humans. A side remark: This would imply ethical considerations that cannot be dismissed or answered easily.

Additionally, for considerable time, emotions have been judged to be a very important feature of the human mind. At ENF, too, their influence of the situated functioning of humans has been emphasised. Therefore, we could be seduced to implement a limbic system. However, this would clearly constitute a confusion of function and method: If – what is often considered as such – emotions have the *function* to mediate between outside and inside world [13, p. 106], it is not said that this cannot be achieved without emotions. Therefore, we may learn about the *purpose* of emotions, but implement something different to achieve the same functionality without necessarily having to rebuild emotions as such.

Speaking of anthropomorphism, we can as well ask the meta-question of why to use humans and not something else, e.g., another kind of mammal, as model for a functional cognitive system. The clear answer is that we know that humans do work, and they (usually) perform very "intelligently" – this means, their variety of accomplishable tasks is enormous. Additionally, through language we have access (although indirect) to the inner world. Recalling Solms' first principle (the evolutionary biological principle), we see that the main *reason* for the human having all his capabilities is his viability (for acting in the world). This said, the question we could ask concerns the viability of an artificial cognitive system – the question where to locate its ecological niche. We argue that this niche is right at the user's side, with other words, the system has to survive our (humans') "evolutionary judgement" whether we want it to "survive", i.e., not to turn it off. From this point of view, the question is not to *model* the mind, but rather to *emulate* the mind (as is the actual subtitle of ENF), which is in our understanding a more realistic view for dealing with a part (function) of the mind. To put it in a nutshell, we would deny Solms' "hard" assumption that we want to ultimately rebuild the human mind. We would, however, not foreclose the possibility that the desired function is not fully achievable without considering the whole mind, i.e., to really model it fully.

Engineer's Constraints

Until now we have motivated that engineers are first and foremost interested in the main goal of robotics, which is building a system that is able to fulfil a function. This is why we would like to focus on those functionalities that seem to be necessary for understanding, interpreting and acting in a scene. In the next section we will list some key functions.

One of the constraints that are imposed by the engineer's approach is synthesising as a whole. Solms mentioned at the beginning of his talk that neuropsychoanalysis is trying to combine the subjective, the objective and the functional approach to the phenomenon called "the human mind". Engineers, however, only have the functional approach at hand, as the others are purely analytical. The subjective (psychoanalytical) view is not applicable at the stage of building the system, the objective (neurological) view is obviously also not feasible. Here lies the main advantage of an interdisciplinary approach with (neuro)psychoanalysts for engineers: They can use the other approaches to better understand the mind and – hopefully – to break this down to the functional level where synthesis can begin. This could – on a very high level – help for a paradigm shift: Not to start from a specific paradigm from artificial intelligence (such as those outlined by E. Barnard at ENF [3, #1, Ch. 2]), but from a purely functional view – i.e., to put the method aside and the high-level goal to the fore.

A very basal – yet for constructing aims very important – point is that we must be very careful in expecting what will happen. It is definitely not likely that the more abstract levels we get and the more complex stuff gets meaning will just arise. To think of a difficult problem like how the mind works in terms of attributing the solution to complexity is seldom correct and hardly useful when synthesising. So we should – even if it was possible – not use the mind as a 1:1 blueprint for designing artificial cognitive systems because we hope for emergence out of greatest possible complexity. Emergence is a notion much discussed and controversial – we do not want to go into detail w.r.t. this debate. To shorten our point of view: Let's stick to the functions we need – and this is what the next section is about.

2.3.3   Main Aspects of a Cognitive System – an Engineer's Perspective

First, let us recall the three basic properties of the mind that Solms mentions in his paper [4]: consciousness, intentionality and free will. In the following, we will try to link these in an analysis from an engineer's (functional) view with the hope to weaken the hard statement that those basic features "[…] do not bode well for Dietrich et al's project" [Part II Section 1.3, point 33].

As outlined in [11] there seem to be the following capabilities that "cognitive systems" need to be able of (this is very likely still not an exhaustive list which is additionally biased by the view on *perceptual* capabilities, as we are mainly concerned with vision for robots to interact with a human and the environment): intentionality, symbol anchoring, abstraction, generalization and prediction. Furthermore, it is needless to say that all these capabilities are, of course, not fully dissociable from each other but rather highly interconnected. In his talk, G. Zucker mentioned that the aim of engineers is to build self-sufficient systems and autonomous robots. There, prediction, embodiment, conflicts and emotions are needed [3, #2, Ch. 5]. With prediction, we totally agree. Anticipation is definitely a function deeply needed as it reflects scene *understanding* and enables adapted actions. Embodiment has become very popular in recent years and it definitely seems that intelligence can hardly be explained when the body (including the specification of its "perceptual channels") is totally ignored. However, when focusing on perception and not on action, we are starting from capabilities that are not necessarily in need of a body. This is a simplification, of course, as recent works have proven the influence of the body on perceptual tasks, cf. [8, 9]. With conflicts, Zucker meant those tensions between the inner and the outer world that enables the need for mediation and therefore the drives that make us do something. Again, it is agreed that this is needed for the whole system, yet for scene and action understanding this does not come to the fore. Concerning the emotions, we already outlined that they are one possible *means* for helping realise those tensions but are not necessarily the only solution. Zucker tackled the problem of a logical description of the world, when he says that it is really hard to nail down all information you have to strict rules (e.g., you cannot foresee every possible exception from the rule). This is now exactly, what we think can be solved or at least eased with the capabilities of generalisation and abstraction as a lot more situations can be covered. Those two abilities are very likely to allow for what the philosopher Sir Karl Popper emphasised in his works: *common sense*.

From a philosophical perspective, we see the tradition of radical constructivism as perfectly suited for orienting oneself when building artificial systems. The reason is simple: Radical constructivism deals with the subjective building of one's knowledge and is totally purged from the metaphysical search for an objective reality. Therefore, we do not first have to find the *grand unified theory* that describes the world but can start with *personal knowledge* that aims at one's own survival. The main points of radical constructivism are (according to [15]): knowledge is actively built up and not passively received, cognition serves hereby the subject's organization of the experiential world and the system's tendency on the whole is guided towards viability. Especially the viability-question is highly related to the discussion of evolution, and has already been tackled in Section 2.3.2. The first two points are covered by the interaction of perception, generalisation and abstraction.

Excursus Into Popular Notions: "Intelligence", "Consciousness", "Cognition"

The notions intelligence, consciousness and cognition have become very fashionable, not only in cognitive science, but also in computer science. Here, we will look how these terms can be understood from a pragmatic and – for engineers – useful perspective. Therefore, we will not review the meanings of these notions in depth as it has been done in the humanities, but rather keep it simple to a manageable synthesising view.

As will be outlined in more detail in the following, the bottom line is that all of those notions must be seen in a gradual sense, just as J. Hawkins states in [5, p. 180] for intelligence: "All mammals, from rats to cats to humans, have a neocortex. They are all intelligent, but to differing degrees."

(A) Consciousness

Consciousness is one of the basic properties of the mind defined by Solms. For him, the mind disappears as such if there is no consciousness. Thereby, consciousness is inherently evaluative and has the purpose of being aware of feeling pleasure or unpleasure [Part II Section 1.3, point 42].

Zucker's statement that we cannot just program the "module consciousness" is right – as we are lacking the exact knowledge how it works and why humans have it in the first place. However, his further implication is misplaced, namely that instead of implementing it we can hope that it will emerge some day. We can not simply expect that there will be emergence, and even if, that the emerged phenomenon really fulfils what we contribute to consciousness. Complexity is not a sufficient explanation for being conscious (although there are in fact philosophical positions that think along this line, cf. [12]). We believe that a more promising approach would be to start from what Solms considers the right level for talking to engineers: What *function* does consciousness have and how can we create something that works this way?

So much for the general considerations: Their bottom line is that the task at hand is to provide the functionality of consciousness. Maybe we should right now really reduce consciousness to the mediation of inner and outer world in the sense of making the system react to tensions between these two. Y. Yovell [4] cites [13] where "[…] consciousness may be defined by its *content* (what we are conscious of) as well as its *level* (how conscious we are)." The first view is covered by the intentionality principle, see below. For now, we would prefer to stick to the latter view, meaning that in the case for artificial systems, we might not define consciousness as the phenomenon that can be found in humans, but – by using the gradual approach – omitting subjectivity and qualia (i.e., the subjective experiential quality of a mental state): Cognitive systems can be more or less conscious – total consciousness finally comprising self-awareness and meta-thinking. The latter is not needed in a functional artificial system – excluding, so to speak, qualia and allowing Zombies, to put it in Chalmers' notion [2]. Such a system would not have subjectivity as we humans know it ("we" implies that none of the readers is a Zombie). Maybe qualia arises anyhow as a *deus ex machina*, rehabilitating anthropomorphism – or maybe not. We claim that the basal function of consciousness, the mediation between inside and outside and hence the functionality of motivating action can be emulated by the intentionality principle (see below). Furthermore, Freud said that consciousness is only a small property of the psyche and does not equal the mind. So let us look at the huge rest!

(B) Intelligence

Intelligence is what engineers really want to achieve in "cognitive technical systems". Consciousness might be a way to achieve it – but the true goal is to build machines that *behave* intelligently. In this phrase, the behaviouristic position is already obvious. Zombies (i.e., non-conscious creatures) may behave exactly the same way as (conscious) humans do. We can likewise look at two kinds of robots: conscious (self-aware) ones and non-conscious ones. If both achieve the same performance (judged by the external third-person perspective), why need consciousness in the first place?

As mentioned in session 2a at ENF ("Getting a grasp" [3]), intelligence can be judged to a large extent when interacting with other agents. Here again, perceptual and understanding capabilities are in the foreground. Still, intelligence as we might say, can be seen firstly gradual and secondly only from the third person perspective. Even when we humans might say to ourselves: "Here, this person seems to be more intelligent than I am" or "In this situation, I behaved unintelligently" – we are always judging our behaviour, our performance, and furthermore we do that in relation to others. This is different to consciousness in this respect that we are conscious of ourselves even without this kind of comparison, and even more: only w.r.t. ourselves. Consciousness and intelligence are therefore the two sides of the same phenomenon, the former from the first person perspective, and the latter from the third person perspective – both are gradual and gradually judged. The point now is that engineers – emphasising on the functional fit (as radical constructivism calls a concept very similar to "viability") of the artificial system to our, human's life-world – are constructing those machines only for the behavioural perspective (at least up to now) as they want the system to behave intelligently.

(C) Cognition

Cognition is one of the most popular terms in recent engineering research. We are arguing for a very careful use of this word as it is overloaded with a lot of different understandings. First of all, cognition, as we understand it, is not necessarily a conscious process (subliminal learning is very likely to be a "cognitive process"). So cognition and consciousness need to be separated. Furthermore, intelligence as stated is only applicable to the third person perspective. We would define cognition now as the superordinate concept that allows for the functional fit. Of course, cognition, consciousness and intelligence are related. If a system (biological or artificial) behaves functionally fitted to its life-world, a third person would call this behaviour intelligent (and in the biological case, the activities can additionally be judged consciously).

The mind plays the crucial role of evaluating and mediating between inner and outer world and generates functional fit through this activity. This might be consciously, not consciously or "less consciously" done (in the sense of awareness) – consciousness has been defined to be gradual. Full self-awareness need not be the key to cognitive behaviour (judged as intelligent by an external viewer). E.g., if the task is to arrive at a specific point beyond a river, the "correct" (viable!) solution is probably not to search for the nearest point (in linear distance) on one's shore but rather to accept a movement away from the goal in order to use a bridge. Using prediction, abstraction and evaluation (and *common sense* in the broadest sense), a hypothetical artificial system should be able to solve this problem without the need for self-awareness. For this task, so to speak, a gradually minor consciousness is way enough; the cognitive activity delivering the needed steps to judge the behaviour as intelligent is nevertheless adequate.

Finally some words concerning the dual-aspect monism: Up to now we argued in line with the notion of the dual-aspect monism which states that consciousness and intelligence are only two different views on the same thing [13, p. 70]. However, even if engineers can use those two notions in a similar understanding (yet maybe to a gradually minor extent), the further implication mentioned during ENF, namely that we can assume that "inside feelings" and "outside behaviours" coincide, makes no sense for engineers: Whereby in humans outside reactions, facial expressions and the like might lead us to further understandings of the inner feelings, qualia, etc. it usually does not help us to understand what goes on in machines if they show a specific behaviour. To put it the other way round: There are so-called embodied conversational agents (e.g., "Max" by Bielefeld University [7]), who are able to express their "feeling" virtually. The underlying principle, however, is the interpolation between hard-coded values of some basic "feelings" and it is obvious that the system does not have any real subjective experience – this means we cannot learn anything beyond what we have programmed the system to have. There is no research outcome in this direction (although there is some in other directions, e.g., for studying the reaction of *humans* to such an agent). With other words: Given the fact that there *is* consciousness, we can be dual-aspect monists. The other way round, however, does not hold: Having one view does not imply the existence of the other – that is the engineer's problem. Engineers *do* have to ask the *how*-question, i.e., Chalmers' hard problem of consciousness [2]. Solms says he is surprised that there is a hard problem of consciousness as this only boils down to a matter of perspective [3, #3, Session 1a] – This is totally understandable from a psychoanalytical view, but it is not from an engineer's (or a philosopher's).

Analytically, it is possible to relate the different views to each other. Philosophers go even further: They question to what extent it is possible that we *feel* something. So the *why*-question is not a question of relating phenomena to each other but rather to question the relation itself. This problem gets very important in systems engineering as it is not clear to what extent it is necessary to have both of the views and if so, then why?

Intentionality

Intentionality is also one of the fundamental properties defined by Solms [4]: "Mental experience is always *about* something, always directed *toward* something. (Freud speaks here of 'wish'.)" Relating consciousness with intentionality also conforms to some philosophies that judge intentionality as the main principle or functionality, e.g., phenomenology or the so-called *theory of objects*. This has been shown to be a useful approach in [11]. Founded by the Austrian philosopher Alexius Meinong, the theory of objects defines intentionality as the central idea, which shall be outlined here as it is especially concerned with perception and establishes the basis for the further notions in this paper: Each experience (which includes each perceptual act) is intentionally directed towards an *object*. Those objects are organised into real and ideal objects, the latter not constituting actual objects but rather anything else (impossible objects, abstract concepts of objects). Furthermore, it is obvious that this intentionality means a directedness of our consciousness towards both, outer (external) and inner (internal) objects.

If, for the time being, we actually reduce consciousness to the notion of intentionality just outlined, we get a manageable and yet important functionality w.r.t. perception and also to attention. As depicted above, this reduction would imply a "minor grade" of consciousness (in comparison with humans). Additionally, note that we totally put aside the qualia-question – we do not have to know whether the system is conscious or not, or whether it "feels" in a specific way when perceiving some object – the main goal is that it does the things it is supposed to.

Symbol Anchoring

Symbol *anchoring* refers to the ability to link meanings of internal structures (symbols) to external objects. The related symbol *grounding* problem has been defined by [4] as the ability to make semantic interpretation of a formal symbol *intrinsic* to the system in contrast to *parasitic* on the meanings in our heads. Those two capabilities are immensely important for a cognitive system to allow for meaningful interaction with the objects in the world. The question, how those meanings arise or can be detected has hence been posed from the early beginnings of artificial intelligence on.

As the viability of an artificial system has been defined as its surviving capabilities in its ecological niche at the human's side, the semantic object-meanings of the system should be more or less similar to the human's phenomenologically spoken, the robot's life-world is congruent with the human's. Generally, such a system should be able to fulfil tasks for humans, i.e., it has to be able to detect similar functions of objects as humans do. Different affordance-based approaches (e.g., recently by [1]) reflect this goal. Approaches of function-based perception try to directly derive functions of objects instead of detecting objects themselves.

Solms' first principle, the evolutionary biological principle seems to be a cornerstone in linking "correct" meanings to objects – practically all animated and unanimated objects we are interacting with have been learned by us. This implies that we have to have a learning principle in our system that learns from its *Umwelt*, to use von Uexkuell's notion [14].

Abstraction

Abstraction and generalization are capabilities that seem to be very important for systems that are confronted with similar, but not identical objects. For us humans it is very simple to immediately "see" an object (including its meaning), even if we have not seen exactly this instance before. E.g., entering a room and using the metal table for putting our coffee mug down, does not even afford conscious thinking – even if we have only seen wooden or glass tables before. This is probably because we store some kind of abstracted representation of learned objects that helps us classify new things.

In order to have a functional robot, we do have to implement this kind of abstract knowledge. A robot that is only trained for a specific living room is unsuited to be sold for every household. Psychologically, this capability is associated with the notion of *human concept learning*. The arising concepts could be loosely connected with the philosophical notion of the thing-in-itself, meaning the constituting substrate of an object. To allow for intentionality towards an inner, abstracted object – which seems to be very important for generalised intelligence – we must provide the system to work with abstract knowledge, breaking it down and connecting it to raw sensory input. To this end, some kind of qualitative description, separated from quantitative sensory input, need to be established (for a first approach, see [10]).

It has already been mentioned that the philosophical tradition of constructivism seems to be a useful approach to the design of cognitive systems. Here, each perception is in fact a re-presentation of the perception itself, meaning that the actual input is processed in a specialized manner, with the system's ultimate goal of functional fit (viability). Again, consciousness, even abstract ontological meaning or understanding, is obsolete; the judgement of the system only depends on its ability to interact with its *Umwelt* in a meaningful way. However, as long as we do not know better, a manual provision of an abstraction learning system seems to be a good means.

Generalisation

In connection with the notion of abstraction there is another capability that seems to be important from an engineer's point of view: generalisation. By that, it is meant that the abstracted knowledge has somehow to be built up during the process of learning, by detecting and abstracting actual appearances to those inner, abstracted objects mentioned in the previous paragraph. For Solms, learning is not one of the basic properties of the mind, but rather in service of them [3, #1, Discussion 1]. However, from an engineer's perspective, it seems that this kind of learning does in fact play a crucial role – at least as long as we are not able to build a memory system like the neocortex. As J. Panksepp mentioned [3, #3, Session 1b], the great power of the neocortex lies in the fact that it starts as a *tabula rasa*, and is then filled by time complying to different constraints. Until we are not able to construct such a generalised "random access memory field", we have to provide some *functionalities* in a different manner, of which generalisation is one.

Tackling the learning of abstracted objects, the question of classification is touched, in context of which L. Goldfarb [3, #3, Session 1c] points out that we are lacking the final class formalism. Maybe – at least that is the hope – a useful class formation occurs automatically if we get to the point where abstraction is powerful enough to detect the real salient features of objects. Because then, new objects might be automatically grouped and class formation will get broader by the time. Such a generalisation technique would be a great step towards general intelligence particularly if it helps us find the right class formalism.

Prediction

The last cardinal function of an artificial cognitive system that we want to address here is prediction. For some, this is the single most important core aspect of intelligence [5]. This seems plausible insofar the connection to *agency* can be established with this capability. Solms explains in [Part II Section 1.3, point 48] that in humans, agency provides the means to inhibit instinctual reactions and therefore to show *free will*. If the artificial system is able to predict the outcome of a specific action, it can likewise inhibit or modify its actions accordingly. This emulates the social brain that works on top of instinctual (default) actions and serves more anticipatory behaviours. Again it does not bother us whether consciousness in the sense of self-awareness is present or not.

P. Palensky [3, #2, Ch. 2] has outlined a very nice idea in his talk: An artificial system should implement something that is comparable to the human's "gut-feeling". This is actually the next step of prediction. Whereas we only mentioned the outcome of an action, here, including more information and – also a currently very important notion – *context*, the system would be better able to assess a given situation. Again, also this is only imaginable when having the system situated in its *Umwelt*.

2.3.4   An Example – Reviewing the Principles

We would like to use a typical example from robotic vision to review the five broad functional principles outlined by Solms in his talk at ENF [3, #1, Ch. 4]:
(1) the evolutionary biological principle
(2) the principle of drives
(3) the principle of feelings/emotions

(4) the principle of motivations

(5) the principle of cognition

Those principles build on top of each other and not all of them have been tackled in this paper yet. As will be shown now, in order to implement those principles, we need the capabilities – especially for the underlying perceptual abilities – that have been mentioned in this paper as being important for artificial cognitive systems. As an example, we want to use a scenario that seems to be of high relevance for a robot: The building-up of knowledge about its environment by using perceptions and actions, i.e., by experimenting with objects[65].

This example shall show that there are some capabilities of robots needed that are not – or not adequately – solved yet. It is an example related to learning, which Solms does not count as a fundamental principle in itself but rather as being in service of the five principles mentioned. However, it seems that this service is deeply required (the totality of common sense knowledge cannot be defined hard-coded); therefore it is a nice example for the long way to go.

For every principle, first the main arguments of Solms from his talk are given, and then their transfer to a technical system is attempted[66].

Evolutionary Biological Principle

According to Solms, the body is governed by the evolutionary biological principle. It is designed to best survive and reproduce, i.e., to maximise the chances of survival in its ecological niche. This comprises also the necessity of interaction with others in order to satisfy the inner needs. So far, there is no "mind" involved.

For an artificial system, the evolutionarily developed layout of the body is "circumvented" by its inventors. We, as designers, judge the necessities for and build up the body of the system suited for its ecological niche. It is, of course, a question in itself whether this is the right starting point at all. As we are also the ones that judge the viability of the system this might not be totally wrong and besides that, we are lacking to a wide extent the possibility of technical prerequisites to build self-assembling robots (for an interesting approach, see [16]). In conjunction with perception, a very important point is that these pre-specifications also comprise (at least to a certain extent) the specific filtering of the sensor data. Of course, not everything that is perceived has relevance and therefore filtering is needed, but – we will see that as we go along – given a constructivist approach, the system should be constrained as little as possible. Constraints should only be applied by the robot's environment (as is the case in the human's). This is especially important for the example of a learning robot. Which perceptual capabilities are needed in order to build up knowledge about the objects in its environment? The possibilities w.r.t. vision range from simple colour blob detection to the extraction of

---

[65] This example is inspired by the European project *XPERO – Learning by Experimentation* [16].

[66] In the following, all references to Solms are taken from [3, #1, Ch. 4] and are not indicated each time.

shapes and pre-defined objects – and from two (monocular vision) to three dimensions (stereo vision, laser).

Drives

For Solms, the mind steps in when the principle of drives comes to work. The role of the mind is the mediation between inner (basic survival and vegetative) needs and outer circumstances. It is located between the visceral body and the muscular, skeletal and perceptual body. Basic drives compel the organism to move out into the world and seek need-satisfaction. Most fundamental is the libidinal drive; the other drives may be classified in different ways.

For our hypothetical system, this would mean that we need to implement both needs and drives into an artificial system, which is actually done by the evolutionary robotics paradigm. For a system like in our example, i.e., that has the goal of finding out more about its environment, this would mean to implement not only basic needs (e.g., the seeking for a power outlet when the battery is running out), but also the need for experimenting with external objects – sometimes termed curiosity drive (a similar drive seems to be in place for a child's playing behaviour).

The latter also reflects the task of the robot, i.e., acquiring knowledge in our example. For any other robotic system, too, the task (predefined or not) is of high relevance for the judgement of its situatedness and further on for its viability.

Feelings/Emotions

As Solms explained, the interoceptive registrations of the oscillation of drive tensions, i.e., of whether drives are satisfied or frustrated, are "felt as *feelings*". They show us how we are doing, i.e., if the drives mentioned above are currently frustrated, we are feeling bad. This is the *function* of feelings. An interesting (social) aspect is tackled when he says that "All values derive ultimately from this."

Furthermore, the notion of consciousness comes into play: first and foremost as interoceptive perception of the state of drive tensions, secondarily also of external perception, in the sense of attributing a specific feeling to an external object. All this is in service of the pleasure/unpleasure principle. It has been found that the basic emotions are universal in all mammalian brains. All of the value systems (from pleasure/unpleasure via basic emotions up to the elaboration in language) have an expressive component that one can see from the outside.

From a technical point of view, those "feelings" would control to which degree the task has been fulfilled. Additionally, symbol anchoring – at least in a primitive sense – steps in at this point. In order to attach a specific feeling to an external object, the need for detecting and labelling an object is important – in the sense of labelling it with a subjective value. Note that this does not necessarily afford the knowledge about what exactly this object is, but at least a rough segmentation into separate objects is required. Perception needs to be capable of this. As (adult) humans we are used to "work" with perfectly segmented data. We immediately know where one object begins and another one ends, which object occludes another one, and much more. This is, however, something you cannot expect from artificial systems (yet).

At this point, the achievement of the (sub-) task(s) that is given via the drives is *evaluated*. Additionally, the external objects get a *specific* significance (*meaning* in the broadest sense) for the system. For our example-experimenter, this means that novel objects or events that are located get a higher ranking because they present a positive feeling about the task pursued: the learning about the environment.

This does not sound constructivist by chance, but to the contrary we are strongly convinced that this is what makes us humans grounded so much into our environment. We have our own subjective valuation of the objects around us (also in the broadest sense). Probably the same could hold for real cognitive robots: They need to build up their own "representation" of the world – yet again implying all the challenging discussions about what to predetermine innately, which was already mentioned briefly.

Motivations

For Solms, the second fundamental property of the mind, namely intentionality, arises with the principle of motivations in addition to consciousness. Motivation allows us to wish (in the Freudian sense) for the repetition of previously experienced satisfactions of needs and conversely for the avoidance of previously experienced dissatisfactions. This implies memory that registers conjunctions between certain feeling states and certain external perceptual states. This also provides further elaboration of the pleasure/unpleasure principle. In that way, *memory extends perception* and allows causal chains.

The intentionality as directedness towards an object is a very good means for pruning the search space. Given our example task, the detection of a novel object and its properties, should motivate the system to further investigate it.

Technically, the storing of the links between certain behaviours and inputs is no big deal. The real problem arises in the storage of the perceptions. Here, the next two functionalities given in this paper get important: abstraction and generalization. We can now extend our robot example to an actual experiment: Finding out that a red cube has some funny property (e.g., making a loud noise when dropped) and therefore raising a pleasure feeling, allows the human child not only to try it again and again with *this* object, but also with *similar* ones, e.g., a red ball. For a robot, this task would be enormously difficult and would bring up a lot of additional problems that are nontrivial: What is a single object? What if the object appears from a different viewpoint, under different lighting conditions, etc.? What means similar object? Before we can think about storing simple causal chains of action and perception, questions concerning the processing of the perceptual input stand in the foreground. But not only common problems of computer vision arise. The need for abstracting and generalising input is a fundamental capability of humans, but is not (yet) of robots. Therefore, what constitutes a specific kind of object (e.g., what makes a table a table) and how to detect these constituting properties are nontrivial questions. The importance can be seen when in need for extrapolating sensory input.

Cognition

The last and highest state of Solms' hierarchy of principles is *cognition* which attaches to *thinking*. By the time, our experience grows and gives us more and more information about the mapping between external perceptions and internal states. This huge amount of data increases the complexity of decisions, giving rise to the occurrence of delays. By experience, it is noticed that short-term pleasure might lead to long-term unpleasure and vice versa, a second reason for such delays of reactions are in place. Finally, also conflicting possibilities of actions or the occurrence of novelties cause delays. All in all, instead of instinctual reactions based on the basic satisfaction system, those reactions might be inhibited and the choice to other, maybe totally different reactions could occur. This is what is termed the *reality principle* and what makes up the *Ego*. This also implies ownership of actions, which eventually constitutes Solms' final fundamental property: agency or free will. To put it in his words: "Agency and free will – perhaps paradoxically – arise out of the capacity *not* to act" [3, #1, Ch. 4].

For an artificial system, this principle is the hardest one, of course. Not only because of the well-known computational complexity problem of deciding, but already when implementing high-level prediction. As mentioned earlier, prediction seems to be very important for cognitive behaviour which is *considered* action; free will therefore boiling down to action decisions after weighing the options. For it is not said that this can only be done consciously (aware), Zombies and therefore artificial systems could be cognitive in this sense.

A. Sloman [3, #1, Ch. 6] pointed out in his talk that for visual perception in humans, there is no building of replicas (as this would not function for impossible objects). Therefore, what cognition does here, is feeding top-down information to visual input. This is very interesting w.r.t. artificial systems as this has not been adhered to yet enough. If sufficient context information and predicted possibilities are provided, a huge amount of the weak hypotheses given by computer vision could be strengthened.

For the experimenting robot, we would need task-dependent and situated prediction of what will happen next. This implies, that the given perceptual input is not only robust enough but additionally tuned to the representations needed for prediction (leading us back to the discussion of abstraction and generalisation). The difficult questions that show up here not only concern the type and quality of knowledge representation but also how to guide visual perception top-down by simultaneously avoiding a "circulus vitiosus": experimenting in order to find something "new" that has already been predefined by the given constraints – an aspect that we already encountered further above when addressing the filtering of perceptual data (first, evolutionary principle).

2.3.5   Open Points

With this paper we did not want to give the impression that if we solve all the implementation problems in order to get the principles work, we will immediately come up with a system as cognitive (or intelligent) as humans. It should rather have shown that such an approach could lead to the first step in the right direction.

An example of what has not been tackled yet is the role of inner speech. It seems – at least for reasoning – that this functionality plays an important role in humans, cf. [13, Ch. 8]. Although for us the focus lies on perception, it is very probable that this capability highly influences the evaluation of the significance of perceived objects. Of course, with tackling inner speech we get quite close to the question of self-awareness again.

Additionally, the depicted "explanation" of a possible implementation of the functionalities is very biased by an engineer's view and therefore highly prone to be a too strong simplification. For example, if we really define feelings as simple "reporting" of how we are doing (in the sense of the mediation between inside and outside), it is doubtable that the same functionality can be achieved as with qualia in humans (which seems to be highly relevant for the "quality of a feeling", cf., e.g., [6]).

A very important discussion during ENF dealt with ethical issues, although not explicitly termed this way. Questions on what "rights" a cognitive robot would have or whether we should implement only "positive" feelings in order not to get an angry robot – which by the way would be senseless if we really want to achieve situated instinctive reactions –, should not be neglected.

2.3.6   Conclusion

So far, we have tried to review the three fundamental properties and five functional principles that Solms presented at ENF in order to find connecting points to engineering sciences. As can be seen, there is still a long way to go until we have the tools ready to implement all of these mind-related functionalities.

Just to name the huge difference between human visual perception and (current) computer vision: Reasoning in the sense of using common sense knowledge is used by humans all the time, whereas is very problematic in computers – how to store all implicit knowledge we humans just "know"? A simple example: Humans would probably not see a white cow flying in the sky but rather know that it is a cow-shaped cloud. Computers would not or, at best, not yet.

Very valuable of an interdisciplinary research such as this seems to be that it helps us engineers not to get lost in details and forget about the conceptual functionalities on top. Hopefully, this approach – to tackle basic implementation issues considering those high-level concepts – could have been adumbrated with this paper.

Although neuropsychoanalysis and technical sciences are totally different, especially w.r.t. their viewpoints – the objective, subjective and functional-analytical approach vs. (only) the functional-synthetic approach –, it seems that a cooperation could be a fruitful inspiration. Or impression, however, is that the inspiration flow is a bit stronger from the psychoanalytic side to the engineer's than the other way round. Maybe some synthetically implemented functionalities will on their part then give a possibility for a deeper understanding of their meaning to the human mind. With other words: Implementing a Zombie could perhaps be very fruitful!

References

[1]   Cakmak M et al (2007) Affordances as a Framework for Robot Control. In: Berthouze L et al (eds) Proc 7th Intl Conf on Epigenetic Robotics, Lund University Cognitive Studies 135
[2]   Chalmers DJ (1996) The Conscious Mind – In Search of a Fundamental Theory. Oxford University Press, USA
[3]   Dietrich D et al (eds) (2007) ENF 2007 – DVDs: Complete Video footage of the conference "ENF – Emulating the mind" (# gives the DVD-number, Ch. the Chapter). Institute of Computer Technology, Vienna University of Technology
[4]   Harnad S (1990) The Symbol Grounding Problem. Physica D 42:335–346
[5]   Hawkins J, Blakeslee S (2004) On Intelligence. Times Books, New York
[6]   Nagel T (1974) What is it like to be a bat? The Philosophical Review LXXXIII, 4:435–450
[7]   Kopp S (2008) Max and the Articulated Communicator Engine. Online at: http://www.techfak.uni-bielefeld.de/. Accessed 30 May 2008
[8]   Proffitt DR et al (1995) Perceiving geographical slant. Psychonomic Bulletin & Review 2(4):409–428
[9]   Proffitt DR et al (2003) The role of effort in perceiving distance. Psychological Science 14(2):106–112
[10] Schlemmer M, Vincze M (2007) On an ontology of spatial relations to discover object concepts in images. In: Zupancic B et al (eds) Proc 6th EUROSIM Congress on Modelling and Simulation, Ljubljana
[11] Schlemmer M, Vincze M, Favre-Bulle B (2007) Modelling the thing-in-itself – a philosophically motivated approach to cognitive robotics, In: Berthouze L et al (eds) Proc 7th Intl Conf on Epigenetic Robotics, Lund University Cognitive Studies 135
[12] Schmidt SJ (1987) Der Radikale Konstruktivismus: Ein neues Paradigma im interdisziplinären Diskurs (Radical constructivism: a new paradigm in interdisciplinary discourse). In: Schmidt SJ (ed) Der Diskurs des Radikalen Konstruktivismus, Suhrkamp, Frankfurt/Main
[13] Solms M, Turnbull O (2004) Das Gehirn und die innere Welt (The brain and the inner world). Patmos, Düsseldorf
[14] von Uexkuell JB (1909) Umwelt und Innenwelt der Tiere (Environment and inner world of animals). J Springer, Berlin
[15] Ziemke T (2001) The Construction of "Reality" in the Robot. Foundations of science 6(1-3):163–233
[16] Zykov V, Mytilinaios E, Adams B, Lipson H (2005) Self-reproducing machines. Nature 435:163-164

2.4 Four Laws of Machine Psychodynamics

Andrzej Buller

Machine psychodynamics represent a paradigm of building brains for robots inspired by Freudian view of mind. A psychodynamic robot's self-development is reinforced by pleasure understood as a measurable quantity that rises when a bodily or psychic tension plummets. It is also proposed that some ambivalence may accelerate a robot's cognitive growth. Mechanisms for pleasure generation and ambivalence jointly make a robot an adventurous creature. The essence of machine psychodynamics is a set of four related laws.

2.4.1 Introduction

The race for human-level machine intelligence is going on. The list of issues concerning a target humanoid robot, proposed by Rodney Brooks, contains: *bodily form, motivation, coherence, self adaptation, development, historical contingencies*, and *inspiration from the brain* [1]. In this paper I concentrate on motivation and coherence. I discuss the two issues from the psychodynamic perspective which emphasizes the perpetual conflict between competing feelings, judgments, or goals, and pays special attention to the intrinsic dynamics of mental events, to which pleasure serves as a universal reinforcer and motivator (cf. [2, p. 15]). I argue that mechanisms built based on psychodynamic ideas may substantially boost a robot's cognitive self-development. The considerations outline a new discipline I call *machine psychodynamics*.

The human being strives after pleasure and seeks to avoid unpleasantness, and so shall a target robot. Let us note that the notion of pleasure (sometimes called joy or happiness) is used in a description of several artificial agents, as, for example, *Cathexis* [3], *Kismet* [4], *ModSAT* [5], *Aryo* [6], *Max* [7], or *Bubbles* [8]. In their case pleasure is a component of a defined state or a function of such a state to be achieved when certain arbitrarily designed conditions are satisfied. Although pleasure defined in this way can reinforce desired behaviors, it is actually not the agent's desire, but still their designer's desire. Machine psychodynamics proposes that desires and pleasures are to emerge in robots' brains. Pleasure is measurable quantity and its volume is closely related to the dynamics of certain bodily or psychic tensions. A bodily tension relates to the degree to which a part of a robot's body deviates from its state of resting. A psychic tension relates to the degree to which a drive (such as boredom, anxiety, fear, or expected pain) deviates from its homeostatic equilibrium. And, what is most essential is that not the state of a low tension, but a *move toward* such a state induces pleasure. Disregarding the question of how well the above idea matches the still unknown truth about the nature of the human psyche, such a view of pleasure is supposed to work well in an artificial mind and to dramatically enhance its ability to self-develop. The supposition is backed by some theoretical considerations and experimental results [9].

The issue of coherence concerns mechanisms owing to which an agent is to cope with contradictory feelings, judgments, or goals. In mainstream AI/robotics, in the case of the simultaneous appearance of conflicting ideas a robot is required to work out, possibly quickly, a rational decision about which of the ideas to implement. Accordingly, their designers employ algorithms for arbitration, action selection, or calculating superpositions of related forces [10, pp. 111-119]. Yet such a machine "self-confidence" looks not too life-like. Note that someone may first love and then hate the same person even when the person has not given him cause for doing so [11, p. 41]. There is empirical evidence that human subjects may abruptly switch from a highly positive evaluation to a highly negative one, or reversely, even if no new data about the object of interest could cause such a switch [12, pp. 97-98]. Hence, machine psychodynamics proposes a mechanism for conflict resolution that allows contradictory ideas to fight against one another in the literal meaning; accordingly, a psychodynamic robot may sometimes hesitate about what to do or abruptly change its mind [13]. I argue that owing to this mechanism and to the specific way of pleasure generation, a psychodynamic robot becomes an adventurous creature.

2.4.2   Pleasure Principle Rewritten

If our goal is to build a robot whose ultimate human-level intelligence is to self-develop in a pleasure-driven manner, we had better keep away from the endless debate on the deep nature of pleasure that, for centuries, has been pending on the border of philosophy and psychology. What we need is only an operational definition of this notion. Some renowned thinkers provide tips for such a definition.

Aristotle in 350 BC wrote:

"We may lay it down that Pleasure is a movement, a movement by which the soul as a whole is consciously brought into its normal state of being; [...] If this is what pleasure is, it is clear that the pleasant is what tends to produce this condition [...] It must therefore be pleasant as a rule to move towards a natural state of being, particularly when a natural process has achieved the complete recovery of that natural state" [14].

Saint Augustine of Hippo in AD 397 wrote:

"The very pleasures [...] men obtain by difficulties. There is no pleasure in caring and drinking unless the pains of hunger and thirst have preceded. Drunkards even eat certain salt meats in order to create a painful thirst – and when the drink allays this, it causes pleasure. It is also the custom that the affianced bride should not be immediately given in marriage so that the husband may not esteem her any less, whom as his betrothed he longed for" [15].

Sigmund Freud in 1920 wrote:

"We have decided to relate pleasure and unpleasure to the quantity of excitation that is present in the mind [...] and to relate them in such a manner that unpleasure corresponds to increase *in the quantity of excitation and pleasure to diminution [...][and] the factor that determines the feeling is probably the amount*

of increase and diminution in the quantity of excitation in a given period of time*"* (emphases his) [16, p. 4].

Mainstream AI/robotics still seems to be blind to the above tips; however, one can meet solutions that almost beg for being supplemented with mechanisms that would facilitate seeking pleasure to be understood in the way Aristotle, St. Augustine, and Freud suggest. As an example, let us consider the motivation system of the MIT *Kismet* [4]. Each of *Kismet*'s drives is represented as a device to which a small steady stream of "activation energy" is being provided. If the volume of accumulated energy exceeds the upper limit of a defined homeostatic range, this means that the robot is under-stimulated. When a satiatory stimulus is provided to the device, the activation energy starts escaping. However, if the energy volume falls below the lower limit of the homeostatic range, we may say that the stimulus is overwhelming. Depending on the state of its drives, *Kismet* seeks an appropriate stimulus or tries to ease itself of the most harmful one. This solution mimics the important homeostatic mechanisms that in the case of animals serve to maintain certain key physiological parameters within healthy limits [4, pp. 108-109]. As for machine psychodynamics, what the paradigm recognizes as a source of pleasure is not the state in which the activation energy is within a homeostatic range, but just *approaching* the state, which can take place only if a robot previously deviated from the state. The more the deviation diminished in a given period of time, the higher the pleasure. A psychodynamic robot is proposed to deliberately seek pleasure through a purposeful "playing" with the deviations. Machine psychodynamics does not propose to resign from the mechanisms for homeostasis maintenance. What machine psychodynamics does propose is to supplement the mechanisms with a machinery that measures how the deviation from the homeostatic equilibrium changes in time and, based on the measurement, generates a pleasure signal whose properties correspond to the tips by Aristotle, St. Augustine, and Freud. The pleasure signal can then be used by the robot as a universal reinforcer of the self-development of various motor skills and cognitive abilities. It can also be noted that deliberate "playing" with the deviations from the homeostatic equilibrium brings us closer to the notion of free will.

Let us now establish order in other related notions. An excessive amount of the activation energy that fills *Kismet*'s drives can be treated as a source of the Freudian unbound excitation. In the case of a deficit of the energy, the satiatory stimulus can be treated as the unbound excitation. In both cases, an "amount" of the excitation positively correlates with the value of deviation from the homeostatic equilibrium. Note that Freud, when in 1938 discussing pleasure-related phenomena, replaced the "amount of unbound excitation" with the word *tension* [17, p. 15]. Hence, tension has been adopted by machine psychodynamics as its key concept – a variable that corresponds to the deviation of a part of the body from its resting state or the deviation of a mental state from a homeostatic equilibrium. Based on all of the above considerations, let us try to formulate a *technical* definition of pleasure to be used in robotics.

Definition of pleasure: Pleasure is a measurable quantity that reinforces certain reactions and behaviours of a creature and constitutes an attractive purpose of actions the creature may plan and undertake.

As it was concluded before, pleasure rises when a related tension plummets. But what happens then? Everyone who can recall an experienced pleasure will surely agree that the feeling more or less abruptly rises and remains for some time after the moment the pleasure-causing stimulus stops being effective; however, the volume of the feeling decays with moderate speed. Hence, the law:

First law of psychodynamics: *Pleasure volume rapidly rises when a related tension plummets, whereas it slowly decays when the tension either rises, remains constant, or diminishes with a relatively low speed.*

Let us consider a mind whose innate feature is a perpetual overwhelming strive to enhance its pleasure record. Let the mind contain a collection of tension-dynamics-driven pleasure generators (working according to the first law of psychodynamics) and a collection of functions mapping certain stimuli onto dynamics of certain tensions. Let us imagine that the mind, in order to enhance its pleasure record, develops for itself not only pleasure-giving reactions to certain stimuli, but discovers and memorizes that a given kind of pleasure will be the strongest if a certain tension is allowed to deviate from its equilibrium to an extreme value, or discovers and memorizes a sequence of actions leading to the acquisition of an efficient tension-discharging stimulus, or discovers and memorizes a successful way of planning such sequences. The above vision outlines the *pleasure principle* in its version dedicated to machine psychodynamics.

2.4.3   Pleasure Applied

The psychodynamic view of pleasure implies that the process of pleasure generation consists of two sub-processes. The first sub-process is the immediate reaction to the plummeting of a tension. The second sub-process is following the reaction with some inertia. Hence the formula that grasps the first law of psychodynamics quantitatively:

$$p' = \lambda(-q') - p/\mathrm{T}$$

where p stands for pleasure volume, q stands for the volume of a related tension, the apostrophe is the symbol of differentiation, λ is a non-decreasing function that returns zero if and only if its argument is lower than a certain threshold, and T is a time constant. Indeed, the higher the T, the stronger the inertia and, accordingly, the slower the decay of pleasure volume. The formula can be implemented based on *integ* – a device which receives a plurality of signals, multiplies each of them by a specified weight, and integrates the weighted sum, where the integral is subject to saturation such that the output value can never go beyond the range [0, 1].

Let the graphical symbol of integ be the one shown in Fig. 2.4.1

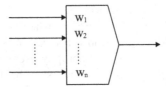

Fig. 2.4.1 Graphical symbol of integ

where w_1, w_2, ..., w_n are weights of inputs. Fig. 2.4.2 shows the scheme of an integ-based pleasure generator. Fig. 2.4.3 shows the answer of the generator to a single act of tension discharge.

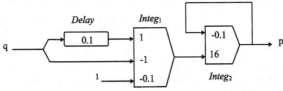

Fig. 2.4.2 Pleasure generator. Delay is a device that returns an incoming signal unchanged after a specified time; Integ is a device that integrates incomong signals, each multiplied by a specified weight while trimming the output value so it never leaves the rang

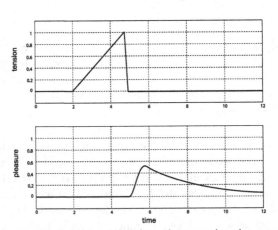

Fig. 2.4.3 Pleasure signal that rises when a tension plummets.

Fig. 2.4.4 shows the effect of pleasure accumulation caused by a series of rhythmic discharges of a tension.

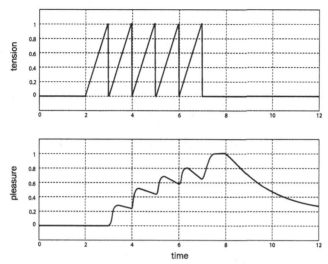

Fig. 2.4.4 Accumulation of pleasure via repetitive discharging of a tension

As an example of a pleasure-generator application, let us consider *psychod* – a creature equipped with a pair of tentacles. The greater the deformation of a tentacle, the higher the tension. Psychod has no innate mechanism for obstacle avoidance; so, initially, upon each encounter with an obstacle, it performs random movements. When one and only one tentacle touches something and then, by accident, stops touching (which means that a tentacle-related tension plummeted), the related pleasure generator produces a signal reinforces circuits that most substantially contributed to the recent movement. In this way, psychod learns, not supervised, to more and more smoothly avoid obstacles, driven only by pleasure.

The recommended mechanism for this kind of learning should be able to suppress random-signal production with increasing strength to eliminate its impact when the objective of the learning is accomplished. However, the mechanism should revert to randomness when the learning does not succeed within a substantial amount of time or when conditions change such that the already learned behavior stops working. Hence, another law:

Second law of psychodynamics: *The degree of randomness of behavior in a given situation is inversely proportional to the progress in learning what to do in such situation.*

The law is not specific to the psychodynamic perspective. Nonetheless, machine psychodynamics has no choice but to add it to its theory, together with yet another law which deals with remembered states and related pleasures and semi-pleasures.

Third law of psychodynamics: *When state Y is a result of behavior B executed in situation X, and Y coincides with the acquisition of pleasure P, then, for situation X, behavior B gets reinforced by the pleasure P, while for other situations be-*

haviors resulting in finding oneself in situation X start being reinforced by a pleasure whose volume is somewhat smaller than the volume of P.

Note that approaching something by oneself is not our first way of object-of-desire-acquisition. Babies first learn a "social way" of handling the environment [18, p. 296]. When something interesting is actually beyond reach, a child produces sounds that sound meaningless, but in fact they express a specific desire. A good caregiver learns how to interpret the sounds – starting from a pure trial-and-error strategy, but, later, reinforced by the child's smiles resulting from providing the truly desired item. Gradually, according to 2nd law of psychodynamics, the strategy fades in favor of better and better guessing. An analogical process happens in the child's mind; however, it results in a purposeful modification of the produced sounds. This effect was observed when *Miao-V* (a simulated mobile robot equipped with a camera, microphone, and speaker) interacted with a human caregiver. Desire related to particular objects rose randomly and the robot produced sounds – initially random, later (according to 2nd and 3rd law of psychodynamics) more or more purposeful. This means *Miao-V*, together with its caregiver, developed a mutually understandable proto-language. When the caregiver was temporarily unavailable, *Miao-V* (also driven by the laws of psychodynamics) started trying to use its mobility potential to approach the object of desire [9].

2.4.4   Constructive Ambivalence

Is the distant object a snake, or only a snake-shaped branch? To select a longer but safer route, or a shorter but riskier one? To fight, or to flee? A conventional robot in the face of such dilemmas tries to quickly work out an explicit decision, whereas a psychodynamic robot may endure ambivalence. In a psychodynamic mind contradictory ideas may coexist and each of them may try to suppress all others; for a domination over rival ideas gives the winning one an influence on the course of things. However, since fortune is fickle, the winning idea may, after a while, lose to a rival one, and after an unpredictable time win again. An intrinsic dynamics of the process may result in irregular switches of judgments Ambivalence may force a robot to develop new methods of judgment and to test their efficiency versus those developed earlier. Also, ambivalence gives a chance to sometimes implement a stupid idea (and, consequently, face an "unnecessary" trouble to cope with), which, as long as the resulting behavior is not too devastating, may give the robot useful knowledge about its own physical and mental capacity.

As an example of a mechanism that facilitates irregular switching let us consider a pair of integs whose outputs are to be interpreted as tension #1 and tension #2 (Fig. 2.4.5). $Integ_1$ receives an external signal r_1, (with weight u_1), a constant signal 1 (with weight -0.2), and the output from $Integ_2$ (with weight -1), whereas $Integ_2$ receives an external signal r_2, (with weight u_2), 1 (with weight -0.2), and the output from $Integ_1$ (with weight -1). The constant signal 1 together with the related negative weight represent a constant leak. Let us assume that weights u_1 and u_2 are variables whose values can be provided as a pair of another external signals.

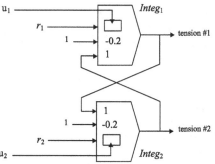

Fig. 2.4.5 A combination of two integs for conflict resolution. u_1 and u_2 are contradictory urges; r_1 and r_2 are instances of a random binary series; **1** is a constant value; the rectangles are slots to be filled with values of weights u_1 and u_2; and tension #1 and tension #2 are associated with urges u_1 and u_2 respectively.

Let us now assume that every two seconds a coin is tossed, and in the case of head, r_1 will equal 1 for the next two seconds, whereas in case of tail, r_1 will equal 0 for the next two seconds. Let the values of r_2 come the same way from separate tossing. What will happen if in the period of interest $u_1 = 0.5$, while $u_2 = 0.31$? An intuitive guess may be that $Integ_2$ would get blocked, while tension #1 would quickly reach the value of 1 and keep it. In fact, $Integ_2$ behaves as if it did not give up and fought bravely (Fig. 2.4.6).

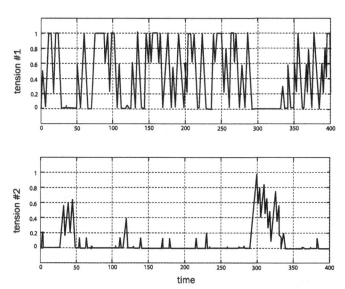

Fig. 2.4.6 A counterintuitive behavior of the circuit from Fig. 2.4.5 for $u_1=0.5$ and $u_2=0.31$. It may happen that a tension that comes from a small urge can suppress, at least for a while, a rival tension – even one that comes from a much stronger urge. Perhaps the underlying property of randomness contributes to the phenomenon of intentionality.

Does this phenomenon not match the event when an individual considers two choices for some time – one that by all means is wise and the second one that is visibly stupid – and, surprisingly, decides for the stupid choice? Events of such kind are not seldom in real life – maybe all of us remember at least a couple of cases of cursing not bad luck but our own stupidity immediately after committing to a decision. Hence the law:

Fourth law of psychodynamics: *A non-zero tension always has a chance to suppress rival tensions – even those that are much stronger.*

As it was experimentally confirmed, the fourth law of psychodynamics may work well even on the level of basic urges. *Miao-1* the robot is a simulated mobile creature, in which two contradictory drives fight for the access to motors. The first drive is hunger related to the state of a battery, and the second one is excitation caused by a toy. A fragment of a related report states:

... Miao punches the ball / it stopped punching and looks toward the battery charger / Miao turns back to the ball and punches it (though not too vigorously) / suddenly it resigns, turns, and slowly approaches the charger / It gets very close to the charger / Miao sadly looks back at the ball, then turns and starts recharging ...

Of course, the word "sadly" is not to be treated literally. The point is that the robot's slow turn and gaze *looked* sad and looked so not because somebody intentionally programmed a masquerade sadness, but because the robot's brain circuitry allowed for a psychodynamic process resulting in such expression [19].

The pair of integs was only a simple illustration of a fight of a tension vs. another tension. In human-level deliberation sophisticated notions fight. Hence, to build machinery for such a fight, we need suitable representations of the notions and, based on them, build (or let a robot's brain develop by itself) an appropriate circuitry. A step in this direction is *MemeStorm* – a grid of processing nodes inhabited by populations of identical pieces of information called *memes* [20]. The populations fight against one another for domination in the grid. A single meme can only contribute to a judgment or belief. When a population of identical memes expels rival populations, it does mean an emergent judgment or belief. It was experimentally confirmed, that when streams of contradictory memes flow to the grid, the dynamics of concluding beliefs resembles the dynamics of tension #1 vs. tension #2 in the experiment with two integs mentioned above [13]. MemeStorm processes propositions, i.e., something more sophisticated than just tensions. MemeStorm nodes can be developed to process multimodal memes [21].

2.4.5 Machine Adventurousness

As Marvin Minsky suggests, one 'secret of creativity' may be to develop the knack of enjoying a sort of unpleasantness that comes from awkward or painful performances [22, p. 278]. Indeed, only an adventurous individual may deliberately select a challenge gain over an easy gain. Psychodynamic robots are adventurous owing to the pleasure principle and susceptibility to ambivalence. And the adventurousness pays. First, adventurousness may facilitate survival. Second, it may accelerate cognitive growth.

As for survival, let us consider a robot's habitat separated from the rest of the environment by an unsafe zone. Assume the supply of vital resources in this habitat started decaying. Let us also assume that the robot's knowledge includes neither the dimensions of the unsafe zone nor what lies beyond the zone. Needless to say, in this situation, if the robot were not psychodynamic, its fate would be sealed. Fortunately, unlike a conventional robot that never engages in a "purposeless" risk, a psychodynamic one may deliberately venture into the unsafe zone – just to increase the expected-fear-related tension and to get pleasure from discharging it. Such a venture may also result from ambivalence related to risk assessment. If the robot does not carry things too far (or is simply lucky), it has a good chance of discovering vital resources just over the unsafe zone. The above vision has been implemented as a simulated environment populated by a colony of rational food-seekers, as well as a colony of pleasure-seekers. The fates of the colonies were such as predicted above – the rational food-seekers became extinct, while the adventurous pleasure-seekers found their new niche [23].

Analogically, the pleasure principle may make a psychodynamic robot "purposelessly" penetrate various areas of its own memories. Unlike a conventional robot that confines the usage of its memories to finding only data that are helpful to solve a problem that is already being faced, its psychodynamic cousin may daydream in the literal meaning. Daydreaming allows it to induce bodily and psychic tensions. In order to magnify dream-grounded pleasures, the robot may embellish facts, design new adventures, or even imagine completely fantastic worlds. It can later try to implement the ideas it has dreamed out. Verily, circuitry that facilitates creativity may emerge just as a result of the robot's strive for more pleasure.

Machine adventurousness may also consist in manipulations on the probability of acquiring a painful strike. Unlike pleasure, pain rises when a tension exceeds a certain level. When the robot bounces into an obstacle, the event will result in a pain-signal. Using an associator, the robot may learn that a portion of pain is usually preceded by a rapid distortion of one of its tentacles. Let an innate mechanism for pain-avoidance react to an expected pain with increased randomness of behavior (2nd law of psychodynamics), which may appear as panic, yet is quite purposeful. If the expected pain accidentally does not come, the pain-expectation-related tension will plummet. Consequently, a related pleasure will be generated and the circuitry that contributed to the accidentally good maneuver will be strengthened. A higher-order control system may use the expected-pain-dedicated tension accumulators as a means of pleasure acquisition. In order to charge the tension accumulators the robot may deliberately undertake a risky action – which may result in severe pain with high probability. Indeed, according to the 4th law of psychodynamics, even a stupid idea may suppress its reasonable alternative. But, when the risky action succeeds and the pain finally does not come, the pain-expectation-related tension will plummet and, according to the 1st law of psychodynamics, the robot will acquire a portion of great pleasure.

2.4.6   Concluding Remarks

Machine Psychodynamics – a paradigm of building brains for robots inspired by Freudian view of mind – admits that a target robot is to develop its cognitive structure by itself. What is novel is that the robot's self-development is reinforced by pleasure understood as a measurable quantity. I proposed four laws of machine psychodynamics. The first of the laws states that pleasure volume rapidly rises when a related tension plummets, whereas it slowly decays in other events. This law forces a robot to develop, all by itself, smarter and smarter methods of changing its relation to the environment or changing the environment itself, just in order to acquire various tension-discharging patterns. The second and third law of psychodynamics provide tips for the construction of such mechanisms.

Machine psychodynamics also proposes that some ambivalence may accelerate a robot's cognitive growth. In the psychodynamic decision-making process contradictory judgments and beliefs fight against one another to dominate a working memory. The winning idea starts being processed toward appropriate action. Nevertheless, owing to a certain counterintuitive property of random series, the actually winning idea may in an unpredictable moment lose to a rival idea. Consequently, the robot may hesitate or abruptly change its mind. These phenomena are reflected in the fourth law of psychodynamics. Mechanisms for pleasure generation and ambivalence jointly make a psychodynamic robot an adventurous creature. The set of four laws have been formulated as a conclusion from several experiments with simulated and physical robots.

Machine psychodynamics does not intend to replace the homeostatic mechanisms employed by mainstream AI/robotics. The psychodynamic approach only proposes to supplement the mainstream solutions with mechanisms for active pleasure-seeking, some deliberate irrationality, and constructive ambivalence. Sadly, a robot designed to deliberately expose itself to inconveniences or dangers may be hardly welcomed by today's corporate investors. The same undoubtedly applies to a robot that displays visible signs of indecisiveness. Nonetheless, I argue that such troublesome properties may be an unavoidable price for the robot's cognitive self-development up to a level beyond that which can be achieved via handcrafting or simulated evolution. Currently, one can notice the dramatic growth of a zoo of artificial "cleaning guys", "speaking" mascots, or human-shaped "reception-desk staff." Yet a robot "speaking" prerecorded sentences is only a masquerade – good to impress naïve folks, but a blind alley as for the dream about human-level machine cognition.

I say, making a machine learn is only a half-success. What is the true point is to make a machine actually want to learn. Machine psychodynamics has up its sleeve a key to the machine's will to learn – the set of four laws one may take into account when designing brains for robots.

References

[1]   R. A. Brooks, "Prospects for Human Level Intelligence for Humanoid Robots," in Proc. First International Symposium on Humanoid Robots (HURO-96), Tokyo, Japan, Oct. 2000.

[2]   D. Westen, Psychology. Mind, Brain, & Culture. New York, NY: J. Wiley & Sons, 1999.
[3]   J. D. Velásquez, "Modeling Emotions and Other Motivations in Synthetic Agents," in Proc. AAAI-97, pp. 10-15.
[4]   C. Breazeal, Designing Sociable Robots. Cambridge, Mass./London, England: A Bradford Book/The MIT Press, 2002.
[5]   E. Henninger, R. M. Jones, and E. Chown, "Behaviors that Emerge from Emotion and Cognition: Implementation and Evaluation of Symbolic/Connectionist Architecture," Proc. Second. International Joint Conference on Autonomous Agents & Multiagent Systems (AAMAS'03), Melbourne, Australia, July 14-18, 2003, pp. 321-328.
[6]   R. Halavati, S. H. Zadeh, and S. B. Shouraki, "Evolution of Pleasure System on Zamin Artificial World," Proc. 15th IASTED International Conference on Modeling and Simulation (MS'04), Marina de Rey, USA, 2003.
[7]   Ch. Becker, S. Knopf, and I. Waschmut, "Simulation the Emotion Dynamics of a Multimodal Conversational Agent," in E. André et al. (Eds.), Affective Dialogue Systems (ADS 2004), LNAI 3068, Berlin: Springer-Verlag, 2004, pp. 154-165.
[8]   P. Palensky, B. Lorenz, and A. Clarici, "Cognitive and Affective Automation: Machines Using the Psychoanalytic Model of the Human Mind," in Proc. first International Engineering & Neuro-Psychoanalysis Forum (ENF'07), Vienna, Austria, July 2007, pp. 49-73.
[9]   J. Liu, A. Buller, and M. Joachimczak, "Self-motivated learning agent: Skill-development in a growing network mediated by pleasure and tensions," Transactions of the Institute of Systems, Control, and Information Engineers, vol. 19/5, pp. 169–176, 2006.
[10]   R. Arkin, Behavior-Based Robotics. Cambridge, Mass: The MIT Press, 1998.
[11]   S. Freud, The Ego and the Id. New York: W. W. Norton & Co., 1960/1923.
[12]   Nowak and R. Vallacher, Dynamical Social Psychology, New York: Guilford Press, 1988.
[13]   Buller, "Mechanisms underlying ambivalence: A psychodynamic model," Estudios de Psicologia, vol. 27/1, pp. 49-66, 2006.
[14]   Aristotle, Rhetoric, Book I, Ch. 11. 350 BC (translation by W. Rhys Roberts).
[15]   St. Augustine, The Confessions, Book VIII, Ch. 3. AD 397 (translation by A. C. Outler).
[16]   S. Freud, Beyond the Pleasure Principle. New York: W. W. Norton & Co., 1961/1920.
[17]   S. Freud, An Outline of Psycho-Analysis. New York: W. W. Norton & Co., 1949/1938.
[18]   M. Minsky, The Society of Mind. New York: Simon & Schuster, 1986.
[19]   Buller, M. Joachimczak, J. Liu, and K. Shimohara "ATR Artificial Brain Project. 2004 Progress Report," Artificial Life and Robotics, vol. 9/4, pp. 197-201, 2005.
[20]   Buller and K. Shimohara "On the dynamics of judgment: Does the butterfly effect take place in human working memory," Artificial Life and Robotics, vol. 5/2, pp. 88-92, 2001.
[21]   Buller, "Operations on Multimodal Records: Towards a Computational Cognitive Linguistics," Technical Report TR-95-027, Intl. Computer Sci. Institute, Berkeley, CA, 1995.
[22]   M. Minsky, The Emotion Machine. New York: Simon & Schuster, 2006.
[23]   Buller, "Machine Psychodynamics: Toward Emergent Thought," Technical Report TR-NIS-005, ATR Network Informatics Labs., Kyoto, March 17, 2006.

2.5 Artificial Group Mind, a Psychoanalytically Founded Thought Experiment

Charlotte Rösener, Tobias Deutsch, Roland Lang, Brit Müller, and Takahiro Yakoh

Intelligent behavior appears to be an overstressed term used in both science and psychoanalysis. Besides controversial definitions of "intelligence", another important issue is to determine whether the mental capabilities of a crowd originates in the mental abilities of the individual or emerges from shared abilities. Psychoanalysis, starting with Freud himself, gives a clear answer to the question regarding human crowds. However the theories have to be enhanced in the case of artificial counterparts. In technology, the capabilities of control systems may vary. In the case of embedded control, models for robotic control are still very limited. The idea to tap the potential of control systems by multiplication in forming decentralized cooperating clusters should resolve current design problems. Smart group behavior should give a solution to the desire for intelligent control, but bears the risk of being unfeasible, reducing the benefit of team work instead of improving. This approach should assist in presenting crucial methods and functions inherent in the single system, as well as for the group, in order to form effective teams guided by psychoanalytical concepts. This paper concentrates on the abilities of groups and its members. It provides a summary of key feature of the Artificial Recognition system (ARS).

2.5.1 Introduction

A core question in various approaches of Artificial Intelligence (AI) is whether or not intelligence can emerge from non intelligent substitutes. The idea of decentralization in control is still challenging, but in nature intelligence does not seem to be an individual matter. A colony of individuals capable of a limited set of actions can respond quickly and efficiently in a group to their environment: the methods evoking these capabilities are called swarm intelligence as has been introduced by G. Beni in [1]. The collective abilities of animals within a colony, crowd or herd showing complex behavior demonstrate the success of collective behavior. Although none of the individuals participating can grasp the big picture, their contribution is crucial for the success of the group. In colonies of animals like ants or honey bees, no one individual seems to be in charge. There is no direct management for central control formulating general commands. Continuing this revolutionary idea, the human crowd is a similar example. However, referring to abilities of human individuals, the human crowd clearly shows the potential risks of multiplying capabilities: The human mind, which is still an unreachable goal for modern control systems, can show a specific lack of mental abilities within a crowd, showing a backward development in behavior.

However, cooperating individuals in teams have become a common operation method in technology and working environments. In science the collective capabilities of entities (software agents or autonomous systems) are used to reach shared goals, which might be infeasible for a single system. So called multi-agent systems (MAS) have gained importance in research during the last few years to solve complex tasks independently [2], [3], [4]. The aim in this context is to determine the crucial capabilities of agents and team members respectively in order to improve the functionality of the team. Efficiency is based on the balance of individual capabilities and their enhancement of the group. The core question dealt with in this work is what level of intelligence has to be inherent in the distributed total system and its autonomous subsystems respectively, which qualities are necessary for a group, which are required for the participating individual to emulate an overall intelligent behavior, and which level of intelligence can be achieved by this method? It is hard to appreciate the complex abilities of a human mind, whose abilities are still far beyond our imagination and exceed the capabilities of engineers in control design. However in biology effective solutions with much simpler concepts can be found. Therefore it is important to determine certain mental abilities which are crucial for effective group behavior.

2.5.2 Challenges in Design

Decentralized designs appear challenging in science as we are not used to thinking in decentralized terms. Although the mind *is* obviously decentralized, this fact appears transparent to the self. In technology an approach based on black boxes is preferred to determine a system by its input and output. Similar to cognitive science, which also concentrates on observable behavior patterns to explain inner functions, this method appears error prone, as a vast number of inner functions have no observable outcome and with all respect to reverse engineering, it is doubtful that high levels of intelligence can be achieved without a fundamental knowledge of basic mental functions. Furthermore, observable, obviously "intelligent" behavior does not necessarily entail that there is any form of intelligence inherent in a system. An excellent example can be found in a chat robot [5] showing the mechanism based on the idea of the Turing test, where a robot pretends to be human in order to evaluate intelligent behavior [6]. A similar concept is the social Fungus Eater of [7], intelligence is supposed to be determined by the observable appearance of intelligent behavior. This however seems to be questionable for the reasons expressed before. In order to achieve a valid form of evaluation, computer science has to turn to disciplines concentrating on inner mechanisms of the human mind, e.g. psychoanalysis. At first, intelligence in the context of the human mind is a main issue which has to be redefined to determine the demands for its artificial counterpart. According to [8, p. 31] "Intelligence endows a system (biological or otherwise) with the ability to improve its likelihood to survive within the real world and, where appropriate, to compete or cooperate successfully with other agents to do so." The definition will be a fundamental basis for this approach, as it contains two main points: the highly diverse actions of human life beyond all simplified conditions of a laboratory setup (programmed or physically constructed) and the crucial aspect of interaction (for cooperation), which is in analogy to the requirements of R. Brooks [9] where intelligence emerges from interactions of the agent with its environment.

Observing one's own mind, we like to think of ourselves as single, monolithic individuals, looking back on a chronology of "obvious" continuity emphasizing this view. Individual history is founded on memorization, recording development through a lifetime. But Freud emphasizes the balance of competitive forces, the Super-Ego, Ego, Id as crucial mechanisms of decision making within the human mind [10] which can be seen as an archetype of agent based decentralized control.

Freud himself described in [11, p. 82] that the collective character of a crowd, transformed from a group of individuals into a psychological crowd. The aggregation of cells form an organism that posses new and different abilities. Following models of psychoanalysis, as described in [10], [12] and [11], the conscious "Ego" is just the tip of the iceberg and cannot legitimate the approach of centralized unified control models at all.

It is accepted in psychology, psychoanalysis and in sociology that a crowd acts differently from its individual members [13], [11]. The so called "group mind" can be described as, "some degree of reciprocal influence between the members of a group" [13, p. 33] this allows "this mental homogeneity" [13, p. 148] based on the mechanisms of affection according to [11, p. 79]. Generally groups share common goals which do not match with the goals of a single participating individual. Psychology and psychoanalysis emphasize that the behavior of crowds appears more instinct driven and therefore evolutionarily behind the capabilities of an individual. However this is not valid for artificially formed crowds, e.g. religious groups, armies, or working teams [11]. The high level organization allows artificial groups to regain individual abilities that would be lost through the accumulation process. But why do some groups appear so much more efficient and goal oriented, while others lack these capabilities? These problems are the object of research in different fields of science. The fundamental issues from a technological point of view in this context are:

What methods, functions and factors are essential to improve the efficiency of groups in terms of its participants?

- How to determine the optimum size of groups?
- Which methods and structures can improve group behavior?
- Is leadership (central control) required and in which context?
- Which role does communication play?
- How does communication influence the efficiency of a group?

These and other concerns shall be discussed within the next Chapters.

2.5.3 The Idea of Artificial Groups

Generally the theory of group minds emphasizes that the psychological group exceeds the simple accumulation of individuals, although aggregations may lead to psychological crowds under special conditions. As described in [11], a certain reduction of individual capabilities seems to be inevitable in forming a group, which has to even out the mental capabilities of its individuals. Beside special forms of highly organized groups, a group mind in general appears evolutionarily behind the capabilities of the individual mind. However certain conditions allow groups to render collective mental processes that are extremely efficient in a group formation. The efficiency and capability of a group itself depends on the efficiency of mechanisms that allow a group to retain the abilities of individual minds in the group.

A. Stupid Masses or Intelligent Groups

McDougall defines in [13, p. 69-71] five principal conditions to categorize types of groups in order to describe the different phenomena within grouped behavior ranging from an unorganized group and simple crowds, to very highly organized groups, whose mental life is far superior to that of the average individual or a mere crowd [13, p. 195]:

- The degree of continuity of a group, which might be either fulfilled through the persistence of the same individuals of the group or in the persistence of roles and positions within the organized group. Highly developed groups exhibit both forms.
- A collectively formed idea about the group, its function and capabilities, composition and relations within the group has to be distributed among the members of the group. This is the source of emotions and actions defining the relations towards other groups.
- The potential interaction with other groups is essential to form a kind of self-knowledge and self-sentiment of a group.
- A body of traditions and customs allow the establishment of relations within the group as well as establishing the group as whole
- The last condition defines the organization of the group itself based on the traditions, which might be invoked by the group itself, but can be maintained by external forces as well.

These conditions allow organized groups to regain a better control of impulses, while they exhibit a coherent, goal oriented behavior. A core question in this context is whether there is a top-down hierarchy or leadership required for achieving superior capabilities in groups. The traditional perspective represented by [11] and [13] opposes the idea of unguided highly organized groups, although Freud admits that this external force does not necessary have to manifest in reality, but can be formal and abstract [11, p. 89]. However the research field of modern swarm intelligence, introduced by G. Beni and J. Wang in 1989 [1], emphasizes the necessity of distributed characteristics of groups, succeeding Minsky's idea of a society of mind, where individual and intelligent behavior is based on massive collective cooperation of simple agents without central coordination [14, p. 20]. The diversity of opinions encouraged by free competition, which can be narrowed in the end, seems to form effective mechanisms to evoke swarm intelligence. This wisdom of the crowd can be also very much referred to the model of Freud [15] where competitive forces establish all mental life. This in turn leads to one of the major design systems found in nature, where "everything is distributed" which does not stop with the individual: just as hundreds of individuals can cooperate in flocks, schools or herds, agents, ideas and synapses can act similarly. Instead of the idea of anticipation or reaction, local imitation and knowledge of proximate synapses or team members can set off an avalanche.

Crowd psychology focuses on the mutual influences of individuals who have only little in common. This stands in contrast to individual psychology, but also sociology. Individual psychology extracts the individual capabilities, needs and methods from its social context with focus on the relationship of the subject to an object, which might act as an ideal, helper or opponent. But the most important influences forming the phenomena of a crowd have to be determined separately. There is an important difference between the anonymous fluctuating crowd without any explicit purpose, e.g. people at a train station or visiting a theatre, and an organized artificial crowd like an army or companies where members cooperate, focusing on a certain goal.

Freud [11] clearly differentiates between the aggregation of people and a psychological crowd by two fundamental relationships based on the existence of affections towards the members within the crowd and towards its leader. The leader itself need not be real, but can also be projected into a virtual person or abstract idea. For Freud [11, p. 89] the idea of leadership is fundamental. This entails the mechanism that can prevent the regression of a crowd towards a more instinct driven behavior. The organization of crowds in combination with any form of leadership (by rule, by ideas or by individuals), allows groups to improve the abilities and the potential of a crowd, similar to the conditions set by McDougall.

The main difference in technology is that there are no human-like abilities yet and the capabilities of the single system can be highly diverse, based on different physics and mental abilities. However McDougall points out that the *similarity* of the mental constitution of individuals is fundamental to allow homogeneity within the group, but this does not necessarily mean that it has to be an exactly humanoid mental apparatus. Therefore there are general tendencies which can be outlined in this case and shall be put to the test.

B. The Psychoanalytical Perspective

The idea of Freud formulating discrete mechanisms building a "mental apparatus" is a great challenge and chance for technological approaches in analytic design. Finding the analogy to mental capabilities allows improvements in the design of automation control. The main focus in this research is to determine important mechanisms between subject and object (other subjects) to organize and reduce to form effective grouped behavior. In analogy to the findings of Freud and McDougall, the following major methods form the premise for superior grouped behavior. The main source of this ability to act as a superior individual is founded on the reciprocal influence of its members [13, p. 33].

According to Freud's individual psychology there are two crucial correlations of the inner world of an individual and the outer world that can describe the relationship between the subject and its external object. Originating from individual psychology, these basic functionalities are crucial to understanding complex relationships within the crowd.

- Projection: The projection describes the ability of the subject to transfer facts, e.g. from subject to object. It is a fundamental mechanism for identification, as it allows the recognition of one's own characteristics in another object. An individual is capable of perceiving the environment through its interests, abilities, customs and affective state, expectations or desires. The subject can therefore compare and equate their own behavior, with the perceived behaviors of other individuals.

- Introjection: Based on psychoanalytical concepts, introjection allows the subject to replicate the behaviors, attributes or other fragments of the surrounding world, especially those of other subjects. This internal mental process should not be mixed up with the intrinsic process of imitation, which leads to similar observable behavior.

In order to describe the relationships between members Freud has defined the following forms of relationships based on individual capabilities.

(A) Identification

Identification can be used in two different meanings. In Freud's theory it describes a psychological process of the subject towards an object, e.g. persons or groups. Identification is crucial in recognizing an object as "what it is", e.g. classifying an object as a stone, furniture, or person. Furthermore it defines "just as if", which does not describe simple imitation: Identification with the characteristics of another person, e.g. the father, a superior, or a teacher, allows the assimilation of the characteristic in the personality of the subject by transformation of oneself in order to become (partly) like the identified person. Identification is not necessarily complete as it extracts distinct characteristics.

Identification is a primary concept establishing affective relationships between individuals. It initializes the psychological "infection" within a group; the ability to put oneself into somebody's position, show compassion and drive any other form of analogy between observable behaviors of group members and one's own individual situation [11, p. 101]. It is a regressive form of an object-relationship using introjections to detect observable mutuality. The important aspect of identification follows from this simple form of compromise and imitation. Because of introjection the subjects assimilate parts of the object, in this case another individual, and develop a new mental structure, giving birth to a unified group mind.

(B) Idealization

Idealization is another mechanism based on identification. It allows the projection of the idealized self towards an external object (individual). The idealized object, e.g. an individual of the group, is enveloped in aspirated perfection, without changing the nature of the object itself. It builds idealized instances standing in context to the self. The self reduces its demands, enduring or even seeking personal restrictions for the benefit of others, the group itself.

(C) Transference

This function is an enhancement of identification, and allows the group to achieve shared goals. It describes all forms of the mechanism of projections, entailing the founding of group identity and regaining of individual characteristics in the group.

The functions described so far allow the description of relationships within the crowd. Identification is a major relationship in allowing the equalization and unification of the homogenous crowd, while idealization defines the relationship towards the leading parts of a group, which can be bound to another individual or abstract. All functions are based on the idea of self awareness. The agent has to visualize its capabilities, physics and localization. It must not only interpret the environment, but also its relationship to it, reflecting itself within its setting. Therefore it must be capable of comparing and recognizing similarities.

As described before, simple crowds focusing solely on these three functions still lack the mental capabilities of the individual. But based on additional levels of organization, transforming a crowd into a so called artificial group [11] can achieve superior behavioral patterns. To fulfill the conditions of McDougall there must be additional mechanisms enhancing the basic relationships of mass psychology. These additionally imposed mechanisms are required to form important criteria of artificial crowds:

- *Rules* represent a form of Super-Ego that does not exist in a regular crowd. They can define relationships and entail behaviors within the group, but also affect group behavior towards other (e.g. competitive) groups or cooperative actions achieving a shared goal. The rules may form a consistent structure, which is necessary for highly organized groups defined in the conditions of McDougall [13].
- *Organization* and *structures* reduce regressive behavior. In combination with rules, the individual of the crowd loses its anonymity and its responsibility increases. Hierarchy structures are an essential requirement according to [13] and [11]. They allow the establishment of roles performed by arbitrary agents, which are interchangeable without any effects towards the structure of the group or adaption of its members.
- Furthermore the *concept of a leader*, physically existing, or idealized as an abstract idea, increases the bond between group members. In combination with the other functions it allows the contribution of individuals in order to reach a shared goal, obeying the rules of the group and the advisor, who takes responsibility for the subordinates.

The need for organizational structures can provide the durability, homogeneity and history of a group that is essential for efficient group behavior.

C. The Technological Perspective

Concepts of autonomous agents and multi-agent systems (MAS) are indispensable for analyzing and designing systems in modern software architectures, improving the potential of modern AI [2]. Among a variety of tools and techniques ranging from comparatively small systems to complex mission critical systems, all designs inherit a shared key abstraction: the agent [16]. Among numerous other definitions found (e.g. [17] and [18, p. 2]) we prefer that of [19] describing an agent as a computational mechanism performing autonomous actions in its environment based on information received from the environment. These concepts are basically applied to the software structure, which appear transparent to its underlying hardware architecture [20]. The approach of multi-agent systems is to determine methods that allow us to build complex systems composed of autonomous agents that operate on local knowledge, but nonetheless can achieve the desired global behaviors [2]. The advantage of MAS is founded on the parallelism of achieving the complex computational schema of the mind out of many smaller processes, each mindless in itself, called agents [14, p. 18]. There are numerous approaches to realising this concept, aiming to arm agents with all the necessary abilities to work in cooperative or competitive environments.

(A) Multi agent systems

Agents can be entities like software or robots which can perform certain tasks independently within their environment, following general concepts rather than obeying actual commands. Sensors, actuators, and communication enable it to interact with the world or with other agents. Autonomous agents are equipped with all necessary resources and abilities to fulfill their tasks. Another important property of agents is the one of flexibility [21]. This means that they have to be reactive, pro-active, and social. To be reactive is the ability to interact with a dynamic environment and to adapt to changes at appropriate points in time. Pro-activeness in this context means the ability to take the initiative when trying to reach a certain goal and not being solely driven by events. The interaction with other agents is a very important issue within MAS and usually left out by artificial intelligence. This topic is referred to as social interaction.

In principle two different possibilities exist for agents working together – benevolent and self interested agents. Benevolent agents help each other whenever they are asked for. They are easy to design as long as it is a closed system, i.e. all agents are benevolent and from the same designer. This approach is also called distributed cooperative problem solver. The main problem is that conflicts cannot be avoided. If one agent is simultaneously asked by two other agents for help, one of the askers cannot be served. Thereafter, agents have to deal with the special situation where help was within reach, but did not respond. In a system with self interested agents this is not a special case; moreover it is the expected case. Here agents help each other on grounds of common interest or on the prospect of future countertrades. A self interested agent may be acting at the expense of other agents.

Thus, there are not only goal conflicts but also conflicts among agents. Cooperation among agents does not necessarily mean that they distribute tasks. Another possibility is sharing of results or information.

A good example and platform for many research groups is robot soccer. As in real soccer, an individual without a team can never win no matter how brilliantly he plays. It is a classic example of an artificial group which has to unite the strengths of its members to achieve superior goals. Acting as a team is crucial to succeed over opposing teams, and – if possible – to score goals. While soccer might be seen as an abstracted and simplified game, many abilities needed for real world applications can be developed within it. For example, in a factory where robots and humans are operating in the same area, techniques for path planning in a hostile non cooperative environment are crucial.

(B) Communication: Imitation as a Secret of Success?

In cooperating environments an agent has to communicate with other agents to complete its tasks. A multi agent system can – similarly to independent agents – communicate and cooperate with other agents or multi agent systems. Communication in order to transmit shared information is a key issue for successful cooperation [22], [23]. These interactions determine the effectiveness of a group, but what are the necessary abilities required for effective communication? This communication can be implemented using speech acts which are theories of how language can be used to change the state of the world. Each utterance of an agent is similar to a physical action and expressed with the intention to change something. This can be for example, passing information, requests, or commands. The complexity of such a language can be very simple – the message with id 1 means help, id 2 means enemy close, id 3 means energy close, etc. More complex agent communication languages (ACL) are for example Knowledge Query and Manipulation Language (KQML) and Foundation for Intelligent Physical Agents (FIPA) ACL [24]. They provide a well defined set of possibilities to express an utterance.

ACLs are usually used either to pass information on a benevolent level or to reach agreements. If two or more parties are arguing about a topic, each party has to deal with the problem of its self interest and reaching a potential mutually beneficial agreement based on common interests. To be able to perform such an act, negotiation capabilities are required. These include a language rich enough to express different positions on a topic, a set of objects/goods/tasks/information to negotiate about, a protocol of how to proceed and a mechanism to define whether an agreement has been reached or not. For simpler negotiations like goods allocations – one agent sells, several agents are potential buyers – auctions can be used. Most common auctions are English auctions, Dutch auctions, first-price sealed-bid auctions, and Vickrey auctions. They differ in the number of rounds, open or sealed bids, and the dominant strategies. They all share the aspect that lies and collusions can occur. For example, shills can inflate the price in an English auction.

Distributed algorithms as we can find in all forms of MAS are usually very limited, without the necessary communication abilities to share common knowledge based on the collective intelligence (COIN) [2, p. 76]. Rather than finding specific plans and rules for each agent, learning how to cooperate best should be the major goal. Still, this is highly dependent on the communication tool provided. In cooperative environments, a key factor is to determine the optimized usage of their communication tool, which exceeds the perceptive capabilities of the individual agent. Especially in cases of cooperative behavior of agents possessing different tools, the exchange of "sufficient" information can be crucial for the efficiency of cooperative behaviors [2, p. 77] In this case the problem of an accurate and sufficient form of communication is a crucial one. The exchange of sensory data in multi-robot coalitions allow a broader perception beyond the sensory abilities of a single system and allows heterogeneous agents to participate [4]. However communication goes beyond sharing sensory data, it can also be used to advise on roles and share tasks. The goals within groups do not necessarily match, as only the sum of these heterogeneous goals may show an achievement in goal allocation. The modeling of ownership, achieving optimized role and task allocation, as can be found in e.g. [22] and [25], show a direct analogy to the rules in artificial human groups. There must be a tradeoff between high transfer performance and the complexity of preprocessed data. The optimum protocol or language cannot be predefined in general as it is a matter of application and embodiment. In nature, imitation, as an observable form of psychological function of identification, appears as a common method. It allows agents to narrow the effective transfer of directly communicated information, but is an evolutionary part of the environment perception process.

2.5.4   The Ability of Progress

Looking at these two major research fields and their discrepancies in research method, intention and outcome, the questions arise not only "What has research learned from natural archetypes? ", but also "how do we learn from nature?"

Many researchers argue that science lacks the ability of individualization in system design. The realistic perspective, which was dominant over the last decades, prefers to describe human experts like a data-processing system similar to computers based on the assumption that the world can be described and formalized under technological premises [26]. As a consequence system designs are driven by technological advances rather than focusing on the needs of their human counterparts.

The origin and the goal of all organisms in biology is their survival [27, p. 19, p. 29]. In analogy to nature, we can use the definition of intelligence by [8, p. 31] as we have used before in Chapter 2.5.2, in the context of a technological approach, as a crucial method to endow an agent and improve the likelihood of its survival. Cooperating intelligent behavior seems to be a key method to extending survival time due to better adaptation to environmental conditions. Not panic but precision is the necessary ingredient of smart group behavior.

An agent must have the following characteristics in order to achieve potentially intelligent cooperative behavior [28]. The agent has to be

- *Autonomous*, so that it may act without human supervision or instruction.
- *Self-sufficient*, in order to be sustained over extended periods similar to the requirements of [8].
- *Embodied and situated*, to act in the real world as a physical system.

In many MAS approaches adaptations are made solely at agent level using a homogenous group of identical agents in order to fulfill tasks in a distributed manner [29]. According to psychoanalytical findings a design solely focusing on individual behavior of the agent itself allows only limited progress. Especially adaptations on system level which consider the cooperative advances on system level allow new design methods and improvement. According to [29] two important criteria have to be satisfied to adjust behavioral structures expanding the idea of McDougall:

- The change of internal states of agents shall be transparent to other agents, as long as there are no grave effects in interaction.
- Agents act locally and their adaption shall not influence the overall system structure, but have an impact on local behaviors of related cooperating agents.

In order to meet these demands, [29] proposes the definition of general roles fulfilled in groups and dynamic role-filling mechanisms that allow interactions in group via roles. However roles themselves are not altered, which allows the replacement of agents without any functional reduction on system level, this is an important method to provide the continuity and stability of highly organized groups that is required by the conditions of McDougall. Furthermore interaction of independent roles allow agents to change and optimize behaviors without redefining interactions in the group as long as they can fulfill the premises of their assigned role. As a consequence communication can be standardized on a higher level.

2.5.5   ARS – A Society Put to the Test

Although the proposed model of a neuro-psychoanalytically-inspired cognitive architecture as described in [30] focuses primarily on individual agents, concepts are included that support and significantly improve multi-agent behavior. The described affective modules and therefore behavior of the system can be seen as a first shift towards intelligent swarm behavior. In [31], an analogy is made according to the optimum level of arousal during the learning phase of humans. The phrase: "No pain, No gain", is applied to athletic training as well as mental training, also indicating that emotions are substantial during the learning of facts, rules, behavior, etc. Within a group of agents, having implemented the proposed neuro-psychoanalytically inspired model, the emotional arousal can be significantly increased and emphasized by feedback to one or more cooperating agents. This feedback can be realized by several communication channels like direct, verbal expressions compared to the humans voice, facial expressions that also indicate the agent's emotional state or observation of the behavior of the cooperating agents. The described Super-Ego-Module, inspired by the Super-Ego described by S. Freud in the second topographical model, is therefore further evidence for multi-agent compatibility of the described model. Without one of the described feedback loops, indicating whether an executed action was beneficial for the desired goal of the group or not, rules within this model can neither be learned nor extended or corrected. Since the Super-Ego can be seen as a storage of rules for social interaction, values and prohibitions, it implicitly represents a guideline for interaction between individuals. Having a look at the higher cognitive layers within the model, an internal simulation of the impact of a planned action within the group can easily be done in a first step, by consulting the rules within the Super-Ego. This simple form of estimating whether the planned action is worth pursuing or not within a situation can be seen as one of the main functionalities within the described module 'Acting-As-If', where strategies for shared long-term goals are planned.

The project ARS (Artificial Recognition System) aims at translating neuro-psychoanalytical terms into technically feasible terms. As a long term goal, building automation systems should operate using a psychoanalytically inspired perception and decision unit. Each room should be represented by an agent, the whole building is a multi-agent system, and people within the building might wear tiny digital personal agents interacting with the building.

As an intermediate step, a simulation platform, described in detail in [32] for autonomous, embodied agents has been built. The agents operate using the described, neuro-psychoanalytically inspired model. They are grouped and have to cooperate within a team to survive as long as possible. This artificial environment is also the test-bed for evoking emotional arousals during the interaction between agents. Simple tasks, like cooperation in collecting energy together instead of collecting it individually have been rewarded and it has been shown that the lifetime of cooperating agents increased. This was not surprising, since it was the predicted behavior, but it was a good chance, to validate the internal emotional variation during such scenarios and improve the model behind it with the help of psychoanalytical advisors. In a further step, the model shall be tested in a multi-agent environment with cooperation between more than two agents.

Currently, three test-cases are defined for the agents [33] – "Ask for dance", "cooperate for energy", and "call for help". They all focus on social reputation. The first one is only performed when no urgent issues have to be considered. Due to social interaction, the mutual social acceptance increases. No other benefit can be gained from this action. Differently the second test-case – an agent detects an energy source which cannot be accessed by one alone and thereafter it asks team mates for help. If one decides to help, different motives could come into play: it needed energy anyway, a benevolent character, the prospect of future help (tit-for-tat). Whatever the true motives, the social reputation of the helper rises. The third test case – "call for help" – focuses on dangerous situations like being attacked by an enemy. The reward for helping is directly proportional to the social level of the caller and the danger of the situation. Thus, if an agent with a high social acceptance is attacked, other agents are motivated to risk damage by helping. All three test-cases together should evoke an emergent social system with reputation as a trade value.

Further test-cases are currently under development. Other than the three described ones, they will focus on emotional reactions and inner conflicts.

One important part of the complex decision unit based on psychoanalysis are the emotions (or affects). A normally acting agent should emotionally react to a certain currently perceived situation within the expected range. If the emotional level differs, this should be explainable by individual experiences made during a single simulation run.

Inner conflicts appear for example – to continue the test-case "call for help" – if an agent receives an emergency call, decides that it would be good for its own social acceptance to help but the drive to ignore the call due to fear is equally strong. In this case the occurring inner conflict – Id competes with Super-Ego – has to be resolved by the Ego.

2.5.6   Criticisms of Modeling

Many systems designs focus solely on an information theoretical structure recreating software architectures on underlying mainstream soft- and hardware systems [20]. [30] shows clearly that there is a main issue left behind in regards to theoretical modeling: The logic gap between the information based "soft" structure and underlying hardware is a core problem in technological designs. Controls are still just models, mathematically transferred and put on the same architecture of the same procedural, serial capabilities like any other conventional system. It entails that a group of physically separated robots which can represent single systems that might work procedurally, simplifies each of them. But as a cooperating group they can form parallel systems, competing and cooperating in similar ways. In analogy to Minsky [14] only a vast physically distributed system can achieve these goals. This goes along with the research embodiment of R. Pfeifer, who emphasizes that the embodiment of agents is crucial for their mental abilities [9], [28] substantiates the third criteria of agents described before. This is an important issue, especially in the case of communication and perception. Pfeifer argues that in the case of sensor-motor coordination, the knowledge of the physical shape and state of the system is crucial for potentially intelligent behavior. As a consequence this would entail the conclusion that human-like intelligence can emerge solely from a humanoid system. This appears doubtful for information research. But the astonishing results of Pfeifer's recent research give much evidence that embodiment is a very capable method of achieving a high potential system acting in the real world. Nevertheless these approaches focus on optimization in special application fields, narrowing a broad concept to specific demands. The results are impressive and provide special solutions to given problems, but may not be seen as general problem solvers.

2.5.7   An Outlook

The project ARS as partly described in [30] met several interesting problems during the development of a neuro-psychoanalytically inspired model. First, different, partly very specialized models within the theory of neuro-psychoanalysis are available. Therefore it was hard, to select the best model for a particular function. In a new approach, only the functional models of the theory of S. Freud – especially the second topographical model – are used to define a basic concept for a decision unit for an embodied, autonomous agent. Second, as described above, the developed simulator was not able to cope with the demands of multi-agent simulation. Therefore, a new simulator is designed to support simulations with crowds of agents, cooperating in groups and not only in couples. This new simulation platform should also be able to contain different versions and types of the developed decision units for comparison and evaluation. This should be the foundation for shifting mechanisms designed for single agent systems to the multi agent system, building artificial groups with superior abilities. In the case of embodiment in robotic systems, this succeeding approach allows achievements in hardware based parallelism and a reduction of processing overheads on the single system. This goes along with the developments in psychoanalysis focusing on individual mental capabilities as well as grouped ones. Agent research has started to focus more and more on autonomous spontaneous group building, leader elections and negotiations, emphasizing grouped capabilities.

References

[1]   G. Beni, J. Wang: "Swarm Intelligence in Cellular Robotic Systems", Proceed. NATO Advanced Workshop on Robots and Biological Systems, 1989.
[2]   J.M. Vidal, Fundamentals of Multiagent Systems, With NetLogo Examples, eBook, 2007.
[3]   L. E. Parker: "Distributed Algorithms for Multi-Robot Observation of Multiple Moving Targets", Autonomous Robots, Volume 12, Number 3, p. 231-255(25), 2002.
[4]   L. E. Parker, F. Tang, "Building Multirobot Coalitions Through Automated Task Solution Synthesis", Proceedings of IEEE, Vol. 94, p.1289-1305, 2006.
[5]   A.L.I.C.E. Chat Robot on Artificial Intelligence Foundation using Artificial Intelligence Markup Language. http://www.alicebot.org/
[6]   M. Turing, "Computing Machinery and Intelligence", Mind, 59, p.433-460, 1950.
[7]   M. Toda: Man, Robot and Society; Martinus Nijhoff Publishing, 1982.
[8]   R. C. Arkin: Behavior-based Robotics, The MIT Press, 1998.
[9]   R.A. Brooks: "Intelligence without reason", A.I. Memo 1293, MIT, 1991.
[10]   S. Freud: "Triebe und Triebschicksale" (Instincts and their vicissitudes), Internationale Zeitschrift für (ärztliche) Psychoanalyse, III, p. 84-100; G.W., X, p. 210-232; S.E. 14: 117-140.
       available: "Triebe und Triebschicksale (1915)", das Ich und das Es. Metapsychologische Schriften (Psychologie), Fischer Taschenbuch Verlag, 1911-1938, 11. unveränderte Auflage 2005, p.79-102, 1915.
[11]   S. Freud: Mass Psychology, originally published 1922, Penguin Classics; Reissue edition (December 2, 2004).
[12]   S. Freud: Vorlesungen zur Einführung in die Psychoanalyse (1916-1917).
[13]   W. McDougall: The Group Mind: A Sketch of the Principles of Collective Psychology, with some Attempt to Apply them to the Interpretation of National Life and Character, New York and London, G.P. Putman's Sons 1920. Reprinted edition by Kessinger Publishing, pp. Xii-418, 2005.
[14]   M. Minsky, The Society in Mind, Simon and Schuster, New York, 1985.

[15] S. Freud, Das Ich und das Es, Internationaler Psychoanalytischer Verlag, Leipzig, Vienna, and Zurich. English translation, The Ego and the Id, Joan Riviere, 1923.

[16] N. Jennings, K. Sycara, M. Wooldridge: "A roadmap of agent research and development", Journal of Autonomous Agents and Multi-Agent Systems, Vol.1, pp.275-306, 1998.

[17] P. Singh; M. Minsky: "An architecture for combining ways to think". Proceedings Int. Conf. on Knowledge Intensive Multi-Agent Systems, Cambridge, MA.; pp. 669-674, 2003.

[18] J. McCarthy: Concepts of logical AI, paper in progress, Stanford University, available (11/2006): http://www-formal.stanford.edu/jmc/concepts-ai.html.

[19] L. Panait, S. Luke "Cooperative Multi-Agent Learning: The State of the Art", Autonomous Agents and Multi-Agent Systems, Volume 11, Number 3 / November, 2005.

[20] W. Jiao, Z. Shi: "A Dynamic Architecture for Multi-Agent Systems", Proceedinsg of Technology of Object-Oriented Languages and Systems, pp. 253-260, 1999.

[21] M. Wooldridge, "An Introduction to MultiAgent Systems", Wiley, 2002.

[22] T. Yakoh, Y.Anzai: "Physical Ownership and Task Reallocation for Multiple Robots with Heterogeneous Goals", Proceedings of International Conference on Intelligent and Cooperative Information Systems, IEEE, p.80-87, 1993.

[23] Y.Uchimura, K.Ohnishi and T.Yakoh, "A controller design method on decentralized systems, Industrial Electronics Society", 30th Annual Conference of IEEE (IECON-2004), Vol.3 p.2961-2966, 2004.

[24] Y. Labrou, T. Finin,Y. Peng: "The current landscape of Agent Communication Languages", Intelligent Systems, Vol. 14, No. 2, IEEE Computer Society, 1999.

[25] Y.Uchimura, K.Ohnishi and T.Yakoh, "Bilateral robot system on the real time network structure", IEEE International Workshop on Advanced Motion Control (AMC-2002), p.63-68, 2002.

[26] Lueg, R. Pfeifer: "Cognition, Situatedness and Situated Design", Proc. 2nd International Conference on Cognitive Technology, p.124-135, 1997.

[27] M. Solms, O. Turnbull: The Brain and the Inner World, Karnac/Other Press, Cathy Miller Foreign Rights Agency, London, England, 2002.

[28] R. Pfeifer, C. Schneider: "Implications of embodiment for robot learning", Proceedings. of Second EUROMICRO workshop on Advanced Mobile Robots, pp. 38-43, 1997.

[29] W. Jiao, M. Zhou, Q. Wang: "Formal Framework for Adaptive Multi-Agent System", Proceedings of International Conference on Intelligent Agent Technology, pp.442-445 , 2003

[30] P. Palensky, B. Lorenz, A. Clarici: "Cognitive and affective automation: Machines using the psychoanalytical model of the human mind", Proceedings of 1st International Engeenering and Neuro-Psychoanalysis Forum, p. 49-73, 2007.

[31] R. Picard: "Affective Computing", MIT, 1997.

[32] T. Deutsch; T. Zia; R. Lang; H. Zeilinger: "A Simulation Platform for Cognitive Agents", *Proceedings of 2008 IEEE International Conference of Industrial Informatics*, 2008, to be published.

[33] T. Deutsch; H. Zeilinger; R. Lang: "Simulation Results for the ARS-PA Model", *Proceedings of 2007 IEEE International Conference of Industrial Informatics*, pp. 1021-1026, 2008.

2.6 Artificial Group Psychodynamics: Emergence of the Collective

Joseph Dodds

Research into computer simulation of group and cultural processes has expanded in recent years [1], including an important recent attempt to incorporate neuropsychoanalytic principles [4]. This paper argues that in order to progress we need to start "taking the group seriously" [13] and utilize psychoanalytic theories of group-level processes. Furthermore, those currently using such psychoanalytic perspectives in a variety of contexts have a lot to gain from computer modelling. This paper aims to elucidate the key elements of three foundational psychoanalytic theories of group dynamics, those of Freud [5], Bion [7] and Jaques [11], with the goal of facilitating future computer-based implementation, and ultimately the formation of a new research field of artificial group psychodynamics.

2.6.1   Introduction

Chao and Rato [4] recently attempted to improve on previous Axelrodian computer simulations of group and cultural processes [1] by implementing an agent-based neuropsychoanalytic model utilizing id-based *homeostatic* mechanisms and Freud's concept of *narcissism of minor differences* [6]. While this approach is highly promising this paper suggests social models based purely on the individual mind-brain without recourse to psychoanalytic theories of groups are necessarily limited. Psychoanalytic social models start from Freud's *Group Psychology and the Analysis of the Ego* [5] and have been developed further by Bion [7], Foulkes [9], Jaques [11], Menzies-Lyth [12], Hinshelwood [14], Young [19], Gould [15], Stacey [16], Dalal [13], Laurence [46] and others in the form of group analysis [37], group analytic psychotherapy, social psychoanalysis, systems psychodynamics, group relations [17] and a range of psychoanalytic social and cultural criticism [71]. This paper intends to elucidate important principles of psychoanalytic group dynamics for the benefit of the ENF community. Developing more formal computer models of these principles will be the subject of a future paper.

2.6.2   Benefits of Such a Project

The potential benefits are substantial. Computer simulations could offer a powerful new research tool for those working in the various areas described above and can act as "philosophical thought experiments" [8], enabling us to test the parameters of our theories in a wide variety of situations, giving us the chance to observe how they 'work' and to follow the emergence of potentially unexpected outcomes. They also help show gaps and errors in our theories, and as with other aspects of the ENF project, the process can help psychoanalysis to gain greater conceptual clarity. Computer scientists in their turn can gain access to the highly detailed and complex conceptual and theoretical work of psychoanalysis, and especially its emphasis on affective and unconscious psychodynamic processes often ignored in other psychologies, and can thus help overcome some of the problems and deadlocks in previous attempts at computer simulations of social processes.

Certain trends in group analysis towards integrating advances in complexity and systems theories [15], [25], [31] may facilitate such a task. Current research into computer and robotic uses of swarm intelligence [24], self-organization [23] and complexity theory [22] could be harnessed towards the emergence of a new field of artificial group psychodynamics. Such research could be used to test group, individual and brain level theories, each level reciprocally supporting the others.

Artificial group psychodynamics could potentially have utility far beyond the scope this paper can study. It is hoped that the simulations would be flexible enough to work in a wide variety of contexts such as aiding the psychoanalytically informed organizational consultant [25], constructing intelligent buildings, mapping goals in conflict resolution, studying cultural dissemination [1], racism [61], revolution, totalitarianism [59], [60], war [41], migration, globalization, terrorism and fundamentalism [18]. Time will tell whether it is an approach worth pursuing.

2.6.3   Brain, Mind or Group?

The ENF 2007 asked *should we model a brain or should we model a mind?* This paper suggests a third option, *should we model the group?* Freud argued that "individual psychology cannot be isolated from group psychology, not simply because one of the functions of the mind is to form relationships with objects, but because the individual's relationship to the object is an integral part of the mind itself. A mind without links to objects is simply not a human mind" [5]. Here it is important to understand 'object' in the psychoanalytic sense, implying a relationship and usually a person. Modern psychoanalysis, especially as developed by the *object relations* school, building on the advances of Klein [68], [69], Fairbairn [67], Winnicott [70], Bion [66] and others, has developed Freud's initial ideas into highly complex theories of internal and external object relations. Internal objects [52] are mental representations of external objects taken into the mind through psychological processes such as introjection, incorporation and internalization, and combined with self-representations including the self-in-relation. The mind is therefore built up, bit by bit, through complex processes of social interaction.

Thus even for the individual, "brain" or "mind" is not enough without "group" and, as Freud wrote, "from the very first individual psychology [...] is at the same time social psychology as well" [5]. One aim of this paper therefore is to recognize the importance of "taking the group seriously" [13] in any psychoanalytic model, not only group models, especially in more relational schools [34], [35] and ecological models of mind and society [28]. Chao and Rato's [4] use of the agent-based modelling system Repast in their research is therefore appropriate as it is designed it "to move beyond the representation of agents as discrete, self-contained entities in favour of a view of social actors as permeable, interleaved, and mutually defining; with cascading, recombinant motives" and to model "belief systems, agents, organizations, and institutions as recursive social constructions" [51].

2.6.4   Emergence and Non-Linear Group Dynamics

Potentially interesting issues in this field are the related phenomena of *emergence* [22], *self-organization* [23], *swarm intelligence* [24] and *artificial life* [27]. These ideas have been used in neuropsychoanalysis [36], psychoanalysis [29], [30], [32] and group analysis [16]. Would simulating the ENF model at the individual agent level be enough to allow for emergence of complex dynamics at the collective level, or would certain group-level phenomena need to be implemented at the design stage? Psychoanalytic understandings of groups are essential in either case as knowing expected outcomes at the collective level helps to fine-tune the model at the individual agent level, and guides the search for potential emergent dynamics. Specific hypotheses for the same initial conditions could then be made by competing psychoanalytic theories and tested to see which best describes the emerging group dynamics. This helps to reciprocally refine both group and individual level models. In addition, complexity theory is ideal for the ENF project of trying to articulate the connections between the complex interacting systems at a range of levels: neurochemical, neuroanatomical, brain, mind, group, society. It is thus invaluable for attempts to connect psychoanalysis, neuroscience, artificial intelligence and group analysis. For recent attempts to combine complexity theory and group analytic perspectives see [15], [16], [25], [58] and section 2.6.8 below.

2.6.5   Psychoanalytic Theories of Groups and Society

Psychoanalytic theories of group and social processes start with Freud's key works on groups and culture [3], [5], [6], [21] but there are many other theories in this area which further research in this area could investigate. For example the Critical Theory of the Frankfurt School [53], [54], [55], [56], the substantial work in cultural studies drawing on the French psychoanalyst Jacques Lacan (e.g. [62]) or the important ideas of the co-founder of group analysis, Foulkes [9], whose theories such as the group's *dynamic matrix* or transferences in the group forming *"continuously re-integrating networks"* [9] are all highly suggestive for artificial group psychodynamics but will not be explored here but be the focus of a future study.

This paper focuses on Freud's contributions and goes on to its extensions in the Kleinian theories of Wilfred Bion [7] and Elliot Jaques [11]. Freud saw groups as an extension of the Oedipal situation of the family (the first group we are a part of). The work of Bion and Jaques extends the Freudian model, and uncovers certain areas it neglected, in particular the more primitive, psychotic aspects of groups and the importance of phantasy and anxiety as studied by Melanie Klein [48]. It is important to stress at this point that just as Freud's psychoanalysis is not only a form of therapy but also a method of studying the mind, so with the psychoanalysis of groups, which have been applied far beyond the clinic. For example in the Tavistock Institute [31], group relations conferences, the Institute of Group Analysis [37], in dialogue with open systems and complexity theories [16], and in journals such as *Free Associations* [39], *Psychoanalysis, Culture, Society* [38], and *Organizational and Social Dynamics: An International Journal for the Integration of Psychoanalytic, Systemic and Group Relations Perspectives* [25].

Computer simulations of these approaches need to take into account complex projective and introjective processes. There need to be ways of putting part of an agent into another, of redistributing internal objects among and between groups (eg. in situations of identification with the leader and corresponding introjections, or in racism where "bad" parts of the group are split off and projected onto a denigrated outgroup.) These processes are very powerful and volatile and should be able to overcome the problem of homogenization found in Axelrodian models [2].

2.6.6 Model 1: Freud's Group Psychology

Key Concepts

(1) *"Groups are bound together by libinidal ties"* [5].

(2) *Emotionality and deindividuation:* Individuals in groups tend "to surrender themselves so unreservedly to their passions and thus to become merged in the group and to lose the sense of the limits of their individuality" [5].

(3) *Leader/father:* In the minds of members the leader "loves all individuals in the group with an equal love. Everything depends upon this delusion; if it were to be dropped then both church and army would dissolve, so far as the external force permitted them to". In the "artificial group" of the army the "Commander in Chief is a father who loves all soldiers equally, and for that reason they are comrades among themselves" [5]. The Church is similarly structured ("Our Father") [26].

(4) *Directionality and volatility of libidinal ties:* "each individual is bound by libidinal ties on the one hand to the leader [...] and on the other hand to the other members of the group" [5]. Libidinal ties holding the group together can dissolve quickly.

(5) Groups have *ambivalent* feelings of love and hate.

(6) Hostile impulses within the group are forbidden and instead *projected* as hatred of an outgroup deflecting hate from inside the group and each individual.

(7) *Love and hate as factors of cohesion:* Love binds in unity while *projection* of intra-group hostility removes it from the group and strengthens group identifications. Each member shares not only good objects (leader, ideals) but also bad objects. These processes can also be viewed with Matte-Blanco's *bi-logic* [13], [50].

(8) *Group narcissism and altruism:* Individual narcissistic interests can be partially reliquished for the good of the group, partly as the whole group is invested with narcissism [5].

(9) *Introjection and identification*: "the earliest expression of an emotional tie with another". The Oedipal boy "would like to grow like [father] and be like him, and take his place everywhere [...] he takes his father as his ideal." Later father's rules are introjected in the form of the Super-Ego and new identifications can arise "with any new perception of a common quality shared with some other person" [5].

(10) *Group members take the leader (or god, idea) as their ego ideal* (Super-Ego subsystem [64],[65]) through *projection* and also *introject* the leader in a two-way process (Fig. 2.6.1). This can lead to ego-impoverishment, loss of individuality, increased risk of being taken over in group processes, reduced ability for thought.

(11) *Group members are identified with one another in their ego due to sharing the same ego ideal*. They therefore love each other as themselves (narcissistically).

(12) *"Narcissism of minor differences"*: Small differences between groups can be magnified, another factor against homogenization [4]. Neighbouring often highly similar groups often have terrible conflicts (English/ French, Freudians/Jungians.)

(13) *Ambivalence towards the leader*: Freud draws on research in ethnography where the King is killed at the end of his reign [3], on religion where the father imago is split into god and devil [57] and the Oedipus complex where the boy wants to be like his father and obey him, but also to take his place and kill him.

(14) *Revival of the primal horde*: Freud hypothesized that civilization began with a primal horde where the brothers ganged together to kill and devour the tyrannical primal father who had kept all women to himself. Then, filled with remorse and guilt, the father became worshipped as a totem ancestor and rules instigated so none could enjoy the position he had vacated, thus instituting the first rules at the origin of culture. Freud believed group life reactivated these ancient conflicts [3].

(15) *Destabilization by (unregulated) love (bypassing the leader)*, especially "asocial" sexual love. Hence the many strict regulations/taboos around sexuality (and marriage needing approval of State and heavenly "fathers".)

(16) *Destabilization by neurosis*: "neurosis has the same disintegrating effect upon the group as being in love" [5]. Freud saw the ultimate origin of our neuroses in the harsh restrictions of civilization [6]. In this model agents would go "neurotic" and abandon group life as a result of internal and external stressors.

(17) *Destabilization by hate*: The group is also in danger of being torn apart by its own aggression, not all of which will be able to be externalized onto an enemy.

(18) *Panic*: Freud describes the situation in war where the leader (ego ideal) is killed. The group then moves into a state of disintegration, a group of disconnected individuals who have "lost their head".

(19) *Multiple identifications blur simplistic ingroup/outgroup distinctions.* Freud claimed multiple identifications and overlapping group ties help reduce inter-group conflict and war, as divisions become less sharp [41].

(20) *Oceanic feeling*: "Mystical" feelings of merger can occur when the ego feels it has expanded to include the whole group, recalling feelings of "at-oneness" with mother before the difficult and painful process of individuation. Promises of overcoming feelings of aloneness, powerlessness and alienation are powerfully seductive [55]. Freud saw omnipotence as a defence against helplessness, and submission to a group as a way to escape the anxiety and struggle of life [6].

Summary of Freud's Group Psychology

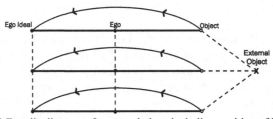

Fig. 2.6.1 Freud's diagram of group relations including position of leader [5]

(1) Group dynamics follow the model of our first group (the family).
(2) Oedipal family dynamics dominate group life.
(3) Ambivalent affects of love and hate are crucial in group dynamics.
(4) Projection, introjection, identification, narcissism.
(5) The role of the leader as the groups ego ideal.
(6) The role of the outgroup (and outgroup within).
(7) Oceanic feeling, loss of individuality/boundaries, group narcissism
(8) Group destabilized by panic, aggression, neurosis, ambivalence to leader, unregulated love, loss of flexibility, cross-group identifications

How to Model Freud's Group Psychology?

Love and hate must be in the system. Love helps bind individuals into groups, hate can break them apart, or strengthen them through mutual hatred of an outgroup. In the model when a certain number of agents come together, the mutually reinforcing ties of "love" which are built up help in group formation, but an outlet for their aggression must also be found for the group not to be torn apart. Group formation should be facilitated by choosing a leader who would play a key role in helping focus the aggression outwards, away both from the leader and other members of the group. This may be modelled by chance interactions between agents resulting in increased binding in libidinal ties, strengthened through repeated contacts in processes of *self-organization* and *clustering.* This would be an unorganized group [21]. At some point, an agent finding itself as a central node in the emerging network, and thus able to bind more of the group to itself, would cross a critical threshold of connectivity and become the "leader". This would result in a phase transition and a major shift in group cohesiveness as individual agents become strongly influenced (but not completely determined) by the leaders actions. The model should also include possibilities, through further self-organization, of alternative "leaders" appearing through libidinal ties being forged outside the individual-leader relation, potentially leading to schisms. Such transitions would probably occur through non-linear processes not directly predictable from the initial set-up. Finally, there should be the possibility of *leaderless groups* [5].

How would projection, introjection and identification be modelled? In the case of the leader-group relationship (Fig. 2.6.1), this could occur through the individual Super-Ego modules becoming *synchronized* with the leader. This should provide tangible benefit, e.g. the reduction in use of energy or levels of anxiety, or an increase in pleasure as the ego feels it has expanded to include the whole group in an "oceanic feeling" [6]. There thus needs to be some initial differences between individual Super-Ego modules, which uniting with a leader removes (or temporarily diminishes) as individuals merge to form a relatively cohesive group. There should also be disadvantages. For although the group becomes more cohesive, reduces anxiety and energy needs, there may be a certain loss of *ecological flexibility* [28], a reduction in the ability to adapt to changes in the environment and to interactions with other groups.

The model should be able to simulate aggression towards an outgroup (war) or an outgroup within (racism). Thus a certain amount of aggression or anxiety which threatens to generate homeostatic difficulties at individual and group levels needs to be projected onto target groups, helping group regulation by displacing intra-group and intra-agent conflict to inter-group conflict. Conflict needs to be modelled inside agents as well as between groups, e.g. between narcissistic drives and the requirements of group living [6], between Id and Super-Ego modules or between neuropsychoanalytic motivational systems such as Panksepp's PANIC, LUST or RAGE systems (social attachment vs. sexual/aggressive drives) [40]. Should homeostatic regulation fail and the conflict tension surpass a given threshold, agents may become "neurotic", breaking group ties or engaging in self-destructive behaviour. As the proportion of such agents increases there will be group-level consequences, possibly including the formation of a marginalized scapegoat group.

In terms of the various threats to group cohesion, some statistical probabilities could be used to determine the likelihood of each outcome. Threats may also lead to a strengthening of the group as individuals seek to bind themselves to the group more fully and give up more autonomy to defend against anxiety (a feature an expanded model using the ideas of Bion [7], [20] and Jaques [11] would emphasize).

Freud's model of group psychology could be implemented in the manifold contexts described above. For example, research on overcoming racism and war might be modelled by allowing for multiple identifications and overlapping group formations which are less dichotomous. Such identifications strengthen the ability of *Eros* (life drive) to bind [6] which as Freud stated in *Why War?* [41], is the main force we have to work against the destructive and seductive power of war and *Thanatos* (death drive). Finally, in terms of the impasse in Axelrod's cultural dissemination model [4], [2], the Freudian model should prevent ending up with purely homogenous cultures as there will always be constant pressures both within and without a "homogenous" group towards change. Overall Freud's theory is fundamentally one of conflict [6].

2.6.7 Melanie Klein: PS↔D and Projective Identification (PI)

Many commentators have pointed out the limitations to Freud's understanding of group dynamics [7], [13], [17]. It is not the place to go into these here. Whatever its faults, it forms the basis in one way or another of all later psychoanalytic research in this area, and is a necessary starting point in any psychoanalytic agent-based simulation of group processes. Artificial group psychodynamics offers the possibility to test alternative models or to put them together in different combinations. Later developments in the psychoanalysis of groups by Bion [7], [20] and Jaques [11], extend Freud's model with ideas derived from Melanie Klein [48]. Klein focused on much earlier mother-infant relationships than Freud and more "primitive" mental processes. She was a founder of *object relations theory* (along with Fairbairn, Balint and Winnicott) [35], which moved the focus away from *drives* and towards *relationships* (with internal and external *objects*), though in contrast to other object relations theorists Klein never lost sight of the importance of drives. Klein described an internal world populated by internal objects in constant complex relations with each other and with external objects. She also argued for the importance of phantasy, anxiety and envy. Artificial group dynamics might be useful in modelling the Kleinian inner world with its complex object relations (see Fig. 2.6.2).

For Bion and Jaques understanding of group dynamics it is Klein's concepts of the *paranoid-schizoid (PS)* and *depressive positions (D)* which are perhaps the most crucial. These are not the same as clinical schizophrenia or depression, or developmental stages (although PS is seen to precede D), as both can be returned to throughout life. Rather they are complex psychological/affective/self/object organizations with characteristic defences and anxieties (see appendix for more on PS and D). PS is characterized by *part-object relationships* (eg. the infant relates to the breast rather than the whole mother, who is not recognized as a separate complete person), paranoid, *persecutory anxieties* and fear of disintegration, and is dominated by the defences of *splitting* (eg. into good/bad breast) and *projective identification* (PI). The latter differs from projection as it involves projecting *into* rather that *onto* the object, subtly pressuring the object to behave in ways conforming with the projection. In PI, parts of the self are felt to reside *inside* the object and can include claustrophobic anxieties of being trapped, phantasies of controlling the object from within, or of evacuating the deadly poison inside. Bion and Jaques see PI as crucial to group dynamics. Robert Young has even claimed that PI "is the most fruitful psychoanalytic concept since the discovery of the unconscious" [42].

D is characterized by *whole-object relationships* and the object is seen to contain both good and bad aspects. This conjunction, though painful, is tolerated. The object is now separate and whole, mourning is now possible (for loss of unity and for the realization that phantasized attacks on the "bad" object also attacked the "good" object as they were in reality the same thing), an urge to repair the imagined damage emerges and the self becomes more integrated. Defences against depressive anxiety are developmentally more advanced and include manic defences. Bion emphasized the volatility of these positions and the way one can move rapidly between them, especially in groups, by putting a double headed arrow between them (PS↔D) [72]. Using complexity theory, PS and D can be seen as *attractors* helping to organize psychic life and can be both normal and pathological. For more information on PS and D see the appendix. These various positions and psychoanalytic configurations can be depicted spatially (Fig. 2.6.2) and computer models of these may be useful to the clinician.

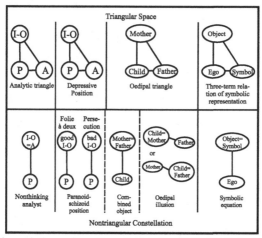

Fig. 2.6.2 Psychoanalytic configurations depicted spatially. P=Patient's experience of self, I-O=Patient's experience of inner object, A=Analyst as object outside patient's inner world. Modified from [45].

2.6.8   Model 2: Bion's Group Analysis

The Binocular Vision

A founder of *group analysis* (with Foulkes [9]), Bion combined ideas from Lewin's field theory of groups with Kleinian object relations. According to Glover, Bion's "account was to offer new ways of bringing psychoanalytic theory to bear on issues outside the consulting room, to illuminate the wider social, political and cultural domain" [43] and is thus of great interest to artificial group dynamics, which can draw on recent attempts to relate Bion's concepts to complexity theory [58] and neuroscience [33]. In *Experiences in Groups* [7] Bion approached the *group-as-a-whole* arguing "No individual however isolated in time and space should be regarded as outside a group or lacking in active manifestations of group psychology", stressing the importance of groups to understanding individuals and echoing Freud's earlier comment on the inseparability of individual and social psychology [5].

Bion claims group and individual psychoanalysis deal with "different facets of the same phenomena", thus providing a "binocular vision" [7]. Bion observed two main trends in groups, the Oedipal dynamics described by Freud [5], and below this powerful, more psychotic aspects involving PI and PS↔D oscillations. These latter dynamics also operate in individuals, but groups allow a "magnified" observation as the group "approximates too closely [...] very primitive phantasies concerning the contents of the mother's body [7]." The more disturbed the group the less understandable it is in terms of Oedipal family patterns. In fact "contact with the emotional life of the group" is "as formidable to the adult as the relationship with the breast is for the infant" [7]. As a result in groups adults tend to resort "to mechanisms described by Klein as typical of the earliest phases of mental life" [7].

Projective Identification and Container-Contained

Bion further developed Klein's concept of Projective Identification (PI), describing the situation where "the analyst feels he is being manipulated into playing a part [...] in somebody else's phantasy" and of the need "to shake one's self out of the numbing feeling of reality" [7]. This is a powerful way to understand how we are pulled into social phantasy systems. Following Bion, Ogden [44] divided PI into 3 stages. First, a phantasy of projecting "part of oneself into another person and of that part taking over the person from within". Second, "pressure is exerted via the interpersonal interaction" so the recipient "experiences pressure to think, feel and behave in a manner congruent with the projection". Finally, the projected feelings are "psychologically processed by the recipient" and "reinternalized by the projector." PI underlies many aspects of psychic functioning, healthy and pathological [42].

These processes are key in intra- and inter-personal interaction and group dynamics. In a healthy *container-contained* ($♀↔♂$) dialectic (e.g. mother-child or analyst-patient) there can be a positive development of this process leading to psychological growth, as the containing and processing *alpha-function* of the container (e.g. analyst, mother) becomes internalized by the projector. The link between ♀↔♂ can be commensal, symbiotic or parasitic [73] and Bion's theory of thinking itself is of a productive union between internal ♀♂ [72]. The ability to have a healthy and creative internal ♀♂ coupling is an achievement connected with D and successful negotiations of the Oedipus complex. In pathological situations, exemplified in destructive group dynamics, both sides become locked into pathological PI-circuits, each projecting and reprojecting split-off aspects of themselves, and unprocessed fragments of experience (*beta elements*) that had been projected into them by the Other. Bion's theories helps us to understand how in these contexts thinking itself is attacked, the ability to think, not just specific thoughts, in a way that resembles schizophrenic *attacks on linking* [74].

Key to "defusing" such destructive group processes is for each side to attempt to take back their projections, to realize the extent to which the "bad" aspects reside in their own self and group, and to which the Others' projections are accurate. This helps *metabolize* or convert *beta-elements* into *alpha-elements* instead of violently (re)projecting them and requires a certain amount of "containing space" in order to manage the difficult and painful feelings that arise. This requires reaching D-level functioning which under intense internal and external pressure may be too difficult to maintain without powerful social containers, as tiny disturbances can result in a massive retreat to PS functioning, and a reigniting of the processes of splitting and pathological PI.

Work Group and the Basic Assumption (BA) Groups

Fig. 2.6.3 Work Group and Basic Assumption Group

One of Bion's most important contributions to understanding groups was his description of the characteristic configurations which they form. He first divided this into two: W*ork group* functioning is the conscious purpose of the group, its "official", reality-based ego-function. This is easily disturbed by more unconscious, primitive behaviour and thought, which Bion calls *basic assumption (BA) groups* (Fig. 2.6.3). These are not different groups, but different aspects of group *functioning,* all groups always contain a mixture of both. Pure Work Groups are impossible and would be emotionally sterile, while pure BA groups would not have enough reality-based functioning to survive for long and would be the group equivalent of severe psychosis. These groups also unite individuals together in different ways: *cooperation* (work group) and *valency* (BA). The latter "requires no training, experience, or mental development. It is instantaneous, inevitable, and instinctive". In addition, "time plays no part in it" and "activities that require an awareness of time [...] tend to arouse feelings of persecution". Finally, "stimuli to development meet with a hostile response" [7].

Connected with BA groups is what Bion called *group mentality*, the "unanimous expression of the will of the group, contributed to by the individual in ways of which he is unaware, influencing him disagreeably whenever he thinks or behaves in a manner at variance with the BA [...] It is thus a machinery of intercommunication" and a foundation "for a successful system of evasion and denial" [7].

The Three BA's: BAD, BAP, BAFF

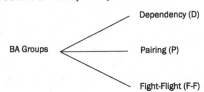

Fig. 2.6.4 BA Groups: Dependency, Pairing, Fight-Flight

Bion identified three main BA constellations which "give form and meaning to the complex and chaotic emotional state that the group unfolds to the investigating prticipant" [7]: Fight-Flight (BAFF, Dependency (BAD) and Pairing (BAP (Fig. 2.6.4). BA's can shift rapidly from one to another or remain long in one BA. Further BA groups have been suggested (e.g. Hopper's Incohesion: *Aggregation/Massification*. BA I:A/M [45]) and artificial group dynamics should also explore these developments. Each BA is a complex "attractor" definable along six dimensions [10]: 1) Sources of anxiety, 2) Affects, 3) Object relations, 4) Major defences, 5) Secondary defences, 6) Adaptive/sophisticated uses [10] (see appendix for details).

Of special interest to neuropsychoanalysis is Bion's postulate that the "valency" of BA groups acts via what he called the "proto-mental system" in which the "physical and psychological or mental are undifferentiated. It is a matrix from which spring the phenomena", a system he designated as *pm* [7]. Bion even foresaw the possibility that neuroscience might one day penetrate this level of group reality: "If, by using a physical approach, we can investigate the physical aspect of the proto-mental system, we may find a way of sampling what the proto-mental system of a group contains at any given time, and from that make the further step that would consist in elaborating a technique for observing the proto-mental counterparts of mental events. Any developments of this nature would make it possible to estimate what the psychological state of a group would be likely to become, because we could investigate it long before it emerged as a basic assumption basically expressed" [7].

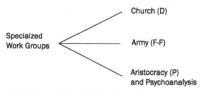

Fig. 2.6.5 Specialized Work Groups

Specialized Work Groups

Taking Freud's examples of Army and Church, Bion claimed that each BA can be to some extent "neutralized" by specialized work groups "budded off" from the main group (Fig. 2.6.5). For example church (BAD), army (BAFF) and aristocracy (BAP). Others have proposed psychoanalysis as an example of a P group [21].

The Leader

The group is "felt as one fragmented individual with another, hidden, in attendance", the leader, who is "the leader is as much the creature of the basic assumption as any other member of the group, and this [...] is to be expected if we envisage identification of the individual with the leader as depending not on introjection alone but on a simultaneous process of projective identification." In fact the leaders personality "renders him peculiarly susceptible to the obliteration of individuality by the basic-assumption group's leadership requirements [...] the leader has no greater freedom to be himself than any other member of the group" [7] (e.g. in BAFF if there is no obvious enemy a paranoid leader may be chosen or a group member is driven into a paranoid state by through collective projective identification of group members).

Complexity at the "Edge" of the BA Group

Stacey [58] recently attempted to articulate Bion's group analysis with complexity theory and asks what would happen "if we move from a membrane metaphor of an organization's boundary to a fractal metaphor in which it is problematic to say what is inside and what is outside"? Self-organization theory focuses "away from regulation at a boundary to the manner in which the system's transformation process transforms itself", "from rational design and regulation to spontaneous self-organizing processes" and thus "the creative potential of disorder" [58]. Stacey argues Work↔BA group interactions create regions of stability and disintegration with potentially creative fractal regions of *bounded instability* at the *edge of chaos* between them [58] (see also Winnicott's *transitional objects* and *potential space* [63].) Computer simulations of such phenomena create unique opportunities to develop these ideas further into "a radically social understanding of individuals" [16].

Summary of Bion's Group Analysis Model

Fig. 2.6.6 Bion's Group Analysis Model

(1) Projective identification (PI), PI-circuits and PS$\leftrightarrow$D
(2) Container$\leftrightarrow$contained. Alpha function$\leftrightarrow$beta elements
(3) Work group$\leftrightarrow$BA group (BAD, BAP, BAFF) and specialized work groups
(4) Leadership in BA groups

2.6.9 Model 3. Social Phantasy Systems as a Defence Against Persecutory and Depressive Anxiety

Key Concepts from the Jaques/Menzies-Lyth Approach

(1) *Individuals make unconscious use of institutions as defence against anxiety.*

(2) *Social phantasy systems (transpersonal defence mechanisms) are constructed to reduce anxiety (PS$\leftrightarrow$D).* In Menzies-Lyth's study of a nursing home these defences operated "structurally and culturally as a kind of depersonalization or elimination of individual distinctiveness" [12] including in part-object functioning (patient as number/illness/body part), effect of uniform/hierarchy/role/routine etc.

(3) *Formation of social phantasy systems:* "When external objects are shared with others and used in common for purposes of projection, phantasy social relationships may be established through projective identification with the common object" [11]. This is implicit in Freud's model (model 1 above).

(4) *Groups dynamics involve interplay of projective/introjective identification*

(5) *Social defences emerge through self-organization of individual defences* with a "reciprocal relationship between social and internal defence mechanisms" [11], continually effecting each other, the emergent global behaviour interacting back to lower levels in a complex, non-linear reiterative process.

(6) *Societies provide institutionalized roles:* Occupants are sanctioned/required to take into themselves projected objects of other members. Over-rigidification of roles may lead to system instability and loss of ecological flexibility [22], [28].

(7) *Manifest/phantasy level functions:* Institutions are "determined and coloured not only by their explicit or consciously agreed and accepted func-

tions, but also manifold unrecognised functions at the phantasy level" [11], corresponding to Bion's work/BA group.

(8) *Social redistribution of internal objects:* Defences against PS anxiety involve putting bad internal objects into particular group members who *absorb* or *deflect* them: "objective fear may be more readily coped with than phantasy persecution. The bad sadistic enemy is fought against, not in the solitary isolation of the unconscious inner world, but in co-operation with comrades-in-arms in real life" [11].

(9) *Scapegoating:* The community is split into good majority/bad minority "consistent with splitting internal objects into good and bad [...] The persecuting groups belief in their own goodness is preserved by heaping contempt upon [...] attacking the scapegoated group". Splitting mechanisms are "reinforced by introjective identification [...] with other members taking part in group-sanctioned attack" [11].

(10) *Unconscious collusion:* Persecuted groups can at the phantasy level seek contempt and suffering to alleviate unconscious guilt, reinforcing defences against depressive anxiety and reinforcing denial to protect internal good objects.

(11) *Identification with the aggressor* [49]: "the phantasy of actually taking the aggressor inside the self in an attempt to control them, then feeling controlled by them and needing to get rid of other, threatened and vulnerable parts of the self into someone else (the new victim)" [75].

(12) *Dynamics of social change* requires restructuring *phantasy* as well as *manifest levels.* Apparent change at the manifest level may conceal lack of change at the phantasy level. Imposed social change provides less opportunity for restructuring of defences. Effective social change may require analysis of common anxieties and unconscious collusions underlying defences of the social phantasy system.

(13) *Social phantasy systems originate* "through *collusive* interaction between individuals to project and reify relevant elements of their psychic defense systems" but for each new individual entering the institution, the social phantasy system is felt as a concrete external object to which they must "react and adapt" [12].

Summary of the Social Phantasy Systems Model

(1) Social phantasy systems transpersonal defences against anxiety (PS↔D)
(2) Individual↔social defence mechanisms
(3) Introjective↔projective identification
(4) Institutionalized roles, hierarchies, routines
(5) Social redistribution of internal objects. Absorption, deflection, reprojection
(6) Manifest↔phantasy social levels and dynamics of social change

2.6.10   Conclusion

It is hoped that the ideas in this preliminary paper, including the three foundational models of Freud, Bion and Jaques, will provide fruitful work in terms of
implementing computer-based psychoanalytic models of groups and social systems, including attempts to integrate these perspectives with complexity theory.
These models, along with others, can be refined, developed, and tested to enable
us to reach a more comprehensive understanding of collective phenomena and allow for the emergence of a new field of artificial group psychodynamics.

Appendix

Correspondences between Bion's Basic Assumption Theory and Klein's developmental positions, modified from L.J. Gould [10].

	Sources of Anxiety	Affects	Object Relations	Major Defences	Minor Defences	Sophisticated /Adaptive Uses
PS	Fear of persecution, fear of destruction of ideal object/self	Anxiety, dread, primitive terror	Part-object and split-object orientation e.g. good/bad breast	Splitting of impulses and objects into positive and negative aspects	Denial, idealization. Projective Identification	Ordering experience, healthy suspicion, forming ideals, ability to act.
BAFF	Fear of persecution by powerful enemies	Anxiety, fear, terror, panic	In/out group mentality. Group members undifferentiated.	Splitting off and projecting outward intragroup anxiety/ aggression	Idealization of leader, denial of aggression	Realistic action, sensitive to danger, loyalty, commitment
D	Recognize dependency, fear aggression may destroy caretaker/cause retaliation	Guilt, despair, depression, envy, greed	Whole object awareness accompanied by dependency on the object	Reparation, manic denial of aggressive impulses which are turned inward	Sublimation Displacement. Inhibition. Repression. Splitting	Impulse control, symbol formation, creative capacity, linking internal states and external behaviour
BAD	Dependency on leader, fear of retaliation or abandonment	Helplessness, emptiness, depression, childishness	Dependent/submissive/hierarchical relation with authority/leader. Leveling of peers	Denial/repression of aggressive impulses toward leader. Idealization of leader	Splitting (idealization of group "believers" and hatred/scapegoating "non-believers"	Appropriate submission to and learning from authority, gratitude, discriminating followership
Oedipus Complex	Fear of exclusion, fear of retaliation	Jealousy, exclusion, loneliness, deprivation	Dyadic/triangular relations. Rivalry/ competition with one parent for favor of other)	Identification with aggressor. Phantasy of combined parent (e.g. phallic mother)	Regression to earlier defences (eg. splitting: idealized/ denigrated parents)	Capacity for passion, mature sexuality, romantic love, reproductive desire
BAP	Recognition of separateness, fear of exclusion	Libidinal excitement, vicarious pleasure, hope	Mobilization/maintenance/ vicarious engagement with the pair. Competition to be favorite child	Idealization/ preservation of pair to sustain hope. Repression of rivalry with pair members	Identification with the pair. Denial of despair. Repression/denial of own sexuality	Recognise pair as source of creativity/ renewal/ change. Realistic future orientation

References

[1]   R. Axelrod, "The Dissemination of Culture", in Journal of Conflict Resolution. Volume 41, Issue 2, April 1997, pp203-206
[2]   Flache and M.W. Macy. "What sustains stable cultural diversity and what undermines it? Axelrod and beyond" http://arxiv.org/ftp/physics/papers/0604/0604201.pdf
[3]   S. Freud, *Totem and Taboo*. Penguin, Great Britain, 1940
[4]   Chao and V. Rato, "A Psychonalitical model for Social Influence in Axelrod Culture Model", personal communication
[5]   S. Freud, "Group Psychology and the Analysis of the Ego (1921)", in S. Freud *Civilization, Society and Religion*. Penguin, London, 1991, pp91-178
[6]   S. Freud, "Civilization and Its Discontents", in S. Freud *Civilization, Society and Religion*. Penguin, London, 1991, pp243-340
[7]   W. Bion, *Experiences In Groups*, Routledge, London, 1990
[8]   D. Dennet, *Freedom Evolves*, Viking, London, 2003
[9]   S. H. Foulkes, *Selected Papers of S.H. Foulkes: Psychoanalysis and Group Analysis*. Edited by Elizabeth Foulkes. Karnac, London, 1990
[10] L. J. Gould, "Correspondences Between Bion's Basic Assumption Theory and Klein's Developmental Positions: An Outline," in *Free Associations* no. 41, July 1997: http://www.human-nature.com/free-associations/bion.html
[11] E. Jaques "Social Systems as a Defence against Persecutory and Depressive Anxiety", in Klein, Melanie *et al. New Directions in Psychoanalysis*. Karnac: Maresfield Reprints, 1977, pp. 478-98
[12] Menzies-Lyth, *Containing Anxiety in Institutions: Selected Essays, vol. 1*. Free Association Books, 1988, pp. 43-88
[13] F. Dalal, *Taking the Group Seriously. Towards a Post-Foulkesian Group Analytic Theory*. Jessica Kingsley, London, 1998
[14] R.D. Hinshelwood: *What Happens in Groups: Psychoanalysis, the individual and the community*. Free Association Books London, 1987
[15] L. Gould, L. Stapley, M. Stein (eds.) *The Systems Psychodynamics of Organizations: Integrating Group Relations, Psychoanalytic, and Open Systems Perspectives*. Karnac, London, 2006
[16] R. Stacey, *Complexity and Group Processes: A Radically Social Understanding of Individuals*, Brunner-Routledge, 2003
[17] D. Armstrong, W.G. Lawrence, R. Young, *Group Relations: An Introduction*. Process Press, London, 1997: http://human-nature.com/rmyoung/papers/paper99.html
[18] R.M. Young, "Psychoanalysis, Fundamentalism and Terrorism". 2002. Available: http://human-nature.com/rmyoung/papers/pap139.html
[19] R.M. Young, "Groups and Institutional Dynamics", 1999: http://human-nature.com/rmyoung/papers/pap118.html
[20] R.M. Young "Distance Learning Units on Bion, Groups and Institutions", 2003: http://human-nature.com/rmyoung/papers/pap146.html
[21] E.S. Person (ed), *On Freud's "Group Psychology and the Analysis of the Ego"*. Analytic Press, New Jersey, 2001
[22] R. Sole and B. Goodwin, *Signs of Life: How Complexity Pervades Biology*. Basic Books, New York, 2000
[23] S. Camazine, J-L. Deneubourg, N. Franks *et al. Self Organization In Biological Systems*. Princeton University Press, New Jersey, 2001
[24] E. Bonabeau, M. Dorigo, G. Theraulaz, *Swarm Intelligence: From Natural To Artificial Systems*. Oxford University, New York, 1999
[25] *Organizational & Social Dynamics*. Vol. 7, No. 2, Autumn 2007
[26] S. Freud, "The Future of an Illusion (1927)", in S. Freud *Civilization, Society and Religion*. Penguin, London, 1991, pp179-241
[27] International Society of Artificial Life: http://www.alife.org
[28] G. Bateson, *Steps Towards an Ecology of Mind*, U. Chicago, 2000
[29] S.R. Palombo, *The Emergent Ego: Complexity and Coevolution in the Psychoanalytic Process*. International Universities Press, 1999
[30] M. Miller, "Chaos, Complexity, Psychoanalysis", *Psychoanalytic Psychology*, 1999

[31] Fraha, "Systems psychodynamics: the formative years of an interdisciplinary field at the Tavistock Institute", in *History of Psychology*, Feb 7, 2004, pp65-84

[32] V. Spruiell, "Deterministic chaos and the sciences of complexity: psychoanalysis in the midst of a general scientific revolution", in *Journal of the American Psychoanalytic Association*, 1993;41(1):3-44:http://www.analysis.com/vs/vs93a.html

[33] R. Kennel, *Bion's Psychoanalysis and Edelman's Neuroscience*, 1997: http://www.sicap.it/~merciai/bion/papers/kenne.htm

[34] S. Mitchell (ed), *Relational Psychoanalysis*. Analytic Press, 2007

[35] R.Klein & H.Bernard, *Handbook of Contemporary Group Psychotherapy: Contributions from Object Relations , Self Psychology & Social Systems Theories*. I.U.P. 1992

[36] NeuroAnalysis: http://neuroanalysis.googlepages.com

[37] Institute of Group Analysis: http://www.groupanalysis.org

[38] *Psychoanalysis, Culture and Society*: http://apcs.rutgers.edu/

[39] *Free Associations*: http://www.human-nature.com/free-associations/

[40] J. Panksepp, *Affective Neuroscience: The Foundations of Human and Animal Emotions*. Oxford University Press, 2004

[41] S. Freud and A. Einstein, "Why War?": http://www.wordit.com/why/why_war.html

[42] Robert Young, "Ambiguous Space: Projective Identification", in *Mental Space*. Process Press, 1994: http://www.human-nature.com/mental/chap7.html

[43] N. Glover, "The Legacy of Wilfred Bion", in *Psychoanalytic Aesthetics: The British School*: http://www.human-nature.com/free-associations/glover/chap4.html

[44] T. Ogden, *Projective Identification and Psychotherapeutic Technique*. Karnac, London, 1992

[45] E. Hopper, "Theoretical and Conceptual Notes Concerning Transference and Countertransference Processes in Groups and by Groups, and the Social Unconscious: Part III", in *Group Analysis*, Volume 39(4):549-559, 2006: http://gaq.sagepub.com

[46] W. G. Laurence, *Social Dreaming*. Karnac, London. 2005

[47] J. Segal, *Melanie Klein*. Sage, London, 1992.

[48] R.D. Hinshelwood, *A Dictionary of Kleinian Thought*. Free Association Books, London 1991

[49] Freud, *The Ego and the Mechanisms of Defence*. Hogarth, London, 1936

[50] Matte-Blanco, *The Unconscious as Infinite Sets: As Essay in Biologic*. Karnac, London, 1998

[51] Repast: http://www.repast.sourceforge.net

[52] T. Ogden, The concept of internal object relations. *International Journal of Psycho-Analysis*, 1983, 64, 227-242

[53] F. Alford, *Melanie Klein and Critical Social Theory*. Yale, 1989

[54] M. Jay, *The Dialectical Imagination*. California University, 1996

[55] E. Fromm, *The Fear of Freedom*. Routledge, 2001

[56] H. Marcuse, *Eros and Civilization*. Routledge, 1998

[57] S. Freud, "A Seventeenth-Century Demonological Neurosis", in S. Freud, *Civilization, Society and Religion*. Penguin, London, 1991, pp377-433

[58] R. Stacey, "Complexity at the Edge of the Basic-Assumption Group", in [15], 91-114

[59] D. Bell, " Anything is Possible and Everything is Permitted": http://www.ipa.org.uk/docs/Fipa-200701-381287-1-Totalit13.doc

[60] W.G. Lawrence, "Totalitarian States of Mind in Institutions". In [17]: http://www.human-nature.com/group/chap2.html

[61] F. Dalal, *Race, Colour and the Process of Racialization: New Perspectives from Group Analysis, Psychoanalysis and Sociology*. Routledge, London, 2002

[62] Seshadri-Crooks, *Desiring Whiteness: A Lacanian Analysis of Race*. Routledge, London. 2000

[63] D.W. Winnicott, *Playing and Reality*. Routledge, London. 1999

[64] S.Freud, *New Introductory Lectures on Psychoanalysis*, Penguin, London, 1991, p66

[65] J. Lapalance and J-B. Pontalis, *The Language of Psychoanalysis*. Karnac, London, 1988, pp144-145

[66] W. Bion, "A theory of thinking". *International Journal of Psycho-Analysis*, 1962., 43, 4-5.

[67] R. Fairbairn, "Synopsis of an object-relations theory of the personality". *International Journal of Psycho-Analysis*, 1963, 44, 224-225

[68] M. Klein, "A contribution to the psychogenesis of manic-depressive states". *International Journal of Psycho-Analysis*, 1935, 16, 262-289

[69] M. Klein, "Mourning and its relation to manic-depressive states". *International Journal of Psycho-Analysis*, 1940. *21*, 125-153

[70] D. Winnicott, "Metapsychological and clinical aspects of regression within the psycho-analytical set-up". In *Through Paediatrics to Psychoanalysis.* Tavistock, London, 1987

[71] Elliot, *Social Theory and Psychoanalysis in Transition: Self and society from Freud to Kristeva.* Karnac, London, 1999

[72] W. Bion, *Elements of Psychoanalysis.* Karnac, London. 1989

[73] W. Bion, *Attention and Interpretation.* Karnac, London. 2007

[74] W. Bion. Attacks on linking. *International Journal of Psycho-Analysis, 40* (5-6), 308. 1959

[75] J. Segal, *Melanie Klein.* Sage, London. 1992

2.7   A Primer of Psychoanalysis for Alan Turing

Robert M. Galatzer-Levy

Contemplating how a psychoanalyst consult might aid an engineer in emulating the mind suggests several possible contributions. One has to do with what evidence would satisfy an analyst that the mind had been successfully emulated, a psychoanalytic version of Turing's test. A second would have to do with what general elements of psychological function are most important to analysts and would therefore have to be adequately emulated. A third would be a detailed critique of particular emulations. In this paper I will explore the first two issues because they permit additional conceptual formulations and point to areas that may be surprising to engineers.

2.7.1   Turing and the Emotions

Alan Turing, a father of artificial intelligence, was such a strange and tragic figure that Hodge's superb biography is aptly subtitled Enigma. Enigma here refers both the German coding device broken by Turing with enormous consequences for the outcome of the Second World War, and to Turing's personality, which puzzled and frustrated many of those around him (Hodges, 1983). Were Turing alive today, he probably would have received a diagnosis of Asperger's Syndrome, a condition characterized by normal to excellent intelligence accompanied by the absence or severe limitation of a spontaneous comprehension of emotional states in oneself or others.

Although people with Asperger's Syndrome are often thought of as being purely rational, like Spock in Star Trek, this characterization is mistaken. They often talk about situations as if they were rational because they lack the means to discuss situations in any other way. As a result they run into difficulty with their own and others' emotionality. This failure to understand others' emotional communications makes them insensitive to others' most experientially urgent communications so they commonly displease or confuse people. Their own emotional responses are often simplified, experienced primarily as impulses toward action, and remain largely disconnected from the rest of psychological life. Like Turing, they commonly are socially isolated in a way that is painful and unclear to themselves, and like him, they often experience ordinary human situations as intrusions into their lives.

In Turing's case, this took a particularly tragic form. Turing was arrested and prosecuted for making contact with a homosexual prostitute. The episode was probability one of his few efforts toward physical intimacy with another person. Homosexual relations were common and well tolerated in British universities during the 1940s but required a fluent comprehension of a complex implicit social-emotional rules which would have posed a far harder problem to the emotionally uncomprehending Turing than the German Enigma Machine. Turning to prostitutes most likely reflected Turing's difficulties in managing complex social situations with important emotional components. It is also likely that, as is the case with many Asperger's patients, Turing experienced sexual impulses as an intense physiological need unrelated to the complex set of emotions that accompany desire in most people.

Turing was profoundly committed to keeping his body in ideal shape. As part of his punishment-treatment for soliciting the prostitute, Turing received hormones that feminized his body. (It is possible that these hormones also interfered with his intellectual function.) The combination of the humiliation of the arrest and the bodily transformation probably caused of his suicide.

Turing's central contribution to artificial intelligence was his essay "Computing Machinery and Intelligence" (Turing, 1950) in which he outlined the "Turing test," a positivistic formulation to the question of whether a machine could think or have a mind, in which machines and people would be said to have equivalent mental functions if an interrogator using "Does the object of interrogation do X" where X is an action or capacity (such as "play chess"). Attempts to approach this standard have largely been frustrated by the technical problems of developing computer programs with adequate linguistic capacities but there seems to be wide agreement that the problem is even more fundamental[67].

It might be thought that pessimism about the usefulness of the Turing test derives primarily from human narcissism, the idea that no mere machine could do what we do. Doubtless this is part of the problem. But the speculation itself points to a central feature of human mentation. Such narcissism is clearly irrational, i.e., the assertion that a mere machine cannot think provides the person asserting it with nothing useful, nothing he or she would be willing to avow wanting and, in fact, interferes with adaptive functioning.

I would like to suggest that the problem with the Turing test is that it focuses on the rational capacities of the object of interrogation and takes little heed of the deep irrationality of human mentation. If one thinks about this in historical context one source of the problem is obvious. An individual like Turing who was deeply unappreciative of irrational and emotional states would find it difficult to conceptualize the problems of human psychology as centering on irrationality and, particularly, emotionality, just as it would be very difficult for a color blind individual to meaningfully include an appreciation of color in a discussion of human vision. (Turing's requirement that communications be limited to teletype so that incidental physical features of the responding sources would not be confused with their fundamental qualities, whatever its other merits, fails to address the problem that such communication involves a much narrower band width than is typical of ordinary human communication and is particularly likely to omit the various means of communicating affect which is generally accomplished to a much greater extent by bodily gesture, especially facial expression and the prosody of speech

---

[67] For the purposes of this paper I am putting to one side the arguments, based in Heidegger's conceptualization of thought, that on a purely philosophical analysis any entity not having a human body and social context are incapable of thought (Winograd and Flores, 1986).

than through words. Although a partial solution of this problem is probably within reach as excellent animations and voice modulations to produce affective expression and prosody are clearly within reach, other aspects of affective communication such as occur through smell and pheromones, which are currently less developed as computer technology would have to be included.) In attempting to emu-emulate the mind it makes sense to recognize that, for better or worse, much (probably most) of mental, life is non-rational and to use what is known about non-rational thought to accomplish this end.

While AI has long recognized the problems of limiting its exploration to rationality (Simon, 1983) considerations have focused primarily on the need for heuristics because complete solutions to problems are impossible within the constraints of computational resources. This approach neither reflects the biological evolution of brains, which are built around and include computational mechanism developed to function in more primitive systems than the modern human brain, nor does it take account of the specifics of the various forms of mentation that have been discovered through psychological and, particularly, psychoanalytic investigation.

Turing made an all too common mistake of assuming that his mind worked like other peoples and equating his mind with his conscious mentation. But, from what we know of Turing, it is fairly clear that he was not only much smarter than most of us but also was far less aware of irrational and emotional elements in his own thinking, much less of unconscious mental processes.

Part of the training of psychoanalysts is to undergo a personal analysis designed to broaden awareness of psychological life and reduce interferences in appreciating the variety of mental life in others. If this analysis goes well it almost always results in profound surprises. Although the particulars of these discovers vary dramatically the analyst in training virtually always discovers elements of personality of which he or she was unaware and which often had the effect of blinding the future analyst to aspects of others' personalities. Although Turing's incomprehension of human emotionality reached the level of significant psychopathology, many individuals attracted to technology have much milder versions of a preference for rationality over emotionality which may make it particularly difficult to focus on the emotional aspects of human psychology. Engineers attempting to emulate the mind would do well to recognize that success in this endeavor will almost certainly depend on becoming quite sophisticate about the mind that is being emulated, i.e., appreciating that naïve introspection, observation and even systematic psychological experimentation are unlikely to yield as rich a picture of human mental life as psychoanalysis and that unless the engineer is either willing to avail himself of psychoanalysis or take on faith some fairly counterintuitive ideas from psychoanalysts, he will likely fail to produce something that works well as a human mind.

2.7.2   *Potential Contributions of Psychoanalysis to Emulating the Mind*

Psychoanalytic investigations have produced a rich picture of psychological function. However, it is difficult to select those portions of psychoanalytic thought that are likely to be useful in emulating the mind from those which might confound the effort.

It would be a mistake to think of psychoanalysis as a unified discipline. Psychoanalysis is better thought of as a group of interrelated ideas, theories, and practices that interact in complex ways. In this, of course, psychoanalysis differs only in degree from other sciences, say chemistry, in which practical applications, e.g., the manufacture of gasoline, is firmly believed by the chemist to be ultimately relatable to the fundamental laws governing properties of substances and solution and chemical combination. But the chemical engineer never relies exclusively on these fundamental law but rather relies on a century of experience in the production of gasoline. The various levels of a discipline interact. Theory leads to new observations and unexplained observation drives the development of new theory.

Throughout the history of psychoanalysis theory and observation have interacted, each feeding on and correcting the other. However, this process has often obscured by a failure to differentiate concepts at different levels of abstraction and a tendency to use terms ambiguously ways that confuse description and theoretical formulation. For example, an analyst might write that a patient had "cathected the Oedipus complex." The statement appears to refer to Freud's theory of psychic energies (cathexis) and a theory of development during a time with the child operates within the framework similar to that described in Sophocles' play *Oedipus Rex*. However, what the analyst actually means is very likely something quite different, e.g., that the associations produced by the patient during analytic hours seem to contain elements that the analyst understands to be fragments or parts of a narrative in which the patient's wish for intimacy with one person is combined with competition for that intimacy with a second individual who is, therefore, experienced as threatening the patient's well being. This sort of confusion makes it difficult even for analysts to be clear about assertions in the psychoanalytic literature and contributes to the difficulty of using psychoanalytic ideas in other disciplines[68].

Following and modifying Robert Waelder's (1960) conceptual hierarchy, we can separate two elements of psychoanalytic theory. One could be described as theory proper, psychoanalysts refer to this as metapsychology and the other as clinical theory. Theory proper, metapsychology, consists of a group of ideas about the functioning of the human psyche that posits the existence of psychological functions in entities such as the ego, psychic energy or intrapsychic forces. It may be regarded as the (not fully developed) basis of a comprehensive theory of mental function which, if properly and adequately instantiated, would result in human-like behavior and which would be capable, in theory, of predicting human behavior given adequate data, much in the same way that classical mechanics, given

---

[68] A related difficulty concerns the problem of inferring the level of evidence supporting statements in the psychoanalytic literature. Analysts tend to describe matters with bold assertions so that it is unclear whether a given statement is based on the analyst's speculation, systematic empirical data or some intermediate basis for belief.

appropriate information about a system's initial state, is able to predict the evolution of the system over time. (We include, of course, in such prediction the possibility of chaotic phenomena as the system will clearly be non-linear.) Occasionally analyst have tried to develop these ideas to the point of making computations based on them (see e.g., (Alexander, 1946; French, 1952; French, 1954)). Such efforts have almost always ended in abandoning the computational effort and using the metapsychological concepts a metaphors to describe clinical phenomena. Metapsychology is modeled on late nineteenth century mechanics and thermodynamics (Galatzer-Levy, 1976). From a metapsychological viewpoint, psychoanalysis is a discipline not unlike classical physics, though obviously much less well developed.

In the mid-1980s metapsychology came under vigorous three pronged attack in the United States after a period of vigorous development following the Second World War (Gill and Holtzman, 1986.) Unlike analogous concepts in physics, metapsychological concepts lead to no new predictions. They tended to obscure psychoanalytic discourse as observations were described in terms of metapsychology in a manner that was often ambiguous. They limited the field of observation as analysts attempted to fit their data into overly narrow metapsychological concepts. Most important, these concepts seldom represented those aspects of psychoanalysis that students of the field found to be most significant, the vivid descriptions of human mental life that arise from the study of the psychoanalysis of individuals. Freud's division of the psyche into three agencies (Id, Ego and Super-Ego) would never have engaged any interest had he not previously shown the richness and complexity of individuals' psychological functioning, in novel-like case histories (Freud, 1905a), (Freud, 1911) or provided tools by which sense could be made of the seemingly senseless content of dreams, slips, and jokes (Freud, 1900; Freud, 1905b; Freud, 1901).

The wide collection of descriptions of human psychological life, the ordering of those observations and their generalization are called the *clinical theory* of psychoanalysis. They describe how people function psychologically, in a manner close to ordinary experience and use language close to ordinary speech. In the view of many psychoanalysts clinical theory constitutes the major contribution of the field. The major controversies within psychoanalysis have concerned the emphasis that should be given to each of the various viewpoints that emerge from the close study of the productions of patients in analysis rather than whether or not any of them is salient. For example, followers of Melanie Klein tend to focus on how people deal with the distress associated with intense and disorganizing aggressive wishes while classical Freudian analysts attend more to the emotional struggles that result from the young child's wish for erotic pleasure and the fears associated with that wish, and "self psychologists" focus on the experience of a cohesive and lively self and the threats to that experience. None of these viewpoints excludes the other though proponents of each feel strongly about the emphasis to be given them.

Let me give an example of the sort explanation that might be given for a human action from a clinical psychoanalytic point of view. How might I begin to explain the writing of the paragraph you are now reading? What follows is an abbreviated and slightly censored version of how an analyst's answer. As I write this I am visiting my son, a high school mathematics teacher, who is occupying an old mansion hidden in a woods not far from my own summer home, which my son has largely taken over, in Massachusetts (a state on the Atlantic Ocean in the U.S.A.). I woke up this morning, earlier than the rest of the house's occupants (my son, another young man who is sharing the house and my wife) having gone to sleep after playing a board game with all of them. I had won. My son and I had a brief but slightly heated and competitive discussion about how best to teach about transfinite numbers. The discussion had started with talking about how best to explain them to my wife. After the game we discussed getting a copy of the Sunday New York Times in the morning. Anticipating that I would be the first one up I volunteered to go buy it. I then felt slightly annoyed and humiliated that they all were convinced that I was likely to get lost in the woods were I to run the errand so they urged me to postpone it so one of the younger men could accompany me. I felt I was being treated as old. To make matters worse the room my wife and I had slept in the night before was cold so my son had provided us with another, warmer room, which, however, required us to sleep in separate beds. My son joked about putting separating us for the night.

When I woke up I had a free hour and chose between several tasks – working on one of three papers (including this one), working on a book that is past due at the publisher, doing some quite remunerative work on legal cases in which I am an expert witness (I thought: but one is about a horrible murder and the other about the death of a mother – not the stuff for a Sunday morning in the woods), finishing reading a book (I thought: no but that will be good for the airplane back home), etc. As I showered the idea for this paragraph came to mind. That would certainly be a good idea, show them (i.e., you the reader) what I actually mean. But why now? What beyond reason led to my choice? The interactions of previous evening stimulated a competitiveness with my son. Before attending medical school I had done graduate work in mathematics and my dissertation would, in fact, have been on Turing machines had I not changed fields. Of course, mathematics has many meanings to me but at this point my thoughts went to my father, a professor of biochemistry, who I can easily see seated at his desk in the evening calculating and graphing the laboratory's data of the day. And then I see his drawing table at the laboratory with the 20 inch long slide rule – obviously a phallic symbol (in highschool "we" all wore shorter but thicker one dangling from our belts). This in turn recalls seeing my father nude and then a vivid (what I now to know to be a screen) memory of a huge fire place in our Massachusetts summer home, which (through analysis) I understood represented a "primal scene" i.e., witnessing my parent's sexual engagement and the excitement and fear associated with that scene. But now I am the professor who writes and though my son is a teacher, I compete with him to teach well. He may try to laugh at me as an incompetent old man who cannot find his way in the woods, cannot explain mathematics, cannot even arrange to sleep in the same bed as his wife but I am the victor because I can write a chapter for a cutting edge book, explain difficult concept, and be victor over both him and my father (my son looks strikingly like my father).

Fully told and explored the psychoanalytic story of how I came to write these paragraphs when I did is even richer than I have described but I think this gives a picture of what an analyst might think of as a partial explanation for a relatively simply decision. We could describe it as a clinical observation or set of clinical observations.

Even those with a cursory knowledge of psychoanalysis will recognized certain themes in this story that lie somewhere between clinical generalizations and clinical theory. Some of these include

(1) Fathers and sons often compete with one another.
(2) This competition commonly includes a slightly disguised sexual element.
(3) People's visions of their place in the world (their identity) is often shaped by their professional competence.
(4) Ideas based on early sexual experiences (such as observing parental intercourse) remain active and emotionally salient over many years and despite changes in the reality of the situation.
(5) Emotional states may be represented in dreams and memories by concrete images (the recalled fire.)
(6) Symbolism (the slide rule for the phallus) plays an important role in allowing one to at the same time address an issue and remain unaware that it is being address.
(7) Experience is commonly organized as narratives, which are overlapping, and whose salient points include not only their content but also what is omitted from their content and the context with which they are told.

The list could be extended much further even based on this simple example.

Any emulation of the human mind that strives to encompass the appreciation that psychoanalysis has developed for the human must, at minimum, include those factors that are generally recognized by psychoanalysts as part of the clinical theory because the clinical theory is the core of what is important in psychoanalysis. While there may be some reward in attempting to carry out the project of constructing a system based on metapsychological concepts the engineer should keep in mind that, in contrast to the clinical theory, metapsychology has seldom proven a fruitful source for the development of significant new ideas and has, at best, served as an awkward metaphoric language in which clinical theory could be described. Theoretical concepts modified to include current knowledge of brain structure and function as, for example, as pursued in the work of Solms and Turnbull (Solms & Turnbull, 2002) have contributed more to clinical psychoanalytic theory than the older metapsychology because looking at psychological activities from the point of view of brain function provides a genuinely new perspective from which to look at the phenomena of psychological life, whereas traditional

374

metapsychology relied on unpromising physicalistic metaphors to reformulate descriptions of phenomena already well described clinically. It viewed things from the same perspective but used less adequate language to describe what was seen.[69]

What follows is a brief description of certain major elements of the clinical theory that must be represented in any device that a psychoanalyst would think of as emulating the human mind well. It is far from exhaustive for the basic technique of the psychoanalyst is much like that of what (Levi-Strauss, 1966) called the "savage mind" using whatever material are available to do the thinking job at hand so that mental functions, as viewed by analysts, have a certain grab-bag quality, in that all sorts of methods are used by the human mind but, at the same time, this far ranging collection of actions must be integrated into a unified whole.

When physiological homeostasis is disrupted or threatened with disruption, actions to maintain homeostasis can become preemptive. The degree of this preemption depends in part on the nature of the physiological process and the time frame within which it must be attended. For example, deprivation of oxygen lasting more than a minute leads to an intense focus on the need to breathe. Eating, defecation and urination become urgent and preoccupying in time frames on the order of 12 to 24 hours. The need for sexual release can be essentially indefinitely postponed. Disruption of bodily integrity, signaled by pain, requires nearly immediate attention. Psychoanalysts discovered that the management of these preemptory needs, especially those that can be to varying extents postponed and those which can be satisfied in a semi-automatic fashion often occurs outside of awareness. The satisfaction of some of these preemptory needs despite their physiological nature can sometimes be partly achieved, at least temporarily, symbolically and through fantasy. Such impulsions toward homeostasis are referred to as "drives."[70] Drives are usefully described in terms of what it is the individual wishes to do (aim), their intensity, and the person (or part of a person or function of person) toward whom the drive is directed (object). For example an infant may experience an intense wish to suck on (aim) its mother's breast (object). Drives may be conscious or unconscious and may be satisfied in disguised ways. A major feature of sexual and aggressive drives is their flexibility with regard to each of their components. Drives can be satisfied, for example, by substituting similar objects for the original object of the drive. Famously, Freud formulated the idea that the mother is the initial object of erotic interest, but both because of her unavailability and the anxieties as-

---

[69] The use of neuroscience be develop better psychoanalytic theories must be differentiated from the use of neuroscience to provide models for the instantiation of those theories, i.e., the engineer's barrowing of techniques developed by nature to build a device that emulates the mind.

[70] The term drive is also used to refer to other tendencies that are psychological preemptive, though the nature and even existence of primary drive unrelated to homeostasis, such as a drive toward destruction or a drive to obtain knowledge is a matter of controversy in psychoanalysis.

sociated with such interest substitute objects are chosen that bear significant resemblance to the mother[71].

The aims of drives can be even more elaborately transformed. This is particularly true because even the negation of a wish may leave its underlying intent more or less intact. For example, the common infantile wish to play with and smear feces can be significantly satisfied in disguised form through an active orderliness if we conceptualize the transformation of the urge through series of (unconscious) thoughts, "I want to play with feces", "I want to be involved with messes by smearing feces", "I want to be involved with messes by *not* smearing feces", "I want to be involved with messes by negating them", "I want to be involved with messes by cleaning them up", "I will live an extraordinarily neat and tidy life" leads to a means of satisfying the underlying urge while at the same time avoiding some of its distressing aspects[72]. This fuzziness's regard to particularly emotionally salient ideas is essential feature of human emotional mentation and is essential to any psychoanalytic model of mind.

2.7.3   *Mental Life Goes on in Several Registers*

The idea that psychological life goes on in multiple registers and that much of psychological life goes on outside of awareness is central to psychoanalytic thought. In particular, some of psychological life is conscious or available to consciousness. Any emulation of mind that is to be psychoanalytically satisfactory must contain consciousness or its equivalent because much of the clinical theory of psychoanalysis concerns how people protect themselves from conscious awareness of distressing ideas as well as the differences in the ways ideas are processed in various aspects of the mind.

The subjective experience of rational thought, so important to economic models of the mind and rational decision theory, plays only a slight role in the psychoanalytic conceptualization of the psyche which tends instead to think of such decision processes as "rationalizations," efforts on the part of the mind to convince itself that it is behaving from clearer and less problematic motives than it in fact is. For example, in the situation described above of the transformation of the wish to smear feces into maintaining an orderly life style the subject would almost certainly point to the material advantages of order or the general merits of orderliness as the "reason" for his behavior. Rough and ready rank orderings (e.g., I like choco-

---

[71] It is important to keep in mind not only that there is looseness and fuzziness in the nature of the object, but also that the original object must be understood in terms of its perception by the subject, i.e. that the object's most salient quality which its substitute must resemble can only be understood and described in terms of the perceiving individual. The person who the child calls mother overlaps only slightly with the woman who the store clerk thinks of as the "next customer" though both might bare the name Harriet Jones.

[72] This example uses the particularly powerful action of negation which invariable evokes that which is negated. Think of the traffic signs that indicate something as forbidden with a red circle and diagonal line over the depiction of what is forbidden or the exercise, "Don't think about an elephant."

late better than I like castor oil) that are commonly used do not integrate into an overall much less quantifiable transitive hierarchy of preferences in the type of mentation that is psychoanalytically interesting. Any model of the mind that emphasizes rationality, in the usual sense of the term, as a major basis of decision processes does not fit well with psychoanalytic experience.

Any satisfactory model of mind must reflect the multiple ways analysts have found information to be processed in various aspects of the mind. For example, similar sounding words and similarly shaped objects are commonly treated as equivalent in such processes as dream construction.

At the same time models that assume a simply hierarchy of agents (Minsky, 1986) do not capture the ongoing coordination, influence and integration of these somewhat separated aspects of mental functioning. An attractive model that goes well with psychoanalytic observation conceptualizes that mind as a hierarchy of mutually influencing coupled oscillators rather than as independent agents (Freeman, 2000) but this idea is not part of current psychoanalytic theory.

The Self and Others

Almost all psychoanalytic conceptualization involve an experience of self (which being an experience is something quite different from the entity having the experience.) This experience becomes distressing insofar as the self is seen as discontinuous in time or space or lacking in vigor (Kohut, 1971.) At the same time, for the vast majority of people involvement with other people is central to psychological life, though psychoanalytic investigation has shown that these other people, are experienced in extremely various ways. To get some idea of this variety consider such representations of monsters in horror films, figures on Greek vases, and portrayals of persons in complex novels. All are examples of mental representations of people.

The study of the fundamental attachment that pulls people toward others, which starts at birth has been one of the triumphs of post-war psychoanalysis (Fonagy, 2001). At this point students of attachment theory have described several characteristic ways in which individuals come to be attached and any emulation of mind must include a fundamental engagement with other minds as a central feature. Since, as Aristotle observed, man is a social animal, any psychoanalytically satisfactory emulation of mind must include interaction with other similar minds and so probably must involve the emulation of a community of minds.

Trauma and Other Forms of Danger

The subjective quality of experiences being "too much", or to use psychoanalytic terminology "traumatic" is another form of preemptive concern. The mind seems to recognize that in response to a sufficiently intense stimulus there is substantial risk of a qualitative shift in function which is experienced as highly dangerous. The risk of such a qualitative shift is often described as a loss of the self or its boundaries. Other forms of danger include the fear of the loss of relations to other people, fear of bodily damage and fear of moral disapproval. The experience of these fears or concern that situations might arise that include them is referred as anxiety and much of psychoanalytic thinking, based as it is in the study of persons in emotional distress, focuses on anxiety and its management.

The means by which individuals manage anxiety are referred to as "mechanism of defense" and this rich collection of psychological actions has been extensively catalogued by psychoanalysts (Freud, 1966). Most mechanisms of defense operating by separating ideas and emotions so that the danger is not apparent. For example, if I feel frightened by the presence of murderous impulses in myself I may simply bar those impulses from awareness (repression) or attribute them to someone else (projection).

The Principle of Multiple Function

A recurring theme of psychoanalytic investigation is that one psychological action may serve multiple ends (Waelder, 1936). It appears to be rare that emotionally significant human actions serve only one function. Consider the example of the compulsively clean and orderly individual discussed above. Being clean and orderly serves multiple functions. It satisfied the urge to be involved with messiness; it protects from the anxiety association with that same urge by denying it; it increases the individual's self esteem and the esteem he receives from other; it commonly leads to financially and practically useful actions. At the same time it may be part of an aggressive act in which the clearly individual makes the life of everyone around her through incessant demands for tidiness and/or it can be a way of being like (identifying with) another person with this same compulsion (Freud, 1905a). Such identification may, in turn, have multiple motivations ranging from keeping a potentially lost loved one with the person making the identification to a self imposed punishment (Schafer, 1968). By bringing all of these goals into a single action the person is not only able to satisfy them all but by satisfying them all he is able to add impulsion to a particular action while at the same time remaining in a position to claim (including to himself) that any one of the goals is not part of his motivation.

A corollary of the principle of multiple function is that any significant pattern of psychological functioning is likely to be difficult to change because it serves has many uses. (Freud, 1914) noted that significant psychological change required repeated engagements of a problem in analytic work and that those instances in which brief interventions appeared transformative usually proved on closer examination to be superficial and inadequate. Insofar as the psychoanalytic human mind is emulated it will have multiple motivations and will transform in depth only gradually

Stories

From approximately age three most people arrange their descriptions of salient events into more or less coherent narratives that encompass aspects of life expe-

378

rience and suggest actions in various situations (Schafer 1992). At least for most people personal narratives have enormous power and the explanation of actions in terms of personal narratives is generally particularly satisfying. The power of such narrative cannot be overestimated and much of the deep sense of nature of life is incorporated in cultural narratives that the individual adopts. Individual's self assessments occur largely through comparisons of their own perceived lives with lives outlined in culturally prescribed narratives, so that, for example, an unmarried woman at age 21 may think herself (and be thought) an "old maid" in one culture, while a woman in the same situation in another culture may be seen as progressing very well. (It is important to remember here that, as with all the psychological functions discussed in the paper, personal narratives may be unconscious, as well as conscious and that their being outside of awareness may intensify their impact on psychological function.)

A particularly important form of narrative might be called the static or "no-change" narrative i.e., the belief that some important aspect of the personality is fixed and unchangeable and so constitutes part of the essence of who the individual is. The power of the static narrative can be seen in the issue of sexual orientation (see (Cohler & Galatzer-Levy, 2000). Until approximately 1900, in Western Civilizations erotic interest in the persons of one's own gender seems to have been regarded as a minor variation in a general erotic interest (similar to the way an interest in hair color might be treated today.) It was not regarded as a stable feature of the personality. A wide ranging shift seems to have occurred a little over a century ago such that same gender sexual orientation came to be viewed as a stable and important feature of the personality. The narrative of sexual orientation shift from one of high fluidity to a static narrative with major consequences for those who experienced same sex arousal including the belief that any other erotic interest was inauthentic, that sexual orientation was unchanging (at least without extremely extensive intervention) and most important that many other aspects of life were fixed by sexual orientation such as gender typical interests. (The "treatment" Alan Turing received was based not only on hatred of homosexuals but on precisely the idea that his sexual interest was of an unchangeable and set him apart from normal men.)

The changes observed in many psychoanalytic treatments are essentially changes in personal narratives that result from bringing them into awareness where they may be transformed by experience. For example, the paradigmatic Freudian narrative, the Oedipus "complex" describes the young boy's fantasy that his sexual interest in his mother will enrage his father and lead to the natural punishment of the boy's castration by the father. This frightening story is repressed (driven from consciousness) but continues to influence the boy's actions and thoughts in that anything suggestive of the erotic interest in mother (e.g., professional success, physical maturation, romantic involvements) may precipitate anxiety and various steps to manage that anxiety (including inhibitions.) Analytic work brings the original narrative and its place in the patients current life into awareness where it is put the test of comparison with material reality (e.g. that the father would not, in fact, castrate his son or that the mother is no longer the object of sexual interest) and hence lose its influence of the boy or man's actions by replacing the old narrative with a far more complex one that among other things provides a history of the old narrative.

Play

The British psychoanalyst, Donald Winnicott (Eigen, 1981; Winnicott, 1971; Winnicott, 1965; Winnicott, 1953; Winnicott, 1968; Winnicott, 1967) building on the child analytic study of play realized the centrality of play and its directives in human psychological function. Winnicott observed that the capacity to rehearse and vary activity in a context without material consequences and separated from the ordinary contexts of action has been transformed from its widespread presence in higher animals to the wide range of human activities such as art, music and literature that are central to the human condition. Analysts have an opportunity to observe serious play closely because one way of thinking about the psychoanalytic process is as a form of play in which the analysand engages in trying out various versions of emotionally salient scenarios in relation to the psychoanalyst. In any case, if the human mind is to be emulated the instantiation would have to include the capacity for play if it is to satisfy an analyst vision of what minds are like.

2.7.4 Conclusion

Had Turing been a psychoanalyst, a highly improbable event given Turing's failure to engage many ordinary human interests, he would have devised a different and far harder test than he did. He could have outlined a group of capacities that he would expect the emulating device to demonstrate. In this paper I have provided an abbreviated and simplified list of the kinds of things an analyst might expect to observe in a successful Turing test. The list emphasizes the ways in which humans differ from traditional "rational actors" even though it can be seen that the various functions described by psychoanalysts have a reason of their own. It might be mentioned that analysts, accustomed as they are to identifying what is peculiar in individuals' psychological productions would make extraordinarily good "Turing testers" (interrogators of the emulation device) both because they are trained to notice peculiarities of psychological function and because they should be quite good at describing what is amiss. Without analytic training sensitive individuals may notice something is wrong but they would often have a hard time trying to describe exactly what it is that is wrong. Running through the list is the centrality of emotional states and affects in determining human action, a feature seldom address in traditional AI. I believe this is because affects are commonly experienced as interferences with cognition rather than as a form of particularly useful rapid cognition (LeDoux, 1996).

It is only by trying to emulate the human mind in its full richness that we can test whether current analytic theories can lead to a synthesis of the type of entity they try to explain (i.e., whether these theories are adequate), provide a tool that would make devices in many ways equivalent to a human mind available to do work and discover more of the computational possibilities of minds.

References

Cohler, B. & Galatzer-Levy, R. (2000). *The Course of Gay and Lesbian Lives: Social and Psychoanalytic Perspectives*. Chicago: University of Chicago Press.

Freeman, W. (2000). *How Brains Make Up Their Minds*. New York: Columbia.

French, T. (1952). The Integration of Behavior: Basic Postulates. Chicago: University of Chicago Press.

French, T. (1954). The Integration of Behavior: The integrative process of dreams. Chicago: University of Chicago Press.

Freud, A. (1966). *The Ego and the Mechanisms of Defense*. Revised Edition ed., New York: International University Press.

Freud, S. (1900). The interpretation of dreams. In J.Strachey (Ed.), *The standard edition of the complete psychological works of Sigmund Freud*. London: Hogarth Press.

Freud, S. (1901). The psychopathology of everyday life. In J.Strachey (Ed.), *The Standard Edition of the Complete Psychological Works of Sigmund Freud*. London: Hogarth Press.

Freud, S. (1905a). Fragment of an analysis of a case of hysteria. In J.Strachey (Ed.), *The Standard Edition of the Complete Psychological Works of Sigmund Freud* London: Hogarth Press.

Freud, S. (1905b). Jokes and their relation to the unconscious. In J.Strachey (Ed.), *The Standard Edition of the Complete Psychological Works of Sigmund Freud*. (London: Hogard Press.

Freud, S. (1911). Psycho-Analytic Notes on an Autobiogrpahical Account of a Case of Paranoia (Dementia Paranoides). In J.Strachey (Ed.), *Standard Edition of the Complete Psychological Wroks of Sigmund Freud* (pp. 3-82). London: Hogarth Press and the Institute of Psycho-Analysis.

Freud, S. (1914). Remembering, repeating and working through: further recommendations on the technique of psychoanalysis II. In J.Strachey (Ed.), *The standard edition of the complete psychological works of Sigmund Freud* (pp. 146-156). London: Hogarth Press.

Hodges, A. (1983). *Alan Turing: The Enigma*. New York: Simon & Schuster.

Levi-Strauss, C. (1966). *TheSsavage Mind*. Chicago: University of Chicago press.

Minsky, M. (1986). *The Society of Mind*. New York: Simon & Schuster.

Schafer, R. (1968). *Aspects of Internalization*. New York: International University Press.

Solms, M. & Turnbull, O. (2002). *The Brain and the Inner World*. New York: Other Press.

Winnicott, D. (1953). Transitional objects and transitional phenomena. In D.W.Winnicott (Ed.), *Collected papers: through pediatrics to psychoanalysis*. (pp. 229-242). New York: Basic Books.

Winnicott, D. (1965). The Maturational Process and the Facilitating Environment. In (New York: International Universities Press.

Winnicott, D. (1971). Playing and Reality. In (New York: Basic Books.

Winnicott, D. W. (1967). The location of cultural experience. *International Journal of Psychoanalysis. 48(3).P 368-72*.

Winnicott, D. W. (1968). Playing: its theoretical status in the clinical situation. *International Journal of Psychoanalysis 49(4).P 591-9*.

2.8   Alexander R. Luria and the Theory of Functional Systems

Wolfgang Jantzen

Luria's neuropsychological theory of functional systems of the brain cannot be traced back to psychoanalysis. Luria's thinking is both influenced by Vygotsky's cultural historical psychology (and by Leontyev's activity theory) and also by Russian physiology (Uchtomsky, Bernstein, Anokhin).

Functional systems are dynamic, self-organizing and autoregulatory central-peripheral organizations of cells, organs and organisms, the activity of which is aimed at achieving adaptive results useful for the system and the organism as a whole. Every functional system is subject of the action of a special type of 'pacemaker'. Correspondingly, Luria's theory models the brain as a space-time regulatory system and the language as a space-time process of activity, based on "sense cores". In Leontyev's theory "sense" represents the emotional, integrating cover of our activity space at any given moment. In ontogenesis sense coincides originally with the emotions.

In our opinion emotions represent Eigen-values or Eigen-behaviour in the organization of the psychic processes. They realize the closure of functional systems. Emotions themselves are multi-oscillatory processes. They mediate between body and perceived as well as anticipated environment at any moment.

2.8.1   Luria and Psychoanalysis

In their summary of neuropsychoanalytical models Dietrich et al (2007) notice, that Luria has developed his dynamic neuropsychology on the basis of Freud's neuropsychological thoughts, particularly on Freud's aphasiology. Luria's collection by the neuropsychoanalysis is not always carried out in such a glaring form – but it is wrong in every case. Of course there was the well-known correspondence between Luria and Freud, stressed by Oliver Sacks (1998, 21) and of course there was his psychoanalytical engagement in the 1920s. But this was still only one of the influences, which had an effect on the scientific ideas of young Luria, who at the time was thinking in rather eclectic fashion. And the psychoanalytical Freud is in many ways a different type than the neuroscientific one (ibid.). For the neuropsychological Freud the article about "Aphasia" from 1891 as well as the "Project for a psychology" (in manuscript 1896) are the most distinguished. Both works remained unnoticed for a long time. The "Project" was published in 1950 for the first time, the work on aphasia had been forgotten for a long time and been reprinted first in 1953. Luria himself had at that time already developed the essential fundamentals of his neuropsychology. Interrupted by his defectological work of the 1950s these main ideas built the base of his great monographs of the 1960s (cf. Braemer & Jantzen 1994, Homskaya 2001). Freud's neurological work was at that time still unknown to Luria. He does not mention the "Aphasia" article in the book "Traumatic Aphasia" (in Russ. 1947, revised 1959; in English 1970). However, he does mention the article for the first time in his later book "Higher cortical functions in man" (in Russ. 1962, in English 1966; Luria 1982). In his lectures Luria has never mentioned Freud in a neuropsychological context (Akhutina; personal communication from 8.3.2008). Akhutina [Ryabova] has been Luria's longstanding assistant and head of the Laboratory of Neuropsychology at M.V. Lomonosov Moscow State University). The fact that Luria as well as Freud in his work about aphasia refer to John Huglings Jackson is no evidence for an orientation to psychoanalysis because Jackson was already known to Vygotsky and Luria due to the works of Henry Head (engl. 1926; translated into Russian soon after that, however, not published) and from Goldstein (Akhutina 2008). Luria is primarily obliged to his early deceased friend and colleague L.S. Vygotsky. He dedicated the book "Higher cortical functions in man" (Luria 1982) to him and published Vygotsky's "neuropsychological testament" (Vygotsky 1997, cf. Akhutina 2002) in one of the first issues of the scientific journal "Neuropsychologia" where he provided a detailed comment (Luria 1965). With this in mind one can follow Akhutina (ibid.) in that Luria works in the traditions of Vygotsky's outline of neuropsychology, which he could only realize in part, however. Also in psychological regard Luria changed his position according to the conception of a "non-classical psychology" justified in Vygotsky's cultural-historical theory and continued in the context of the psychological activity theory by Leontyev. In agreement with Vygotsky he writes: "Scientific psychology develops where natural and social sciences meet; because the social forms of life force the brain to work in a new way, they let evolve qualitatively new functional systems. Exactly these systems are the subject of psychology" (Luria 2002, p. 58). Within the senseful and systemic development of psychic processes there is no mentioning any more of psychoanalysis or psychoanalytic categories such as Id, Ego and Super-Ego, but instead space-time interrelations inside functional systems of the psyche or the personality are cited. The positions of trying to separate the works of Vygotsky, Luria and Leontyev, and especially to see Leontyev in discontinuity to Vygotsky (Kozulin 1996, Keiler 1997), became totally untenable with the increasing accessibility of the complete works of these three authors ("troika") (cf. Jantzen 2003, 2008). In the background of these theories we also have to consider N.A. Bernstein's (1967, 1987) physiology of movements. Bernstein ideas have been of great influence not only to Leontyev's activity theory but also to Luria's neuropsychology. Moreover, clear mutual interrelations exist between Bernstein and Vygotsky. Bernstein particularly stresses to be obliged to Vygotsky's conception of the chronogenetic and dynamic localization of psychic processes (Feigenberg 2005). In much the same way as there are tight theoretical relations between Vygotsky and Luria and between Vygotsky and Leontyev, there are also tight relations between Luria and Leontyev. Luria takes up Leontyev's conception of a nature historically and cultural historically well-founded general psychology in a systematic way and develops it further.

A Spinozian foundation is common to this psychology. Openly this is carried out in Vygotsky (Vygotsky 1996, cf. Jantzen 2002) and Leontyev's work (2006, 25-32). But through the latter it also found the way into the works of Luria and

Bernstein (it is worth mentioning that the famous Spinozian philosopher Deborin had also been a member of the Moscow Psychological Institute just like these four people in the 1920s). This foundation obtains further deepening by the close cooperation of Leontyev with the philosopher Evald Ilyenkov in the 1960s and 1970s (Jantzen & Siebert 2003). This creates clear homologies to psychoanalysis, which in its neuropsychological foundation can also be regarded as an attempt of a Spinozian psychology (Jantzen 1989, 1994). However, it neither allows to subsume cultural historical and activity theory under psychoanalysis nor psychoanalysis under these theories. The theories of Vygotsky, Luria, Leontyev and Bernstein have all in common the creation of a space-time system (functional system) of psychic processes with its neurophysiological and neuropsychological bases and connections. However, the theoretical reconstruction is, among other reasons, difficult because the problem of the emotions is not systematically discussed in either Luria's or Bernstein's theory. This topic is dealt with by Vygotsky especially in his late work (cf. Jantzen 2008). Emotions and cognitions develop in the human ontogenesis in different age stages and can be observed in the development of experience/emotional experience (Russ.: pereshivanie, Germ. Erleben). In General Psychology Leont'ev has treated this problem in detail and in neuropsychological aspects he agreed with Simonov's neurobiology of emotions (Leont'ev 1998). Our own work shows clear perspectives for reconstruction, unification and further development of the complete theory. A general theory of functional systems is in the centre of this reconsideration (cf. Jantzen 1990, VII). For this theory the already mentioned body of work is of vital importance and on top of that Anokhin's physiological systems theory needs to be mentioned as well as Uchtomskij's postrelativistic theory of neurophysiological processes. Although almost entirely unknown in the West, Uchtomskij's theory has played an enormous role in the development of Russian neurophysiology and neuropsychology.

2.8.2   General Theory of Functional Systems

Anokhin (1974) himself uses the idea of the functional system from the level of the cell up to the level of the integral organism. "Functional systems are dynamic, selforganizing and autoregulatory central-peripheral organizations the activity of which is aimed at achieving adaptive results useful for the system and the organism as a whole" (Sudakov 2007). Anokhin's concept is different from other systems concepts by two central aspects: (1) Systems forming factor is the "beneficial or adaptive effect"; (2) functional systems possess a dynamic, differentiated, and operational architecture with feedback information about the action results. Many of these assumptions were summarized in models by Anokhin, others appear merely in text passages and require a systematic reconstruction, and others are indicated but not developed in detail. Fig. 2.8.1 takes up Anokhin's model (1974, p. 253) and differentiates it according to his own assumptions as well as under citation of Simonov's emotion theory (1982, 1986).

384

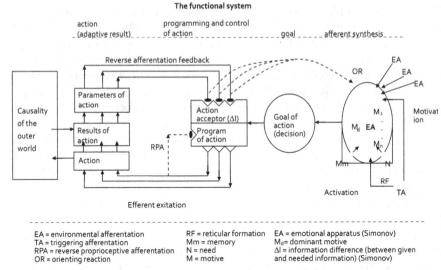

Fig. 2.8.1 Anokhin's model enhanced by Simonov's emotion theory

Afferent synthesis is the initial stage of the purposeful activity. The "pre-triggering integration" proceeds at the front of it. While actively drifting in the present each (human) being judges its current state in the world in the given system organism-environment at any moment. This judgment is done on the basis of external (environmental afferentation) as well as internal conditions (somatic afferents, memory, need). The four main components of the afferent synthesis are: (1) motivational excitation, (2) the sum total of situational afferentations, (3) triggering afferentation and (4) utilization of memory mechanisms (Anokhin 1974, p. 215). The formation of the *dominating motive* presupposes the mechanism of the *dominant* introduced by Uchtomskij (2004) into physiology. A dominant is a functional system with a "system history". It realizes the activity, which is directed towards the satisfaction of the currently dominating need (hunger, thirst, relationship etc.) and reproduces within its functional rhythm the rhythm of external influences. On elementary psychophysiological levels this concept corresponds to the one of the instinct as an elementary unity of emotional valuation, of anticipation, and of cognition. In relation to the conditioned reflex the dominant is genetically primary (Vygotsky 1998, p. 224). It is its inner prerequisite. As a spacetime-structure (a chronotope; Uchtomskij) it integrates the activity for the development and learning of the subject at any given time and with that enables differentiations in dynamic flexible activities. It is a functional system as such (Anokhin 1974, p. 459). But its inner structure as a component of the afferent synthesis remains undefined. And therefore the nature of the dominant motivation also remains undefined (ibid., p. 450). "Dominance is a physiological means by which the functional systems are manifested in the adaptive effects of the organism by changing the levels of excitability" (ibid., p. 458).

Before the afferent synthesis and the decision making dominants can exist in latent (subdominant) form. Correspondingly I have modified Anokhin's graphic model of the afferent synthesis.

Symmetrically to Mm (memory) I've used the term N (need) instead of motive (M) in Anokhin's original graphic (1974, pp. 241). This corresponds to Leontyevs (1978) suggestion, to understand physical demands of the organism and also of the brain and should be distinguished from psychological needs (developed by experiences). Both are assigned to past/present of the organism and should be distinguished from the dimension of possible future. Motives arise only from this and in this possible future. The need relevant side of the activity (in Russ. "dejatel'nost"; in German "Tätigkeit") transforms itself into the motive of the activity through an object of the possible satisfaction of needs. This object is signalled by its goal (Jantzen 2003). I speak of motives only from the beginning of the process of the assessment of different alternatives (subdominant motives M1 - Mn) mediated by the emotional apparatus (EA) – which is in continuation of Anokhin's considerations and taking into account Simonov's psychophysiological theory of the emotions. At the end of this process comes the decision about the possible action. This decision is carried out on the basis of available information (ΔI) at this time and is taken simultaneously with the initiation of the dominating motive (Md). The setting up of an action programme and an action acceptor related to a "beneficial or adaptive effect" is therefore in general conducted by need, is motivated, senseful and systemic (cf. Jantzen 1990, p. 57, fig. 10). The attempt to relate Anokhin's theory of the functional system and Simonov's needinformation- theory of the emotions to each other needs to take certain facts into account: Anokhin distinguishes spaces of the *past-present* (memory, need in the form of latent dominants) from thoseof the *fluent present* (pre-trigger integration, afferent synthesis until making the decision; reverse afferentation feedback related to the steering of the action and the results of the action) from those of the *possible future* (motive of action, goal of action, action programme). By taking up the parameter *time* he treats the spacetime of psychophysiological processes in a post-relativistic way (cf. also Chauvet 1996, chpt. 5). "Every functional system is subject of the action of a special type of *'pacemaker'* having a high energy charge. This is the only possible explanation for he the fact that the numerous components of the functional system which are often very remote from each other, are unified into a single, integrate whole and continuously maintain his unity" (Anokhin 1974, p. 528).

This is obviously is a general quality of living systems, because under stress every nerve cell can take the role of an oscillator (Kryzhanovsky 1986, 77), able to return into the rest condition again by hyperactivity and the generation of corresponding action models. At this point we have taken up Simonov's (1982, 1986) emotion theory (Jantzen 1990, Feuser & Jantzen 1994). Emotions are a function of the needs N and the information difference ΔI, i.e. the difference between the necessary information (expected) and the existing information before the action is started: $E = f(N, \Delta I)$. However, Simonov's considerations do not include the just discussed temporal transformations. The needs themselves are not yet determined by the object (goal) of the activity because this is formed only in the afferent synthesis. In accordance with Simonov's formula emotions only refer to the past/present, i.e. the perception of the situation in the pre-trigger integration.

At modeling possible future in the afferent synthesis ("anticipatory reflection" of the reality; Anokhin 1974) previous experiences must be considered. Possible need-satisfying actions must be weighed up. By doing so the strength and eventually the direction of the emotion changes compared with the one in the bare *perception*. In the process of the *assessment* the needs become transformed into distinguished subdominant goal-motives. Due to the comparison of these different goal-motives done under the aspect of the expected emotional benefit the transition from assessment to *decision* is carried out – the dominating motive has been selected (cf. Klix 1980, p. 51, p. 57). The emotional assessment remains permanently in the present but this present changes. The present changes because alternative actions from the past are included in the present and reflected in relation to the future. This creates a senseful and systemic space (a "dominant") within which the actions are carried out until the "beneficial or adaptive effect" is obtained. Consequently, by taking the decision a new pacemaker of activity is established. This is the emotional fluctuation (oscillation) itself between satisfaction of needs and success of the action in the context of the motivated activity within the given functional system of activity (chronotop). This activity itself is embedded into motives of a higher order, which arise level by level in the ontogenetic development (Leont'ev 1978). At the programming of partial actions the pacemakers of the movement coordination (Turvey 1990) are multi-oscillatory coupled with the entire pacemaker of the whole activity. In this way the entire motive remains unchanged. Action acceptor and action program not only become afferentated from the usefulness of action by realizing the goal but also from its need satisfying power by realizing the motive. The other parts of Fig. 2.8.1 are self-explanatory. In addition, it is important that there is a permanent reverse feedback of the steering mechanism of the action (action acceptor/program of action) via proprioceptive afferents. This kind of reverse afferentation has to be distinguished from the one that announces the results. If the results do not correspond to the expected reverse afferents a renewed cycle starts mediated by orientation reactions. On the basis of a renewed afferent synthesis a new discrete "system quantum" (Sudakov 2007) comes into being.

2.8.3 Luria's Theory of Brain Functions and Language

In continuation of Anokhin's theory of functional systems and taking into consideration the traditions of the cultural historical and activity theory and also the Russian physiology (Uchtomskij, Pavlov etc), Luria, in his monograph "Human brain and psychological processes" (1966), develops a general theory of functional systems of the human brain which gets its final theoretical setting in the three block theory of the brain functions (Luria 1973, pp. 43).

Every functional system has memory, which it realizes permanently in the context of its afferent synthesis (physical afferentation and environmental afferentation). This also applies to the brain as a whole – as a functional system of functional systems. At the same time every functional system has the ability to program and assess actions related to a "beneficial or adaptive effect". These qualities are reflected in Luria's separation of a (second) functional unit for the receiving, analyzing and storing information in the parietal, temporal and occipital area of the human brain (1973, pp. 67), its afferent system, and of a (third) functional unit for programming, regulation and verification of activity in the frontal area (ibid. pp. 79), its efferent system. Luria therefore considers activity at the highest level of the brain as the interaction of afferent synthesis and formation of actions and thus at all other levels, too. In this functional organization of the brain the action acceptor is the mediating unit i.e. the frontal, language mediated formation of a "model of the future" (Feigenberg 1998) on the basis of activation and motivation. The foundation of these processes is a corresponding (first) unity of brainstem, mid-brain and orbitalfrontal cortex as a unit for regulating *tone and waking and mental states* (Luria 1973, pp. 44). The concept of tone used here corresponds to Bernstein's definition of tone as "the centrally controlled tuning of all functional parameters of every muscle element and its effectory nerve fibre" (1987, p. 209).

This first unit projects into the two others and secures both the non-specific activation as well as the specific one (and with that the motivation). This conception of the brain as a functional system makes it possible to think about brain functions as a space-timesystem, as a chronotopical connection of both the whole and parts (Uchtomskij). Some other authors have tried to subdivide brain functions into functional blocks just like Luria. Although, they have failed to notice this aspect of chronotopical organization (cf. Roth 1987, p. 225; Edelman und Tononi 2000) Edelman (1987), taking up Bernstein's ideas, assumes that the process of the global mapping of the consciousness arises from actions based on the selection of "degenerated" gestures. Moreover, Edelman & Tononi (2000, pp. 42) assume that on the one side a strong functional differentiation between three different "topological brain arrangements" (connected in parallel) exist and on the other side there is the thalamo-cortical system producing consciousness by means of a "reentry" network. These three different topological brain arrangements are (1) the time producing and regulating cerebellum, (2) the space producing and regulating hippocampus and (3) the space-temporal movement segments producing basal ganglia. Both assumptions suggest that it is not only appropriate but mandatory to accept Luria's solution of the whole brain as a functional system with a space-time-structure (chronotop). Luria does not only suppose such a space-time-structure at the level of neuropsychology but also at the level of psychology itself. Language organizes itself between time pole and space pole of the brain (frontally or post-centrally) as a unity of processes which are *syntagmatic* (structuring the sentence form) and *paradigmatic* (structuring the sentence contents logically and hierarchically). These processes are involved in understanding and speaking language in a different way and can be damaged in form of aphasias by brain injuries (Luria 1976; 1982). Every speech production presupposes motive and intention (1982, lecture 10). And the understanding of any language communication requires the evaluation of "sense cores" (ibid. lecture 13). But Luria uses "sense" in a rather cognitivistic way. In Leontyev's work, in continuation of Vygotsky's considerations of the dimension of emotional experience (Russian: pereshivanie; German: Erleben), sense (Russian: smysl) has received a fundamentally more exact emotional-psychological definition. Unlike the common notion of "sense" in this new version the cognitive content appears as "meaning". For this reasons particularly the concept "sense" must be certainly going beyond Luria.

2.8.4 Sense and Emotions

"Sense is not obtained by the meaning but through life", so Leontyev (1979, p. 262). In ontogenesis sense coincides originally with the emotions. Sense expresses itself through meaning. By finding a meaning sense represents the emotional, integrating cover of our activity space at any given time. In this respect it is conceived in a way homologous to the psychoanalytical idea of the libidinous object cathexis. Sense integrates, however, not only the activity given in form of the dominating motive; sense in principle integrates all life events. In the Spinozian basic idea, sense is the result of the transformation of the affects by affects, or, according to Vygotsky, the transformation of the emotions by emotions. This transformation is based on the objectivity of the world increasingly acquired by the meanings. "The deep nature of the psychic sensory images lies in their objectivity, in that they have their origin in processes of activity connecting the subject in a practical way with the external objective world" (Leont'ev 1978, p. 84). In the ontogenetic development the transformation of the consciousness takes place with special clarity within developmental crises (Vygotsky 1998, pp. 187). The "crux of the issue" is the relation between emotion and intellect (Vygotsky 1993, p. 230). "Things do not change because we think about them but the affect and the functions linked to them change according to how they are recognized. They then are in a different connection to the consciousness and to another affect; and this consequently alters their relationship to the whole and to its unity" (Vygotsky 1993, p. 236).

Fig. 2.8.2 represents the interrelations within the functional system of the senseful and systemic psychological processes in accordance with Leontyev's theory (cf. Jantzen 2006, 2008).

In the broadest sense activity means to do something (Russian "aktiv'nost"; German "Aktivität") [2]. In Leontyev's theory *activity* has a more narrow meaning and describes the *need relevant side* of the activity (Russian "dejatel'nost"; German "Tätigkeit") [1]. Activity in Leontyev's sense [1] always arises when the possible object of a need is realized in the psychic process as an object. An *action*, however, is the objective and goal centred side of the activity [1]. Actions always take up former, automated actions, too, these are *operations*. Consequently the activity [1] is a process, which expresses a need. It exists, however, only in the form of the action. Action therefore is senseful as well as objective. The other parts of Fig. 2.8.2 are self-explanatory. However, what are the emotions? Obviously emotions represent Eigen-values or Eigenbehaviour (von Foerster 1987, Varela 1987) in the organization of the psychic processes because they realize the closure of the functional system.

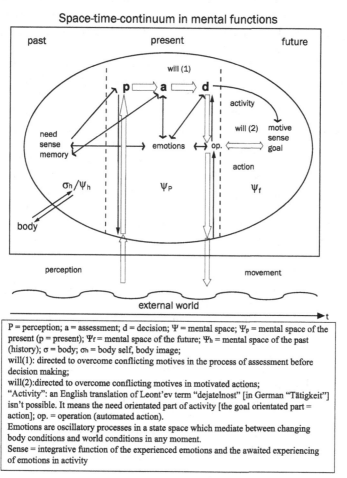

Fig. 2.8.2 Interrelations within the functional system

2.8.5 Some Remarks on the Nature of the Emotions

In accordance with the quoted considerations emotions realize themselves strictly in the window of the fluent present. In the process of assessment and decision they are replaced by accompanying emotions of possible activities in the form of those actions that provide a greater emotional valence for the subject.

The kernel of this mediation, and that's our in the mean time worked out hypothesis for an independent theory of the emotions, is to regard emotions themselves as multioscillatory processes. Strictly connected to the perceived or remembered present they mediate between body and environment at any given moment. They are the base of our self assessment processes (cf. Jantzen 1990, chpt. 7 and 8, 1994, 1999, Jantzen & Meyer 2008). As oscillating structures they are embedded in fluctuation processes by hormones and neurotransmitters (Vincent 1990). They are connected by coherence or interference to the various neural perturbations (cf. Maturana & Varela 1992) at the periphery of the body just like at the periphery of the world. In accordance with the topology of these different fluctuation processes perturbations are caused by the given periphery or result from interferences between the peripheries. In the brain the inner organs are represented differently than the extremities, the representation of visual processes is different from hearing, from smelling etc. The "system time" is realized by the emotions. Because the different fluctuation processes refer to the "system time" and because of the "system time" (the emotions) being coupled to the perturbations coherence can be established once again. This happens either by reassessment of the perception or by environment-related actions. As an anticipating formation of coherence emotions ensure the decision in the process of assessment between different possible alternative actions (= process of motive formation). They are multioscillatory entities of "the thinking body in the world" (Il'enkov 1994), which form coherence at any moment or suspend it in favour of a future coherence between body and world. They can be taken along and strengthened by an outer source of resonance corresponding to them.

Research about early attachment shows exactly this. Mother and baby "synchronize [...] the intensity of their affective behavior in fractions of seconds" (Schore 2003, p. 56). This "synchronicity of affects" is based on resonance processes caused by attuning by the mother in the mother child dyad. In these resonance processes transitions frequently happen from a condition of quiet mutual attention to a condition, which is more intensive and positively affective. The ability to control these processes of resonance formation and one's own positive excitement in a way that they can be reverted to a resting state is the central developmental achievement in the second year of life (Schore 1994). Trevarthen & Reddy (2006) stress the role of music and art just like Wallon (1984) has done for the social organization of the emotions. In agreement with our assumptions Trevarthen understands emotions as transition balanced states. They "signify transitional equilibria in the whole subject's changing motives and the accompanying internal autonomic processes, and in the expression of outwardly directed actions and interests that the motives are generating" (Trevarthen 2005, p. 68). In their ontogenetic development, via initially intersubjective actions, children increasingly use their emotions to interfere with their emotions. The emotional regulation then shifts from interpsychic to intrapsychic processes like Vygotsky assumed.

These considerations demonstrate shortcomings of the emotion theories worked out by the neuropsychoanalysis until now. And from this point of view the speculations about emotions coming from the field of Artificial Intelligence are, in my opinion, totally untenable. Panksepp distinguishes a number of "core emotional feelings" in his various different works reflecting "the neurodynamic attractor landscapes of a variety of extended transdisencephalic, limbic emotional action systems" (2005, p. 30). But emotion itself is only an "umbrella" concept for him (ibid. 32). Schore (1994) suggests a model of the emotions quite different from Panksepp's considerations. But also in his model questions about the nature of the emotions remain open. However, the question must be:

What is an emotion? As far as I see it our own theory is the only one, which is able to handle the shown contradictions. Moreover, it is in agreement with a second order cybernetics. It is essentially based on the traditions of Vygotsky, Luria and, Leontyev. In a number of questions it is in clear proximity to neuropsychoanalysis. However, for other questions an open scientific discussion is mandatory.

References

[1] Akhutina, Tatiana: Foundations of neuropsychology. In: Robbins, Dorothy; Stetsenko, Anna (Eds.): Voices within Vygotsky's non-classical psychology. Past, present, future. New York, Nova Science, 2002, pp. 27-44

[2] Anokhin, P.K.: Biology and neurophysiology of the conditioned reflex and its role in adaptive behavior. Oxford: Pergamon, 1974

[3] Bernstein, N.A.: The co-ordination and regulation of movements. Oxford: Pergamon, 1967

[4] Bernstein, N.A.: Bewegungsphysiologie. Leipzig, Barth, 1987, 2. Aufl.

[5] Braemer, Gudrun & Jantzen, W.: Bibliographie der Arbeiten von A.R. Lurija. In: Jantzen,

[6] W. (Hrsg.): Die neuronalen Verstrickungen des Bewußtseins. Zur Aktualität von A. R. Lurijas Neuropsychologie. Münster: LIT, 1994, pp. 267-345

[7] Chauvet, G.A.: Theoretical systems in biology. Vol. III. New York: Pergamon 1996

[8] Dietrich, D. et al.: Considering a technical realization of a neuro-psychoanalytical modelof the mind. In: Dietrich, D. et al. (Eds.): Conference Proceedings ENF 2007. 1st International Engineering & Neuro-Psychoanalysis Forum. Wien, 2007, pp. 13-20

[9] Edelman, G.; Tononi, G.: A universe of consciousness. New York: Basic Books 2000

[10] Edelman, G.M.: Neural Darwinism - a theory of neuronal group selection. New York: Basic Books, 1987

[11] Feigenberg, J.M.: The model of future in motor control. In: Latash, M. L. (Ed.): Progress in motor control. Vol. 1.: Bernstein's traditions in movement studies. Champaign: IL, Human Kinetics, 1998, pp. 89-104

[12] Feigenberg, J.: Bernstein und Vygotsky. (Translation of Feigenberg, J.: N. Bernšteijn: From reflex to the model of the future. Moskau: Smysl, 2005, pp. 47-51 [russ.]) In: Mitteilungen der Luria Gesellschaft, 12 (1), pp. 4-8, 2005

[13] Feuser, G; Jantzen, W.: Die Entstehung des Sinns in der Weltgeschichte. In: Jantzen, W.: Am Anfang war der Sinn. Marburg: BdWi-Verlag, 1994, pp. 79-113

[14] Foerster, H. von: Erkenntnistheorien und Selbstorganisation. In: Schmidt, S.J. (Hrsg.): Der Diskurs des Radikalen Konstruktivismus. Frankfurt/M.: Suhrkamp, pp. 133-158, 1987

[15] Freud, S.: Project for a scientific psychology. In: The standard edition of the complete works of Sigmund Freud. Vol. 1. London: Imago, pp. 295-391, 1950

[16] Freud, S.: On aphasia. [Translation of: "Zur Auffassung der Aphasien". Leipzig: Deuticke 1891] London: Imago, 1953

[17] Head, H.: Aphasia and kindred disorders of speech. Vol. I - II. Cambridge: UP, 1926

[18] Homskaya, Evgenia D.: Alexander Romanovich Luria. A scientific biography. New York: Kluver Academic/Plenum Publishers, 2001

[19] Il'enkov, E.V.: Dialektik des Ideellen. Münster: LIT, 1994

[20] Jantzen, W.: Freud und Leontjew oder: Die Aufhebung der Psychoanalyse im Marxismus. In: Jahrbuch für Psychopathologie und Psychotherapie, 9, pp. 44-68, 1989

[21] Jantzen, W.: Allgemeine Behindertenpädagogik Bd. 2. Neurowissenschaftliche Grundlagen, Diagnostik, Pädagogik und Therapie. Weinheim: Beltz, 1990

[22] Jantzen, W.: Am Anfang war der Sinn. Marburg: BdWi-Verlag, 1994

[23] Jantzen, W.: Freud neu marxistisch lesen. In: Jantzen, W.: Am Anfang war der Sinn. Marburg: BdWi-Verlag, pp. 24-34, 1994

[24] Jantzen, W.: Transempirische Räume – Sinn und Bedeutung in Lebenszusammenhängen. In: H.J. Fischbeck (Hrsg.): Leben in Gefahr? Von der Erkenntnis des Lebens zu einer neuen Ethik des Lebens. Neukirchen-Vluyn: Neukirchener Verlag, pp. 123-144, 1999

[25] Jantzen, W.: The Spinozist programme for psychology: An attempt to reconstruct Vygotsky's methodology of psychological materialism in view of his theories of motions. In: Robbins, Dorothy; Stetsenko, Anna (Eds.): Voices within Vygotsky's nonclassical psychology. Past, present, future. New York: Nova Science, 2002, pp. 101-112

[26] Jantzen, W.: A.N. Leont'ev und das Problem der Raumzeit in den psychischen Prozessen. In: Jantzen, W.; Siebert, B. (Hrsg.): Ein Diamant schleift den anderen – Evald Vasil'evic Il'enkov und die Tätigkeitstheorie. Berlin: Lehmanns Media, 2003, pp. 400 - 462

[27] Jantzen, W.: Die Konzeption des Willens im Werk von Vygotsky und ihre Weiterführung bei Leont'ev. In: Mitteilungen der Luria Gesellschaft, 13 (2) pp. 23-56, 2006

[28] Jantzen, W.: Kulturhistorische Psychologie heute – Methodologische Erkundungen zu L.S. Vygotsky. Berlin: Lehmanns Media, 2008

[29] Jantzen, W. & Meyer, Dagmar: Isolation und Entwicklungspsychopathologie. In: Feuser, G. & Herz, Birgit (Hrsg.): Emotionen und Persönlichkeit. Band 10 von "Behinderung Bildung, Partizipation" Enzyklopädisches Handbuch der Behindertenpädagogik. Stuttgart: Kohlhammer, 2008 i.V.

[30] Jantzen, W.; Siebert, B. (Hrsg.): Ein Diamant schleift den anderen – Evald Vasil'evic Il'enkov und die Tätigkeitstheorie. Berlin: Lehmanns Media, 2003

[31] Keiler, P.: Feuerbach, Wygotski & Co. Studien zur Grundlegung einer Psychologie des gesellschaftlichen Menschen. Berlin: Argument, 1997

[32] Klix, F.: Erwachendes Denken. Berlin: DVdW, 1980

[33] Kozulin, A.: Commentary. In: Human Development, 39, pp. 328-329, 1996

[34] Kryzhanovsky, G.N.: Central nervous system pathology. New York: Consultants Bureau, 1986

[35] Leontyev, A.N.: Activity, consciousness and personality. Englewood Cliffs N.J.: Prentice Hall, 1978

[36] Leontyev, A.N.: Tätigkeit, Bewusstsein, Persönlichkeit. Berlin: Volk und Wissen, 1979

[37] Leontyev, A.N.: Problems of the development of the mind. Moscow: Progress, 1981

[38] Leontyev, A.N.: Bedürfnisse, Motive, Emotionen. In: Mitteilungen der Luria- Gesellschaft, 5 (1) pp. 4-32, 1998

[39] Leontyev, A. N.: Frühe Schriften. Band II. (Hrsg.: G. Rückriem,), Berlin: Lehmanns Media, 2006

[40] Luria, A.R.: L.S. Vygotsky and the problem of functional localization. In: Neuropsychologia, 3, pp. 387-392, 1965

[41] Luria, A.R.: Human brain and psychological processes. New York: Harper & Row, 1966

[42] Luria, A.R.: Traumatic aphasia. The Hague: Mouton, 1970

[43] Luria, A.R.: The working brain. Harmondsworth/Middlesex, Penguin, 1973

[44] Luria, A.R.: Basic problems of neurolinguistics. The Hague: Mouton, 1976

[45] Lurija, A.R.: Introduccion evolutionista a la psychologia. Barcelona: Editorial Fontanella 1977

[46] Luria, A.R.: Higher cortical functions in man. New York: Basic Books, 2nd. ed., 1980

[47] Luria, A.R.: Language and cognition. Washington D.C.: Winston, 1982

[48] Lurija, A.R.: Lectures about General Psychology, (Russ.). Moscow (http://www.psychology.ru/library) 2000

[49] Lurija, A.R.: Die Stellung der Psychologie unter den Sozial- und den Biowissenschaften. In: Jantzen W. (Hrsg.): Kulturhistorische Humanwissenschaft. Berlin, Pro Business, 2002

[50] Maturana, H. & Varela, F.: The tree of knowledge. The biological roots of human understanding. Boston: Shambhala, rev. ed., 1992

[51] Panksepp, J.: Affective consciousness: Core emotional feelings in animals and humans. In: Consciousness and Cognition, 14, pp. 30-80, 2005

[52] Roth, G.: Erkenntnis und Realität. Das reale Gehirn und seine Wirklichkeit. In: Schmidt, S.J. (Hrsg.): Der Diskurs des Radikalen Konstruktivismus. Frankfurt/M.: Suhrkamp, pp. 229-255, 1987

[53] Sacks, O.: Sigmund Freud: The other road. In: Guttmann, G.; Scholz-Strasser, Inge (Hrsg.): Freud and the neurosciences. From brain research to the unconscious. Wien (Öster. Akad. d. Wiss.) 1998, pp. 11-22

[54] Schore, A.N.: Affect regulation and the origin of the self. The neurobiology of emotional development. Hillsdale/N.J.: LEA, 1994

[55] Schore, A.: Zur Neurobiologie der Bindung zwischen Mutter und Kind. In: Keller, Heidi (Hrsg.): Handbuch der Kleinkindforschung. Bern: Huber, 3. Aufl. pp. 49-80, 2003

[56] Simonov, P.V.: Höhere Nerventätigkeit des Menschen. Motivationelle und emotionale Aspekte. Berlin: Volk und Gesundheit, 1982

[57] Simonov, P.V.: The emotional brain. Physiology, neuroanatomy, psychology and emotion. New York: Plenum 1986

[58] Sudakov, K.V.: The theory of functional systems as a basic of scientific ways to cope with emotional stress. http://www.rsias.ru/eng/scntfc_sudakov/ (22.08.07) Moscow 2007

[59] Trevarthen, C. & Reddy, V.: Consciousness in infants. In: Velman, M. & Schneider, Susan (Eds.): A companion to consciousness. Oxford: Blackwell 2006

[60] Trevarthen, C.: Stepping away from the mirror: Pride and shame in adventures of companionship. In: Carter, C.S. (Ed.): Attachment and bonding: A new synthesis. Dahlem Workshop Report 92. Cambridge/M.: MIT Press, pp. 55-84, 2005

[61] Turvey, M.T.: Coordination. In: American Psychologist, 45, (8), pp. 938-953, 1990 Uchtomskij, A.A.: Die Dominante als Arbeitsprinzip der Nervenzentren. In: Mitteilungen der Luria-Gesellschaft, 11 (1/2), pp. 25-38, 2004

[62] Varela, F.: Autonomie und Autopoiese. In: Schmidt, S.J. (Hrsg.): Der Diskurs des Radikalen Konstruktivismus. Frankfurt/M.: Suhrkamp, pp. 119-132, 1987

[63] Vincent, J.-D.: The biology of emotions. Oxford: Blackwell, 1990

[64] Vygotsky, L.S.: The problem of mental retardation. In: The fundamentals of defectology. The collected works of L.S. Vygotsky, Vol. 2, ed. by R.W. Rieber & A.S. Carton. New York: Plenum-Press, pp. 220-240, 1993

[65] Vygotsky, L.S.: Psychology and the theory of localization of mental functions. In: Problems of the theory and history of psychology. The collected works of L.S. Vygotsky, Vol. 3, ed. by R.W. Rieber & A.S. Carton. New York: Plenum, pp. 139-144, 1997

[66] Vygotsky, L.S.: Infancy. In: Child psychology. The collected works of L.S. Vygotsky, Vol. 5, ed. by R.W. Rieber. New York: Plenum, pp. 207-241, 1998

[67] Vygotsky, L.S.: The teaching about emotions. Historical-psychological studies. In: Scientific legacy. The collected works of L.S. Vygotsky, Vol. 6, ed. by R.W. Rieber. New York: Plenum, pp. 71-235, 1999

[68] Wallon, H.: The emotions. In: Voyat, G. (Ed.): The World of Henri Wallon. New York: Jason Aronson, pp. 147-164, 1984

2.9   A Mind for Resolving the Interior-Exterior Distinctions

Ariane Bazan

During the ENF conference day on July 23ʳᵈ 2007, both Peter Palensky and Gerhard Zucker (né Pratl) presented schematic simulation modules of the mind in which one detail struck me. While the brain was conceived as embedded in a body and likewise, this body was conceived as present in a context or an environment, the mind-module was built so as to make, at the level of the perception entry point, an a priori distinction between external and internal signals, i.e. between signals coming from the environment or from the body respectively. In Pratl's model (flashed for only one second), these entry points were even anatomically distinct.

I think it is worth mentioning that this distinction in a simulation module a priori solves an essential biological problem, which is the distinction between interior and exterior. Freud (1895; 1915) proposes the following line of thought: when a stimulus arrives at an irritable membrane, the first priority of this membrane will be to get rid of this stimulus, which by increasing the potential energy of the system threatens its stability. In a crucial step towards the emergence of a living system, the closure of this membrane gives rise to a dramatic distinction: the distinction between an internal milieu and an external context. In this new configuration, when a stimulus, coming either from within or without, arrives at the level of the membrane, the main survival strategy for the organism will again be to get rid of the increased energy so as to safeguard the stability of the system. In 'Instincts and their vicissitudes' Freud (1915, p. 119) proposes: *"Let us imagine ourselves in the situation of an almost entirely helpless living organism, as yet unorientated in the world, which is receiving stimuli in its nervous substance. [...] On the one hand, it will be aware of stimuli which can be avoided by muscular action (flight); these it ascribes to an external world. On the other hand, it will also be aware of stimuli against which such action is of no avail and whose character of constant pressure persists in spite of it; these stimuli are the signs of an internal world, the evidence of instinctual needs. The perceptual substance of the living organism will thus have found in the efficacy of its muscular activity a basis for distinguishing between an 'outside' and an 'inside'."* Freud then proposes that the organism will distinguish the interior from the exterior stimuli on the basis of the *efficiency of its motor system* to make the stimulation stop. Indeed, if the stimulus comes from the external environment, the primitive organism can easily stop the stimulation by fleeing. On the other hand, fleeing does not influence the internal source of stimulation. The 'efficiency of the motor system' therefore is the efficiency of the flight at this primitive level.

Modern sensorimotor models have described an important physiological mechanism which precisely monitors the efficiency of the proper motor system (e.g. Blakemore et al., 1999; see Fig. 2.9.1). In this model, the organism (or its 'mind') is informed of the movement efficiency by a message that is directly derived from the motor command aimed at the peripheral motor elements (e.g. the muscles and the joints). This information message, which is a copy of the given command feeding forward to the central nervous system, is called an 'efference copy'. The efference copy is the signal by which the organism knows that the peripheral changes, it is about to sense at the proprioceptive or somatosensory levels, are in fact a consequence of the movement of which the organism itself is the agent.

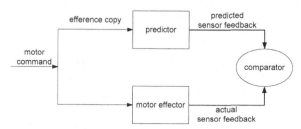

Fig. 2.9.1 An intended forward model makes predictions of the sensory feedback based on the motor commands sent (Blakemore et al., 1999). These predictions are then compared to the actual sensory feedback at the level of a comparator.

Therefore, when a stimulus arriving on the membrane (of the 'perceptual substance') – i.e. an incoming stimulus – is influenced by a movement of the organism and the organism is informed on the proper agency of this movement, the movement is then *efficient at the level of the stimulus* and a 'perception' in the true sense of the word (and as apposed to a sensation or an interoception) is constituted. Perception then means the ability to situate the 'content' of the stimulus in an exterior space (e.g. Lenay, 2006), i.e. in the world outside. In other words still, the organism has been able to determine the exterior origin of the stimulus.

It seems obvious that the distinction between interior and exterior is of critical importance for the survival of the organism. In particular, Freud (1895, p. 325) indicates that, at the highest level of complexity, a human organism must be able to distinguish between a perception, a memory, a fantasy etc. Indeed, acting on a fantasy image, with no external reality, not only leads to disappointment but might lead to downright damage to the organism. It is for the solution of this problem – the distinction between memory and perception – that Freud (1895, p. 325) supposes the ω ('omega')-neuron system. These ω neurons have a particular status: even though they are "activated along with perception" and "behave like organs of perception" (Freud, 1895: 309), their discharge direction is efferent, i.e. in the direction of motility (Freud, 1895, p. 311). In other words, the ω-neurons are – in agreement with the then current line of thought lead by von Helmholtz (1878) – true motor neurons actively implied in perception. Freud (1895) there, like von Helmholtz, holds an *enactive* view of perception. More precisely it is, in Freud's (1895, p. 421) literal terms, the 'Abfuhrnachrichten', the 'efference messages' of these ω-neurons which deliver the information that the active content element coincides with a *movement* of perception and is therefore indeed informative of the external world – and therefore is not merely a wish or a fantasy informative of the internal world. Elsewhere, I argue that that these 'Abfuhrnachrichten', which Freud (1895, p. 325) tellingly calls the 'indications of reality', might precisely correspond with the 'efference copies' in the modern sensorimotor models (Bazan, 2007; Bazan & Van de Vijver, *in press*).

Earlier, Shevrin (1998) had already proposed that the very function of consciousness is indeed to distinguish the origin of content elements – perceptions, thoughts, fantasies, memories –, i.e. that this ability to distinguish the origin is precisely the added value of consciousness as compared to mere unconscious processing. Indeed, at the level of the unconscious, the mind works exclusively with content elements, disregarding their origin, and there is a blurring of memories with wishes and external information. This blurring can be observed in all instances where unconscious processing is accessible: free association, dreaming, symptom formation and psychotic delusions and hallucinations. In other words, one line of thought in metapsychology supposes that the emergence of the characteristic added value of conscious processing (as contrasted to unconscious processing) is precisely a consequence of the problem of distinguishing external from internal stimulation, i.e. that a conscious organization of the mind emerged from the necessity to resolve this distinction. For this reason, the fact that this distinction is a priori in the proposed simulation modules might dramatically falsify the starting conditions for the emergence of a mind-like apparatus.

This might be all the more true since, on a more general level, the evolution of life might really be thought of as the complexifying of the interior-exterior problem. Indeed, Freud (1915) had left us at the level of a primitive organism arising from the closure of a simple membrane. In the further evolution of invertebrate animal[73] life, all evolutionary milestones can be considered as a result of a complexifying of the interior-exterior problem. The most primitive organisms simply consist of a cell lining of an interior cavity (e.g. sponges), with a very simple distinction between interior and exterior. A first milestone in natural evolution is the emergence of gastrulation, a movement during the embryological development whereby part of this outer membrane or *ectoderm* moves inwards and forms an inner closed system, delineated by *endoderm*. This inner closed system then will form the gut.

Gut formation (Stainier, 2005) is one of the first outcomes of multicellularity. Bringing cells together allowed the organism the opportunity to make specialized cell types, with selective pressure leading to the emergence of an outer protective

---

[73] The distinction between animal and plant life is based upon the composition of the outer cell membrane: the cell membrane of plants is structured by sugars or carbohydrates, which render them stiff. Animal outer cell membranes have no sugars, such that their cells remain flexible. As a result, animal cells (and organisms) can move while plant cells (and organisms) cannot. This might again suggest that membrane status in relation to movement crucially 'sets the stage' for the eventual emergence of a mind.

coat (the ectoderm) and an inner layer (the endoderm) involved in food absorption. This innovation, which in evolution starts with the Cnidarians (including the corals, sea anemones, jellyfish, sea pens, sea pansies and sea wasps), brings about a major complication of the inner outer-problem: first, there is the exterior environment, which is outside the outer membrane or the ectoderm; second, there is an interior milieu, inside the cavity delineated by the endoderm (namely, the gut); third, there is an 'in between' cavity inside the ectoderm but outside the endoderm.

A second milestone, then, is the formation, some 40 million years ago, within this 'in between cavity' of a third germ layer, the *mesoderm* when some of the cells migrating inward from the endoderm form an additional layer between the endoderm and the ectoderm. A third critical innovation emerges when a cavity arises in within this mesoderm, called a *coelom*.

Organs formed inside a coelom can move, grow, and develop while fluid cushions and protects them from shocks. Animals having such a body cavity are called 'coelomate animals'. Spoon worms and molluscs are among the first coelomates but this key innovation led to the evolution of nearly all complex animals. One can see that this embryological innovation has, again, complicated the inner-outer problem: indeed, a new 'inside' arises, which is of another kind than the inside gut cavity delineated by the endoderm. The gut, indeed, communicates directly with the outside world by both a mouth and an anus, such that it is an interior space in which exterior milieu can directly penetrate. It is an inner cavity that at two points merges with the outer space. The coelom cavity, by contrast, now is an inner space that is entirely interior and entirely enclosed by mesoderm. The interior-exterior problem, there, is dramatically more complex than in Freud's simple closed membrane: e.g. when a stimulus arrives at the level of the 'mind' it might be important to distinguish if this stimulus originated either from the external environment, from the incorporated external environment, or from the enclosed internal environment.

One of the most dramatic milestones in the evolution of animal life is the emergence of an outer skeleton, operated by 'voluntary' striated muscles, in the lancet fish, the first vertebrate animal. This innovation resets the stage for the interior-exterior problem altogether. Indeed, as Mark Solms indicated in his talk at the ENF, it might be argued that in vertebrates it is as if two bodies, each with their proper cohesion, are continuously operating in parallel: an 'internal' body, inherited from the invertebrates and where smooth muscles and glands are the predominant effectors, and an 'external' body, emerging in the vertebrates and where the 'new' voluntary striated muscles, operating the outer skeleton, are the predominant effectors. The internal body is the source of major stimulus production – such as respiratory needs, hunger, thirst, sexual tensions etc. – which threaten the stability of the organism. This internal body, thereby, produces demands for work upon the external body: the external body is *driven* to interact with the external world so as to ultimately make the internal stimulation stop. At the same time, the external body is also a source of major stimulus production – namely those arising from the perception of the outside world. It thus probably has to manage its own stability in terms of keeping the potential energy at a non-threatening low level. It is as if a vertebrate's mind, then, faces the problem of having to manage two stimuli-metabolisms, each with their proper inner-outer complexity, as well as to attune both metabolisms reciprocally (i.e. the inner with the outer body and vice versa). I agree with Mark Solms that consciousness might critically serve for these multiple adjustments of the interior-exterior problem.

References

Bazan A (2007). An attempt towards an integrative comparison of psychoanalytical and sensorimotor control theories of action. *Attention and Performance XXII*. Oxford University Press, New York, pp. 319-338.

Bazan A, Van de Vijver G. (*in press*). L'objet d'une science neuro-psychanalytique. Questions épistémologiques et mise à l'épreuve. In: L. Ouss, B. Golse (éds.), *La Neuro-Psychanalyse*. Odile Jacob.

Blakemore,S.- J., Frith, C. D. and Wolpert, D. M. (1999). Spatio-Temporal Prediction Modulates the Perception of Self-ProducedStimuli *Journal of Cognitive Neuroscience* 11: 551-559.

Freud, S. (1895/1950). *Entwurf einer Psychologie*. G.W., Nachtragsband, pp. 375-477.

Freud, S. (1895/1966) Project for a scientific psychology (Stratchey, J., trans.). In Standard Edition I, pp. 281–397/410. Hogarth Press, London. (Original publication in 1950.)

Freud, S. (1915). Instincts and their vicissitudes. *Standard Edition* XIV: 117-140.

Helmholtz, H. (1878/1971) The facts of perception. In Kahl, R. (ed.), *Selected Writings of Hermann von Helmholtz*.Wesleyan University Press,Middletown, CT.

Lenay, C. (2006) Enaction, externalisme et suppléance perceptive. *Intellectica*, 43, 27–52.

Shevrin, H. (1998) Why do we need to be conscious? A psychoanalytic answer. In Barone, D. F. Hersen, M. and Van Hasselt,V. B. (eds), Advanced Personality, Chapter 10. Plenum Press, New York.

Stainier, D.Y.R. (2005). No organ left behind: tales of gut development and evolution. *Science* 307: 1902-1904.

2.10 The Vision, Revisited

The Mental Apparatus for Complex Automation Systems – A Combined Computer Scientific and Neuropsychoanalytical Approach

Dietmar Dietrich, Mihaela Ulieru[74], Dietmar Bruckner, Georg Fodor

The INDIN/ENF [16] heralded the start of a paradigm shift in the design of intelligent systems. As the emerging community of interest further investigates ways to move forward on this path, we herewith underline how, if several boundary conditions are met, the hypotheses and suggestions which were postulated at the first meeting, can be backed by scientific and technological advances tested in concrete applications and projects.

2.10.1 Motivation

Any paradigm shift must be based on substantial arguments backed by scientific evidence that serves as hypotheses for further research to either support or dismiss the initial arguments. Our initial claim is that, in order to break through the current limitations to designing human-like intelligent technology, scientists in the area of artificial intelligence need to pursue new paths to capture the essence of the human mind in its entirety, and to simultaneously retain scientific consistency. Based on the Forum deliberations (see also the DVD of the [16]) we provide, as a strong foundation for the emerging community pioneering these efforts, examples in this regard to substantiate these claims by evaluating the status quo and at the same time setting out directions for future research. To achieve our goal we will need concerted interdisciplinary effort bringing together engineers and psychoanalysts in innovative projects and working at the junction of a paradigm shift that stretches the boundaries of both disciplines in creating a new platform for automation science capable of tackling the high degree of complexity that has become pervasive in all areas of automation (with security, safety, geriatric care, energy management, and building automation as important examples) as well as the whole information technology revolution driven world [10]. In approaching complexity it is natural to look at nature – of which we are living proof, for successful ways to intelligent evolution. What follows is our hypothesis for further discussion.

2.10.2 Challenges in Automation

Historically, starting from the trick fountains that measured the time during court hearings in ancient Egypt and followed by the "marvels" like Leonardo da Vinci's timepiece, which possessed various clever mechanical processes to measure time in a continuous fashion, automation was concerned with executing

---

[74] This work was supported by the HarrisonMcCain Foundation.

400

processes without the direct intervention of humans[75]. The momentum created by the advances in automation led to the industrial revolution – when it became possible to using large amounts of energy for animating devices or controlling processes. Progress in electric generator and propulsion technology resulted in electronic control becoming pervasive. What characterized automation at that time was the mechanical-electrical dichotomy manifested on the physical plane – energy was used to supply the power to either alter something physically or chemically (like move objects or synthesize substances), or to direct or control a process. As with da Vinci's timepieces – within this dichotomy – the flow of energy is directly coupled to the flow of information.

The computer substantially changed these structures and enabled the separation of the information flow from the energy flow, Fig. 2.10.1. The current time to be displayed by watches at train stations, for example, is computed and controlled by a computer. The computer calculates the necessary control signal – the information – for the motor to maneuver the clock's hands to display the current time.

In former times steam engines were mechanically controlled by a device operated by centrifugal force, which – just like in da Vinci's timepiece – was controlled by mechanical information flows. In contrast, today's control devices, of which the most current ones are networked embedded control systems, are composed of a nano-scale computing and peripheral devices that control the actuators via information signals flowing through networks, e. g. so-called eNetworks.

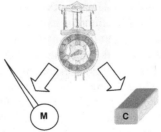

Fig. 2.10.1 Separation of energy flow and information flow
bottom left the moving mechanical part (M: Motor),
bottom right the "intelligence" behind (C: Computer)

This principle seems to be obvious in retrospect, but evolution also needed time to take this step forward. As an example one can compare the amoeba with a bug as shown in [8]. The bug utilizes a dedicated apparatus for controlling itself – the nervous system. If in a next step one compares the information processing architecture of a bug with that of a human, further development in automation is clearly noticeable: data about the processes have to be acquired with increasingly diverse

---

[75] Automation mainly aims at taking the monotonic work load from humans, increasing precision, saving energy, and controlling processes that exceed human ability to handle.

and precise sensors, and the performance needs to be steadily enhanced for data processing – which exactly mirrors the current developments in automation. As the complexity of systems increases with the ubiquity of communication infrastructures, the number of sensors is increasing dramatically, thanks to micro- and nanotechnology and more and more information is being processed and provided (saved) for other tasks. Communications systems not only connect sensors and actuators with their controllers, but also whole networks are interconnected resulting in systems of systems connected by networks of networks with hybrid characteristics integrated via a unified information communications technologies (ICT) infrastructure, uniting all former standalone processes. This enabled Computer Integrated Manufacturing (CIM), among other things, an aspiration of automation engineers for 25 years, to become a reality, while opening perspectives that were previously unimaginable – like the deployment of systems that merge the physical and the virtual into the novel Cyber-Physical Ecosystems that are being applied pervasively in many areas, from safety and security to green electricity distribution, vehicular technologies, homecare and building automation [12].

This implies increased complexity. It is usual today that modern buildings have several thousand embedded systems (computers) installed in order to control the process "building" [7]. Similarly in modern automobiles hundreds of embedded systems perform their functions, which would be impossible to maintain efficiently or operate safely without the help of automation. Or if the low-energy and passive houses are to be introduced over a wide area, energy has to be controlled adequately. A parallel with the way humans consume and store energy is a very good example in this respect. Without the nervous system we would not have such an efficient energy household that reacts dynamically to changes and challenges from the outside. And energy control is just a small part of the 'human system' process that is necessary for us to survive. This sheds light on the future directions for automation of which we will focus on two essentials: the separation of energy and information flow and the requirement to acquire as precise data as possible to allow efficient process control. However using the human body as source of inspiration is just the tip of the iceberg. In spite of the enormous investment in developing innovative solutions for intelligent systems, Artificial Intelligence failed in general to emulate substantial capabilities of natural systems, let alone human abilities, which therefore cannot yet be used by engineers in creating devices that would display similar capabilities. Natural adaptation of a device to a particular human, including the emotional status and instant assessment of situations, anticipating needs or learning from experience are still technologically in their infancy. Such capabilities would provide automation with a new quality of artificial natural-like intelligence that would transcend the traditional incremental improvement solution through a disruptive shift in paradigm with its origins in the deepest possible understanding of nature and its most intimate principles and solutions for information processing. Psychoanalysis offers the knowledge and hypotheses for such a radical shift towards embedding human-like experience into machines[76].

2.10.3 Functional Model vs. Behavioral Model

Although the discussion whether to use a behavioral or a function model is very controversial in psychology, it was not articulated at the INDIN/ENF (see the DVD of the [16]).

While both methods have advantages and disadvantages, depending on the area of application, from the automation perspective the results are usually mathematical formulas or can be evaluated by practical experiments [3]. The goal is to design a model which can be used for the development of a device intelligent enough to perform complex tasks, such as, for example, to be able to recognize a dangerous situation as described in [31, p. 57]: A small child enters the kitchen and a hot pot is found on the stove. No adult is present.

To design an automated device that would be able to protect the child it is not sufficient to know the behavior of the process, but it is essential to define its functions. Therefore, in the example with the child the important feature of the control device is to recognize and handle the complex relationships in order to anticipate any possible danger. The key aspect here is complexity. It is practically impossible to control such a complex situation by instructing the device what to do at each step – thus the traditional AI methods that would involve, for example, either the composition of tables with the information, specifying what has to happen in which case or to use rule-based algorithms which define the behavior as a reaction to several circumstances (see also [27]). However, such behavior-based descriptions can only be seen as a simplified option for solutions to tasks which do not meet the needs of complex real-life situations, be they as simple as in the example considered. To approach this, one has to take the next step by analyzing the system and splitting it up into its functional entities. Normally, functionalizing results in an infinitely more complex process which is not feasible to pursue, however it is the only way to approach reality[77].

---

[76] This may be further used as inspiration in the design of large scale adaptive systems by using eNetworks as the nervous system of the ICT controlled 'ecosystems' such as, e.g. 'energy webs' - energy networks capable of adapting energy production and distribution based on natural user demand; holistic security ecosystems animated by network enabled operations bringing together ad-hoc first responders and networked devices/weapons to respond timely to unexpected crises; hazard free transportation (automotive networks for aerospace and avionics) and eHealth – homecare and telecare, disaster response and pandemic mitigation [34].

[77] This was one of the crucial points that prompted Dietrich to decide using psychoanalysis in 1999 and not another scientific direction, like behavioural psychology for example [8]. Behavioural psychology is able to explain things in an easier fashion, but the question is whether this kind of modelling is close to reality. On the other hand, the disadvantages of psychoanalysis are its enormous complexity and the fact that it invokes opposition in many people against its way of thinking.

Considering another example leads to the similar conclusions: It is hardly possible to describe the behavior of a PC due to its complexity. However, such a description would not even be useful if one wished to use it to build another PC. Instead, considering the structure of its functions and their mutual influences – hence, a functional model – would rather do the job.

From this we can conclude that while with a behavioral description it is possible to analyze or verify how a process behaves and under which circumstances, in order to design one, a *functional model* is required. Once this functional model is in place, it may be evaluated using the behavioral model of the system at hand.

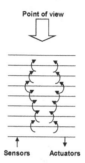

Fig. 2.10.2 Hierarchical model of the brain

In the sequel we will elaborate on why the behavioral model will not lead to success when attempting the design of complex automation systems. Considering the hierarchical model of [23] – if it is seen to correspond well with reality, it is important to recognize that the knowledge of the highest (ternary) level – was very limited in his time. Extrapolating his approach, we can define a structure as depicted in Fig. 2.10.2. Additionally, psychoanalysis assumes that the functions in the higher levels are organized as a partial hierarchy, which leaves some degree of autonomy to the distributed entities to communicate according to their needs (heterarchy).

It is general knowledge that the human body has numerous control structures, both mentally and physically. According to Fig. 2.10.2 this means that all levels are interconnected and have influences and feedbacks on and to each other. The control structures within the levels take on stable states which are dependent on their boundary conditions. In this structure, while observing the behavior of the upper level, it is possible that one of the lower levels is altered which leads to significant change in this level. On the other hand the lower control structures manage to keep the system stable – which cannot be considered while observing the upper behavior. One main reason is because the middle layers are not yet known and described, there is just a vague description of their functions [10], [11], [23].

In this sense it is clear that e.g. statistical analysis of multiple nested control loops, which are additionally mutually connected and non-linear, cannot help in

404

finding the functional model thereof. This has to be achieved in another way. All this leads to psychoanalysis.

2.10.4   Psychoanalysis

Artificial Intelligence and psychoanalysis were mentioned together for the first time in [32; p. 241] and the question was raised as to whether there should be co-operation between the two fields.

At the WFCS [8], Dietrich for the first time presented a model based on consi-dering cooperation between those disciplines and worked out by his team in Vien-na. These were rudimentary considerations [10], however today we need to ques-tion why these considerations came up at exactly that time and if they make any sense at all.

Ms. Turkle [32] points out the contradictions between the disciplines and per-ceptively analyzes the root of the clash between the two branches of science. In spite of the mutual prejudices, engineers have increasingly embraced the vocabu-lary of psychologists, without building the necessary foundation that would ensure proper use. In their attempt to apply ideas from psychology engineers built a theo-retical patchwork based on superficial knowledge – given that, in the engineering approach it didn't matter whether it matched the original theoretical psychology (as per [2, p.44]). One characteristic of the engineering method was to build a ma-thematical formalism, which in such cases led to useful results (see also [20], HBM[78]). Given that the task was to produce a solution to some concrete technical problem rather than to build a model of the human mental apparatus – the task proved useful (as confirmed by HBM).

Aside from these solutions, no significant progress on modeling the mental ap-paratus is worthy of note [15]. We attribute this to the impossibility that engineers would have in educating themselves in the psychoanalytical field. One has to con-sider that the academical training of psychoanalysts itself takes – according to WAK[79] – longer than the average education of an engineer. Thus it makes a lot of sense to bring the communities together in a common endeavor when researching the functions of the human mental apparatus in an interdisciplinary fashion. Even then, it is essential to avoid inherent confusion that may result from the unavoida-ble clash between the two so different sciences. Therefore it is suggested that a start be made by constructing a unitary model which is consistent to only one school, rather than 'mixing-and-matching' insights and concepts from various schools, or contradicting schools from the humanities as per [2], where, in spite of the dauntingly colorful mixture of terms and theories would in no way account for a functional model of the human mental apparatus if put together without further research into their compatibility. A design procedure based on such considerations could potentially help to construct a particular robot, which could, at most, copy several typical behavioral patterns, it is questionable whether we would consider

---

[78] http://www.seas.upenn.edu/~barryg/HBMR.html

[79] Wiener Arbeitskreis für Psychoanalyse; http://www.psychoanalyse.org/

them 'human-like'. To clarify the bias here, one has to distinguish between what a robot is and can do autonomously and what a human observer projects into what the observer perceives. Moreover, we need to distinguish between human-like appearance (Fig. 2.10.3), human-like behavior (as per [2]) and human-like *thinking* – which we claim to be a function performed by the human mental apparatus, which definitely cannot be assumed for the two objects[80] presented in Fig. 2.10.3.

Fig. 2.10.3 Are these human-like? (Source of left picture: Honda)

Of course a natural question now is: Why is psychoanalysis considered promising for robot and automation technology? And why is the time ripe for it right now?

In Section 2.10.2 of this article was dealt with why statistical models from behavioral psychology cannot provide design policies. Furthermore, the humanities – the science disciplines concerned with – among other things, also – this issue – provide no other convincing and exhaustive functional psychological models than the one originally postulated by Sigmund Freud[81], making it quite straight forward to use that. While other mental models provide widely accepted models of human behavior, psychoanalysis is deeply rooted in the functions of human mental control loops as a major drive in their actions. It is therefore the goal of the authors to found a community to build an automated system capable of encapsulating this powerful mechanism.

With the advent of pervasive information and communication technologies society is undergoing a radical transformation, from a command economy to an eNetworked ecosystem characteristic of the eSociety. The pervasiveness of networks linking large, sophisticated knowledge repositories managed by intelligent agents – is becoming more and more a universal "brain", capable of finding and answering almost any question. Linked by eNetworks, global enterprises and businesses merge seamlessly into a forever growing open market economy in

---

[80] The ASIMO robot is a technical masterpiece with previously unseen motor abilities. However, it is still far way from human-like consciousness.

[81] The authors could not find any other functional model during many years of studying literature and talking to experts in the humanities.

406

which dynamic adaptation and seamless evolution are equivalent to survival. The Global Collaborative Ecosystem is becoming increasingly hybrid [35].

Therefore, there is currently an urgent need for strategies to deal with the large degree of complexity that the information technology revolution will bring, as envisaged some time ago by the authors [7], [8], [9], [34]. It is in this spirit that we want to use recent results obtained on this path to design solutions for intelligent automation [31], [29], [28], [24].

2.10.5 Neuropsychoanalytically Inspired Model

The clash between psychoanalysts and engineers however goes far beyond their different jargon - into, their different way of thinking and work methods. After all most psychoanalysts work in therapy rather than in theoretical model building, thus for them the natural scientific point of view is of minor relevance. The clash can, at most, result in an acceptable compromise, as the perspectives are so diametrically opposed that there is little chance of converting either side. For example, engineers often refuse to accept scientific findings if they don't seem logical to them (e.g. engineers would go as far as to integrate a state machine in a bionic model – where it does not belong at all [18]). On the other hand psychoanalysts often refuse to give up their mechanistic way of thinking (Freud's imagination of mental energy is still part of the daily vocabulary of psychoanalysis, yet it contradicts the above mentioned requirement of separating energy flow from information flow).

While the jargon may be dealt with via a learning process, we consider a larger obstacle to be the essential divergence of the two schools of thought. In this respect there are many potential traps and pitfalls, of which we will underline two.

(1) While the engineering approach is based on the formulation of a unitary model that is valid for all systems of the kind the model was built for, psychoanalysts instantiate the universal model for each particular individual case. Thus, while for engineering the hypothesis is universal and valid in all cases, for psychoanalysts an explanation or hypothesis can only hold true in one instance, for a particular individual patient. While, for example, to a psychoanalyst the topographical model of Freud (Id, Ego, Super-Ego) is not required to match the unconscious-conscious model on all points, such an inconsistency would be unacceptable to an engineer.

(2) The second example: In computer science one makes a distinction between hardware, software, and application[82]. An application behaves according to the underlying software on which it is built (which comprises the algorithms and the programming) and on the hardware on which it runs. Hence, a behavioral description can be formulated for the application – as a function of the particular software and hardware. Looking at the neuron in a similar manner, as a small, particular computer [14] (as was proven already in 1976 in [21]) – the human brain can be regarded as a heavily distributed

---

[82] The application is the utilization view, i.e. the functionality provided to the user.

computer system. With this definition the functional distinction of hardware, software, applications and their behavior can also be applied to psychoanalysis. This leads us to conclude that not only is research on the information theoretical aspects of psychoanalytic and neuroscientific concepts done with mechanistic methods, but psychoanalysis also refuses to distinguish between the possible descriptions[83].

So, although full of promise, we anticipate the path to using functional models inspired by psychoanalysis for intelligent systems to be a difficult and stony one.

The new Bionic Approach

When considering a top down design, psychoanalysis offers two possible models, so-called topographical models. The first model distinguishes between the system unconscious and perception/consciousness, the second model describes the Ego, Id and Super-Ego. These are two models describing specific phenomena. They overlap in large areas, but seem to be very hard to unify, as the sustained efforts invested in this thus far have proven. To facilitate the task of unification we propose an engineering approach that starts with simple theoretical constructs, leaving aside initially the unconscious-conscious model, which cannot be easily integrated into the Id-Ego- Super-Ego model without leaving too many unanswered questions.

As postulated in the respective presentations at the ENF, we suggest that the second topographical model initially be modularized into its functional entities, while clearly defining and specifying the interfaces between them. A question raised at the ENF (see also the video of the [16]), but not discussed, was the question whether the psychoanalytic model should be integrated – in the sense of Mark Solms – in the neurological model that distinguishes core consciousness and extended consciousness.

The theory of psychoanalysis needs to be experienced by oneself by going through a psychoanalysis in order to internalize it, which makes it difficult to grasp by traditionally trained engineers. However, many areas are based on the Freudian topographical model which is very well functionally organized, so one can envision ways towards a common approach by e.g. augmenting the Freudian topographical model (of the mental apparatus) with the neurological model of Alexander Luria [23] (of the brain) as a functional description of the perception system, which would result in a holistic model to which computer engineers can relate. Hence, this modular approach can be further subdivided into functions which can be researched separately while they still interact with the rest of the model in an orchestrated, harmonious manner as stated in [6, p. 154].

Now let us see how one can tackle the concept of consciousness with these basic considerations.

---

[83] The authors are aware that up to now there was no necessity, since Freud turned away from neurological observations.

Core Consciousness

The representation field constitutes the main part of the core consciousness module [16]. The inputs thereof are data (images and scenarios, which last just moments in time [6, p. 29]), which are obtained via symbolic relations between sensor values. These inputs are associated with other images and scenarios from memory and weighted by the emotions. The result generates new evaluations that are further imprinted on the newly stored images and scenarios. Over time everything is matched within core consciousness and therefore generates (emotionally weighted) reactions [6, p. 29] - there are no other functions performed by the core consciousness module, even time or history does not play a role here (past and future are not taken into account - aka are not perceived)[84]. Only the 'here and now' is of interest. The being is equipped with this 'core' functionality to react efficiently and optimally to external circumstances.

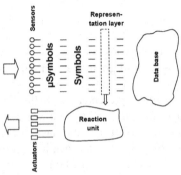

Fig. 2.10.4 Information flow from sensors to the „data base".

The core consciousness module depicted in Fig. 2.10.4 – does not possess feelings or consciousness in the everyday sense – thus the reaction unit is not very large. The main part of the core consciousness module consists of the perceptions data base which stores previously seen images and scenarios. It may be assumed that this data base is very small or does not even exist in the case of primitive creatures taking into account their relatively small number of sensors. Such creatures are considered to be "hard-wired" or pre-programmed through their genes, although biology still has to confirm this assumption. The structure of the symbolization network, from the sensors to the database, compresses (and does not extend, as might have been anticipated) [4], [26], [36], and [3] while the structures of the outputs underlie the exact opposite.

Thinking of modern robots with capabilities like dancing or riding a bicycle, their intelligence might be rather limited compared to primitive creatures. The ca-

---

[84] E.g. a worm does not know about yesterday or tomorrow, it is just aware of the action about to happen now.

pabilities of even the most intelligent robot should not be overestimated in spite of the enormous sophistication of the control algorithms required to perform these functions. A bee might possess more intelligence, but, as mentioned above, this is still an open question for biology.

So, what is the advantage of a core consciousness module compared to the usual control loop algorithms? Core consciousness can be seen as lying above the basic control loops in creatures' bodies like the reflex loop. Just like a control algorithm, core consciousness provides a mechanism that decides how the body should react (in automation such a system is called real-time system). However, unlike in many technical systems, the data is not being analyzed directly. Instead perception is essentially based on the comparison of incoming symbolic data which are associated with inner scenarios and images and directed to the evaluation system – as per Fig. 2.10.4.

On Embodiment

With these considerations in mind we will attempt to answer a fundamental question: Do we need embodiment when trying to simulate or eventually emulate the mental apparatus? Considering the mental apparatus to be evolution's solution as the control device for living creatures, with the ultimate purpose of survival of the individual (body) which is imminently related to the survival of the whole species, a natural result is that *all* capabilities of the mental apparatus aim at fulfilling needs of the body (or the social entity). In this light intelligence modeled on the mental apparatus cannot be thought of without the needs of the respective body. With respect to modeling such a 'mental apparatus' for technical devices, the body can be defined as a functional entity and interfaces between 'body' and 'mind' can be specified. Such interfaces and closely related functions can be structured hierarchically [23].

Extended Consciousness

For the functionality of the extended consciousness according to [6, p. 195] the principles are basically the same as before, but have to be considered in a more complex context in which the mental apparatus additionally takes past and future into consideration, which makes the *control system* "mental apparatus" a complex, multiple nested control loop containing several underlying control loops[85], as depicted in Fig. 2.10.5. Extended consciousness takes into account previous perceptions and evaluations and *relates them to oneself* in finding feeling-based[86] decisions that further match previous experiences. Feelings in this definition are again evaluations, but in contrast to the emotions identified in core consciousness, feelings always evaluate in relation to the impact on oneself. Recognition or evalua-

---

[85] Fig. 2.10.5 is a very coarse depiction, which can be refined or extended depending on the aspects one wants to enhance it (e.g. with the hormones system of humans).

[86] English language just knows the term *emotional*, although in this case feelings are involved.

410

tion is not available to core consciousness. (A bee for example, of which can be assumed to possess only core consciousness and no extended consciousness, does not feel pain, it just perceives it.).

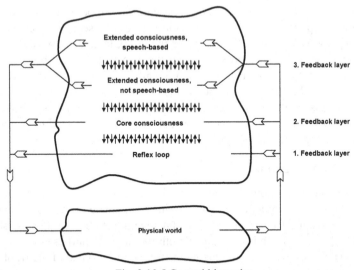

Fig. 2.10.5 Control hierarchy

In this top level control loop two representation layers, independent of the one held by the core consciousness, are assumed: Humans have a mental representation of the physical world. This includes both, objects outside the body and the body itself. These two together form the outside world, while the mental representation thereof is (part of) the inner world[87]. Additionally, humans have a representation (imagination) of themselves, the *self*. The self is also part of the inner world. In humans consciousness seems to originate in the relationships between the self and recognized objects from the outer world. Experiences about this relationship create feelings. The major problem with modeling this concept is that the self is not a fixed representation, but dependent on various influences [30, p. 279], like the state of the homeostasis, psychic condition, outer influences, etc. The self is in an ongoing flow and will never arrive at a state again that it has lived through once already.

With this, another hypothesis can be postulated: In robotics researchers are inclined towards modeling state machines [1]. For the mental apparatus, this approach has to be rejected, because, in light of the above one can never reach the exact same state a second time.

---

[87] The terms *outside* and *inside* world are seen in relation to the mental apparatus. Therefore, the body is outside, too.

The extended consciousness module has even more capabilities. Humans can not just look back in the past, but are also able to plan the future, to think and try out [6, p. 197]. This means that based on experiences creatures with extended consciousness think of alternatives in some kind of movie-like imagination in order to find the most advantageous reaction to a given situation.

Thinking

This trial-action is what is called *thinking*. How this works in detail, is as yet unknown. According to Solms [30, p. 209] this process is related to dreams, in which the sensory input is just generated as desired and real sensor values from outside are filtered. This raises the following questions for the computer scientist: do these trial-actions run in parallel, or do they run sequentially? How are they initiated? How are they terminated? What is the basis for decisions? Is the evaluation based on feelings[88]?

It is of the essence that these 'trial actions' follow pathways that were learned in the past. This means that current perceptions are continually overlaid with patterns from childhood – all unconscious. The way adults perceive the world is very different from the way children do, the adult world is perceived partially, and with much bias. If one wants to fully understand what someone else means, all the interpretations will be rooted in interpretations made by oneself acting-as-if, one's own learned pattern, which obviously cannot fully comply with the ones of the person to be understood. Interpretations of the behavior of another person can principally be no more than estimates. We have to accept that everybody lives in their own world. Ultimately, these considerations lead to a definition of the range of the functions of language: it shall transport the subjective imaginations of the speaker (including sensations from sensory modes) to allow the listener to interpret what the speaker desires to communicate. Since this is in principle not always possible – because of the different past and therefore the different allocation of terms – many interpretations remain fragmentary.

The presented concepts of reflex arcs, core consciousness and extended consciousness of humans reveal to the control and automation engineer a hierarchical concept of nested feedback loops (depicted in Fig. 2.10.5), in which every layer needs the lower layer, and every higher layer logically contains a higher functionality – and reacts slower – following control theory in this respect. Going into more detail one will soon recognize that the triple pack in Fig. 2.10.5 needs to be further subdivided. Consequently, each layer with enhanced "intelligence" capabilities for ever more complex processing will have a correspondingly lower speed of response.

---

[88] The term *feeling*, which according to Damasio clearly describes an evaluation instance, may not be confused with the colloquial term feeling – which has quite diffuse meanings.

2.10.6 Consequences and Conclusions

The ENF Workshop has ignited an unprecedented dialogue between engineering and neuro-psychoanalysis that not only showcased the enormous progress in brain sciences over the last decades but also demonstrated that psychoanalysis can serve as a strong foundation for the development of extremely useful engineering models.

In spite of the usual refusal from the mainstream engineering school of thought – we can already easily dismiss several erroneous attempts at solving the new, complex problems with paradigms from the old engineering school of thought - still sucking large amounts of funding into dead-end research. As a generic example consider Fig. 2.10.4, derived from the work of Solms and Damasio, showing that imaging technology-based studies of the brain can only provide an analysis of simple tasks involving neurons nearby sensor data, such as e.g. if one decides to go right or left, or to raise a hand, etc. – when a mechanism similar to a device driver[89] – in the language of computer engineers – has to be executed by the mental apparatus – which, can thus be localized in the brain for such simple 'decisions'.

However, if more complex considerations are undertaken in higher cognitive areas, such as speaking, feeling or thinking, respectively planning, this involves the whole brain – which was already postulated by Sigmund Freud 100 years ago [17]. The idea of finding concrete neurons from which thoughts can be read contradicts neuro-psychoanalytical models, being as unfeasible as the idea of finding the behavior of a complex system by analyzing the behavior of its parts [34].

Another misconception that our work dismisses on similar grounds is the idea that tomography can help investigating consciousness. Even if it had the resolution to depict the impulses of single neurons or synapses - imaging can only be of auxiliary help - while often lacking scientifically sustainable interpretations, especially if not restricted to the lower neurological layers. A trivial example from computer science easily makes our point: it is like attempting to analyze the functionality of the text processing program WORD (by Microsoft) by scanning myriads of transistors ($\approx$ synapses) in the computer hardware. (And let's take note that a computer is much (!) simpler in its design than the human brain ...). Any simple function of WORD – such as inserting a letter – triggers bunches of transistors thus leaving the researcher puzzled in its attempt to explain the complex functionality by looking at the behavior of the elementary components (transistors). The fundamental bias here is that one cannot achieve complex functionality by instructing every elementary element exactly what to do at every step. The attempt to find which transistors account for the respective function is an impossibility. Signals would be viewed as a whirlpool of events. Such a synthesis reveals though that below the application software (WORD in our case) lie the operating system,

---

[89] A device driver in computer science is a program that allows other parts of the software access to hardware.

device drivers, etc. Thus to understand how WORD is processed in the computer (aka complex thought process in the brain), one first has to understand the operating system – which luckily enough we are not required to do after leaving the university undergraduate benches. The idea here is that to synthesize behavior one has to scale the system by clustering the components accounting for various functions – which can be done at several levels of resolution [33]. Such considerations - also with philosophical implications – make it clear why humans will never be able to objectively recognize reality, because we always think in models via the association of images and scenarios which we already learned. We live in a subjective world, the physical world will always be only accessible to us indirectly through our models.

Another point worth mentioning here is the quite popular view in brain research that the brain does not work like a computer – which is in fact not based on any scientific evidence. Even if one is to look at it from the most popularly accepted viewpoint, as a device that processes information [13] [19] [25] – a computer is certainly very similar to the brain!

A brief look at the history of computers further confirms this: The Z3 built by Zuse in 1941 is today globally accepted as the very first computer. It was a fully automated calculator for binary floating point arithmetic. From the perspective of automation the Z3 is a device with one single task: to process incoming information and use the results to control processes. Recalling now the process clock from Fig. 2.10.1 we would talk about a device that provides the mechanical power to operate the clockhand and a computer to control how that has to be done. Based on this preamble, a computer can be defined as a system which processes information. Since '41, there have been many attempts to enhance the computer's functionality, starting with analogue calculators and threshold systems to the nowadays attractive quantum computer. However, if we do not apportion the brain metaphysical phenomena, it is very hard to find a principal difference in task and function between the brain and a computer.

If we are to truly achieve the breakthrough in computing that will enable the processing of complex tasks required for 21st Century problems [5] we will have to break through the current paradigms and redefine computing and its abstractions [22] in terms that will enable it to conquer the new challenges. The ENF community strongly believes that the psychoanalytical perspective which is convinced that allowing a natural scientific top-down description of the functions of the mental apparatus– is one way towards achieving this breakthrough. Thus we are determined to capitalize on the achievements of psychoanalysis by building the new computing abstractions for the next generation (fifth generation of Artificial Intelligence) intelligent machines. Such abstractions would – in addition to the future human-friendly technology and environments – be extremely useful in the design of the much needed 'intent analysis' systems [37] capable of capturing e.g. the intent of a human – being the information used for increasing comfort, security, safety, etc.

With this said – we, the ENF community are ready to redefine artificial intelligence, cognitive science and brain research by bringing the insights of psychoanalysis into the engineering world.

References

[1] Argall B., Browning B., Veloso M.: Learning to Select State Machines using Expert Advice on an Autonomous Robot; In Proceedings of ICRA'2007, Rome, Italy, April 2007

[2] Breazeal, C. L.: Designing Sociable Robots; The MIT Press; 2002

[3] Bruckner D.: Probabilistic Models in Building Automation: Recognizing Scenarios with Statistical Methods, Dissertation Thesis, Vienna University of Technology, 2007

[4] Burgstaller W.: *Interpretation of Situations in Buildings,* Dissertation Thesis, Vienna University of Technology, 2008

[5] Systems Thinking: Coping with 21st Century Problems, CRC Press, ISBN 978-1-4200-5491-0

[6] Damasio, A. R.: The Feeling of What Happens: Body, Emotions and the Making of Consciousness; List Verlag (Econ Ullstein List Verlag GmbH); 1999

[7] Dietrich, D.; Loy, D.; Schweinzer, H.-J.: LON-Technologie; Verteilte Systeme in der Anwendung; Hüthig Verlag, 2. Auflage, 1999

[8] Dietrich, D.: "Evolution potentials for fieldbus systems", IEEE International Workshop on Factory Communication Systems WFCS 2000; Portugal (2000)

[9] Dietrich, D.; Kastner, W.; Sauter T. (ed.): EIB Gebäudebussystem; Hüthig, Heidelberg, 2000

[10] Dietrich, D.; Kastner, W.; Schweinzer, H.: Wahrnehmungsbewusstsein in der Automation - ein neuer bionischer Dekansatz; at, 52. Jahrgang – März 2004, pp. 107-116

[11] Dornes, M.: Der kompetente Säugling - Die präverbale Entwicklung des Menschen; Geist und Psyche Fischer; Fischer Taschenbuch Verlag, 2001

[12] Doursat and Ulieru: Design by Emergence: The Challenge of Engineering Large Scale eNetworked Systems, IEEE Transactions on Systems, Man and Cybernetics - Part A, Special Issue on Cyber-Physical Ecosystems (to appear in fall 2008)

[13] Duden - Informatik; Dudenverlag, Mannheim, Wien, Zürich, 1989

[14] Eccles, J. C.: The Understanding of the Brain; McGraw-Hill Book Company, New York; 1973

[15] Eccles, J. C.: How the Self Controls its Brain; Springer Verlag Berlin, Heidelberg, 1994

[16] Conference Proceedings ENF 2007; 1st International Engineering & Neuro-Psdychoanaysis Forum der IEEE/INDIN 2007; Dietrich, Fodor, Pratl, Solms; Vienna; July, 2007
DVD: to buy from Dietmar Dietrich: http://www.ict.tuwien.ac.at

[17] Freud, S.: Zur Auffassung der Aphasien; Eine kritische Studie; Herausgegeben von Paul Vogel; Psychologie Fischer; Fischer Taschenbuch Verlag, 2001

[18] Goldfarb L., Gay D., Golubitsky O., Korkin D., Scrimger I.: What is a structural representation? A proposal for an event-based representational formalism;
http://www.cs.unb.ca/ ~goldfarb/ETSbook/ETS6.pdf; 2007

[19] Hayes, J. P.: Computer Architecture and Organization (2nd ed.). New York, NY: McGraw-Hill Publishing Company, 1988

[20] Johns, M.; Silvermann, B. G.: How Emotions and Personality Effect the Utility of Alternative Decisions: A Terrorist Target Selection Case Study; 2001

[21] Lange, D.: Aufbau zweier Neuronenmodelle mit modernen Bauelementen nach Küpfmüller-Jenik zur Nachbildung gegenseitiger Erregungs- und Hemmvorgänge bei neuronaler Verkopplung über exzitatorische beziehungsweise inhibitorische Synapsen; Diplomarbeit; 1976

[22] Lee, A.E, Computing Foundations and Practice for Cyber-Physical Systems: A Preliminary Report, UC, Berkeley, UCB/EECS-2007-72, May, 2007

[23] Luria, A. R.: Working Brain: An Introduction to Neuropsychology. London, UK: Penguin Books Ltd., 1973.

[24] Palensky, B.: *From Neuro-Psychoanalysis to Cognitive and Affective Automation Systems,* Dissertation Thesis, Vienna University of Technology, 2008

[25] Patterson, D. A.; Hennessy, J. L.: Computer Organization And Design - The Hardware/Software Interface; Morgan Kaufmann Publishers; San Mateo, California, 1994

[26] Pratl (Zucker) G.: *Processing and Symbolization of Ambient Sensor Data,* Dissertation Thesis, Vienna University of Technology, 2006

[27] G. Pratl, D. Dietrich, G. Hancke, W. Penzhorn: "A New Model for Autonomous, Networked Control Systems"; IEEE Trans. on Industrial Informatics, Volume 1, Issue 3 (2007), S. 21 - 32.

[28] Rösener Ch.: *Adaptive Behavior Arbitration for Mobile Service Robots in Building Automation*, Dissertation Thesis, Vienna University of Technology, 2007

[29] Russ G.: *Situation Dependent Behavior in Building Automation*, Dissertation Thesis, Vienna University of Technology, 2003

[30] Solms, M.; Turnbull, O.: The Brain and the Inner World: An Introduction to the Neuroscience of Subjective Experience. New York, NY: Karnac Books Ltd., 2002

[31] Tamarit C.: *Automation System Perception – First Step towards Perceptive Awareness*, Dissertation Thesis, Vienna University of Technology, 2003

[32] Turkle, Sh.: Artificial Intelligence and Psychoanalysis: A New Alliance; editor: S. R. Graubard; MIT Press, Cambridge, MA; 1988

[33] Ulieru M. and Este R., "The Holonic Enterprise and Theory of Emergence", *International Journal Cybernetics and Human Knowing* (Imprint Academic), Vol. 11, No. 1, 2004, pp. 79-99.

[34] Ulieru M. Evolving the 'DNA blueprint' of eNetwork middleware to Control Resilient and Efficient Cyber-Physical Ecosystems, Invited paper at BIONETICS 2007 - 2nd International Conference on Bio-Inspired Models of Network, Information, and Computing Systems, 2007.

[35] Ulieru M. and Verdon J. "IT Revolutions in the Industry: From the Command Economy to the eNetworked Industrial Ecosystem" - *Position Paper at the 1st International Workshop on Industrial Ecosystems*, IEEE International Conference on Industrial Informatics, Korea, 2008.

[36] Velik R.: *A Bionic Model for Human-like Machine Perception*, Dissertation Thesis, Vienna University of Technology, 2008.

[37] Werther, G., Profiling 'Change Processes' as a Strategic Analysis Tool, Competitive Intelligence Magazine, January-March, 2000

Part IV Explanations for Engineers and Psychoanalysts

List of Authors in Alphabetic Order

Dietmar	Bruckner	Vienna University of Technology, Austria
Tobias	Deutsch	Vienna University of Technology, Austria
Dietmar	Dietrich	Vienna University of Technology, Austria
Georg	Fodor	Vienna Psychoanalytic Society, Austria; University of Cape Town, South Africa
Roland	Lang	Vienna University of Technology, Austria
Brit	Müller	Vienna Circle of Psychoanalysis, Austria; Vienna University of Technology, Austria
Anna	Tmej	Vienna Psychoanalytic Society, Austria; Vienna University of Technology, Austria
Rosemarie	Velik	Vienna University of Technology, Austria
Heimo	Zeilinger	Vienna University of Technology, Austria
Gerhard	Zucker (né Pratl)	Vienna University of Technology, Austria

Actuator (engineering)
 ... is the counterpart to sensor. Used to interact with the environment by causing an action. Actuators are mainly technical, a biological counterpart is, for example, the muscle.

Ad-hoc sensor network (engineering)
 ... is a network of computer nodes that is equipped with sensors and is able to connect dynamically with other nodes. Nodes can change their position, therefore a permanent update of the communication paths is necessary.

Affect (psychoanalysis)
 ... represents one of the two drive representations (see presentation) in the psychic apparatus. *Affect* is the qualitative form of drive energy in all its variations.

Agent, Software ~ (engineering)
... is a software program, capable of acting on behalf of a user or another program. Most important attributes of an agent are persistence, autonomous, flexible (hence reactive, proactive, and social), adaptive, distributed, and mobile. The more attributes that are implanted, the closer the agent gets to being a so called smart agent (or intelligent agent).

Agent, embodied, autonomous ~ (engineering)
... extends the attributes of software agents by embodiment, situatedness, and self-sufficiency. Thus, this agent has to have a defined body (real or simulated). The main focus in embodied agents is the agent's interaction with the world. A very important design principle is the ecological balance - the internal processing capabilities have to match the complexity of the perceivable data.

Aphasia (neuropsychology)
... is a language disorder as a result of a neurologic disease. There are aphasias affecting different aspects of language like comprehension, or speech, or language in a global way.

Application (engineering)
... is a specific process, for example, surveillance of a building, controlling the temperature in a room or controlling the gait of a robot.

ARS
... is an abbreviation for Artificial Recognition System, a research project at the Institute of Computer Technology, Vienna University of Technology, Austria. The project integrates neuropsychoanalytic principles into technical models. By initiating cooperation between psychoanalysts and engineers the project aims at creating a functional model of the human psychic apparatus that can be implemented in a technical system.

Artificial Intelligence (engineering)
... is a technical research area that studies and applies the principles of intelligence of beings.

Artificial Life (engineering)
... is a simulation of embodied autonomous agents. They are placed within a simulated virtual environment. Artificial life simulations are closely related to social simulations. The focus of such simulations lies within the understanding of the information processing and the social interactions of the agents.

Association (psychoanalysis)
... describes any connection between psychic contents. This connection may be *conscious* or *unconscious*.

418

Association (engineering)
... describes the task of combining elements of perceptions (e.g. symbols) with other elements of perception or with symbols from memory to create new symbols.

Attention (psychoanalysis)
... describes the direction of the focus of perception on certain environmental conditions or internal processes.

Automation (engineering)
... is a general term for transferring work that has been done by animals or humans to machines. Engineers create machines, which make use of electrical, mechanical or other power to take over tasks (e. g. transport restructuring of material).

Automation communication systems (engineering)
Modern automation enables communication between machines. Today communication is based on protocols, which are often organized in a hierarchically layered system.

Behavior (psychology)
... is the sum of all actions and responses an organism performs as a result of internal or external stimuli. Behaviorism is a science dealing with the observation of responses to certain stimuli.

Binary data (engineering)
... represents data with only two symbols, for example "0" and "1". Data like letters, numbers, audio or video streams can be expressed by sequences of these two symbols. The decimal number 9 can be represented by the sequence "1001".

Bionics (engineering)
... translates biological principles into technical principles. The main idea is that evolution forced living organisms including fauna and flora to become highly optimized in order to survive. The principles and actual solutions those living organisms developed are transferred into technical systems.

Body (computer science)
A body is situated in a world and perceives the environment through its own sensors. Using its own actuators it can move within the environment and is able to alter it to some extent. Additionally, it has to have the ability to maintain all the necessary resources on its own. The body can be real or simulated.

Bottom-up Design (engineering)

... refers to a design method where design is based on existing solutions. The functionality of the existing solution is enhanced piece-wise in order to finally gain a new system capable of solving other problems or committing other tasks. See also *Top-down design.*

Building automation (engineering)

A building automation system is a control system for a commercial building. Modern systems consist of a network of computerized nodes – fieldbus nodes – designed to monitor and control mechanical, lightning, security or entertainment systems in a building. The functionality of the whole system is called building automation. The most common applications today are control systems like heating, ventilation, air condition, or access control systems. See also Home automation.

Cathexis (psychoanalysis)

... describes the allocation of psychic energy (see psychic energy) to drive representatives (see presentations, affects).

Chip design (engineering)

... refers to the process of designing an integrated circuit. An integrated circuit consists of miniaturized electronic components built into an electrical network on a monolithic semiconductor substrate. Chip design is about designing a single chip solution with a lot of integrated functionality – being either digital data processing or analog or as mixed signals.

Cognitive science (engineering)

... is an interdisciplinary field that employs conclusions from biology, neuroscience, psychology, psychoanalysis, philosophy, and computer science to study the mind, especially its cognitive functions, such as thinking, problem solving, planning, and, more generally, information processing functions.

Computer (engineering)

... is a unit that manipulates data.

Computer algorithm (engineering)

... is a set of instructions to solve a computational or mathematical problem. The formal definition says: It is an effective method, which computes a solution from a defined set of input variables into a defined set of output variables. The computation performs a sequence of well-defined instructions. Effective means: for each input (within the allowed set) there will be a result within a finite time frame, whereas efficient also incorporates the necessary time.

Conflict (psychoanalysis)

... evolves from contradictory drive demands located in the Ego.

Consciousness (psychoanalysis)
... is a quality of mental life, consisting of anything the individual is currently attending to and subjectively aware of, e.g. *affects*.

Controller (control engineering)
In control theory the controller is a functional unit in the control loop that is described in mathematical algorithms.

Controller = Control unit (computer engineering)
The controller in a control system that manipulates data provided by sensors to monitor a process and uses the actuators to affect that process.

Decision function (psychoanalysis)
In psychoanalysis, the decision function is an Ego function responsible for conscious and unconscious decisions according to reality demands.

Desire (psychoanalysis)
... is used synonymously with the term *wish*.

Desire plan (engineering)
... is a known or calculated sequence of actions and events that lead to a fulfilled desire.

Digital data (engineering)
... is data represented in a digital format, in contrast to analog data, whereby the value range is not continuous.

Distributed system (engineering)
In a distributed system the components of control are distributed in the process. This is in contrast to a central control system.

Diverse sensors (engineering)
... are sensors of different types.

Drive (psychoanalysis)
... is responsible for converting physiological stimuli to psychic information. It is activated by a 'bodily need' and initializes a demand which needs to get discharged by the use of an object.

Dust, smart ~ (engineering)
... defines an approach for merging sensor, controller, communication devices and an energy source to one single chip. The devices should nearly be the size of dust particles and create ad-hoc networks.

Ego (psychoanalysis)

... is one of the three agencies defined in the second topographical model. It synthesizes psychic processes and takes up a mediation function between environment, Id, and Super-Ego.

Ego ideal (psychoanalysis)

... represents the ideal presentation of the Self and is included as a function of the Super-Ego.

Embedded system (engineering)

... is a special purpose computer system for a specific process. The computer is part of the process.

Emergence (computer science)

... is the spontaneous development of phenomena or structures at the macro level of a system based on the interaction of its elements. Emergent attributes of a system are not directly represented by the isolated attributes of the elements of the system.

Emotion (psychoanalysis)

... is the psychic evaluation of contents of perception based upon memory traces.

Energy (engineering)

... is the capability of a physical system to perform work. The law of conservation of energy postulates that within a closed system the total energy remains constant. Forms of energy are heat, kinetic, mechanical, electrical, and others.

Energy (psychoanalysis)

... was hypothesized by Freud to drive the operations of the mental apparatus and to define the relevance of representations. This usage of the term is not related to the physical term.

Environment (engineering)

In nature, the environment is the sum of all living and non-living things which occur naturally on earth. A simulated environment contains – similarly to the natural environment – all dynamic and static things that are needed (e.g. for an artificial life simulation). This could be a set of agents and energy sources (dynamic) and some rocks and artifacts (static).

Feeling (neuropsychology)

The comparison of an individual's inner and outer world leads to feelings, which represent subjective awareness.

Fieldbus (engineering)
... is an industrial real-time communication system which connects many devices like sensors and actuators with controllers. They are used to automate complex industrial systems like manufacturing assembly lines.

Fly-by-wire (engineering)
... is an aircraft flight control system where the mechanical operating mechanisms are replaced by a fieldbus. Hence, the pilots steering movements are digitized using sensors, transmitted via the communication system to the corresponding actuators for flaps.

Focus of attention (psychoanalysis)
... is the allocation of – limited – resources of the mind to selectively concentrate on a certain conscious perception or a special thought.

Free association (psychoanalysis)
In the course of a psychoanalytic session, the patient is asked to report everything that comes to their mind, thereby producing so-called free associations. (See also *association.*)

Function (electrical engineering)
... as seen from an engineer's viewpoint in contrast to behavior; it is an extension of the mathematical definition. A function is the mathematical, textual or graphical description of a (physical or virtual) object that operates according to specifications. In electrical engineering the term function is mainly use to refer to single elements of a bigger unit. Behavior on the other hand describes the temporal course of a function.

Hardware (engineering)
Within computers, hardware consists of all physical parts including the digital circuits within the processor.

Home automation (engineering)
In Europe, home automation corresponds to building automation, but is only related to private homes. In the USA and Japan, home automation furthermore includes functions of brown goods (living room) and white goods (kitchen).

Human-like
Human characteristics applied to animals or inanimate objects.

Id (psychoanalysis)
One of the three agencies defined in the second topographical model. As defined by Freud, the Id is the first agency to exist developmentally. The Id and its

contents (e.g. *drive* representatives) are *unconscious* and thus governed by the *primary process*.

Image (engineering)

... is a modality-dependent matrix (collection) of characteristic features from perception at a particular point in time (e. g. optic, haptic, etc.).

Image, mental ~, perceived ~ (ARS)

A predefined template, that describes a typical set of perceived sensory data (e.g. object X in a distance of Y units straight ahead). A mental image is a computational object that contains all sensory data, which a system perceives in one single moment in time.

Implementation (engineering)

... is the realization of a model/specification. In computer science, the implementation of software is the work of programming.

Information (engineering)

... is knowledge of circumstances and course of a perceived reality.

Inhibition (psychoanalysis)

... suppresses drive demands and therefore permits the execution of Ego functions.

Intelligent environment (engineering)

... consists of cooperating electronic enhanced devices providing automatic functionality.

Interface (computer engineering)

... is a definition of the communication between two modules. E.g. the definition of a graphical user interface includes the communication between a human operator and a computer.

Knowledge base (computer science)

... is storage in a computational system that includes known facts, collects new ones, organizes them and makes them accessible.

Linear approximation (engineering)

... is an approximation of a non-linear, mathematical function by a linear one. In the area of engineering, systems cannot be precisely described by a linear function. However, using a linear approximation of the system characteristics can be enough to describe the system between two limits that include the normal operational range. This approach reduces the complexity of the model of the described system and simplifies further calculations.

Machine perception (engineering)
... concerns the building of machines that sense and interpret the environment.

Maxwell's equations (engineering)
... is a set of four partial differential equations with secondary conditions, describing the interrelationship between the electric and magnetic fields.

Mechanisms of defense (psychoanalysis)
... describes the processes that decide whether drive demands get access to the pre-consciousness and consciousness or remain unconscious. In addition mechanisms of defense define the manner of realization in the psychic apparatus. They are functions of the *Ego* (see second topographical model).

Memory (engineering)
... is a unit in a computational system that is able to store data that can be accessed by the system.

Memory (psychoanalysis)
... has a subjective quality due to the subjective quality of the original experience which can only be attempted to be reconstructed.

Memory (psychology)
... is the ability of an organism to store, retain and retrieve information.

Memory trace (psychoanalysis)
... describes as a psycho-physiological concept the modality of representing memories in the mind. Memory traces are stored in different systems. They are preserved permanently untill they are reactivated, once they have been assigned.

Model, first topographical ~
... describes the first of two topographical models by Sigmund Freud that divides the psychic apparatus into three agencies – unconscious, preconscious, conscious.

Model, second topographical ~
... describes the second of two topographical models by Sigmund Freud that divides the psychic apparatus into three functional agencies – Id, Ego, Super-Ego. Also known as structural model.

Module (computer science)
... is a unit containing functions with closely related relevance. It has a well defined interfaces to other modules.

Motility (psychoanalysis)
… describes the function of executing movements.

Multi agent system (engineering)
… is a system of interacting (software) agents. They need the ability to coope-rate, coordinate and negotiate with each other and human operators.

Network (engineering)
… is a wired or wireless connection of computer nodes for the purpose of communication between those nodes.

Network node
… is a computational unit (e. g. an embedded system) in a computer network that can receive, store, manipulate, and transmit data. In this sense, a network node is, for example, a sensor, actuator or controller.

Neuro-symbol (engineering)
… describes the basic processing units of neuro-symbolic networks. Neuro-symbols combine characteristics of neural and symbolic information processing and are to be seen in the first and second abstraction layer of Luria's three-layer model.

Neurology
… is a medical science that specializes in the nervous system of the human body.

Neuropsychoanalysis
… is the research field concerned with the correlation of psychoanalytic and neuroscientific terms and concepts.

npsa
… is the International Neuropsychoanalysis Society. The website address is www.neuropsa.org.uk/npsa.

Palsy, cerebral (neurology)
… is a motility disturbance as a consequence of damage to the brain.

Pedagogy
… is the science of education and development of humans.

Perception-consciousness (psychoanalysis)
… is defined as a function of the first topographical model and serves as a pool for received information from the inner and the outer world and assigns psychic energy to it.

Pre-consciousness (psychoanalysis)
... is defined as a function of the first topographical model and serves as a pool of presentations and affects which are not conscious at a given moment but can become conscious.

Presentation, thing ~ (psychoanalysis)
... corresponds to sensorial characteristics of an object and is mainly processed in the Id.

Process (engineering)
... defines interdependent operations in a technical system that converts, transports, and stores matter, energy, or information. The process is based on functional units that define the behavior of the process.

Process, primary (psychoanalysis)
... represents the modality which handles unorganized and contradictory processes located in the Id. See also Secondary process.

Process (psychoanalysis)
See *secondary process* and *primary process*.

Process control (engineering)
... describes methods, architecture, and algorithms that manage a process.

Process, secondary (psychoanalysis)
... comprises pre-conscious and conscious. In contrast to the primary process the satisfaction of the pleasure principal is inhibited. The secondary process is characterized by rational thoughts and conscious actions.

Propulsion technology (engineering)
... is technology to create motion. Machines can move items like wheels or arms by using, for example, electrical, mechanical or pneumatic power.

Psyche (psychoanalysis)
In ancient Greek, *psyche* refers to a concept of the self, including mind and soul. In Freud's structural theory, the psyche is composed of three components: the Ego, the Super-Ego and the Id. The forces of these three agencies determine behavior and subjective experience as well as inner-psychic conflicts.

Psychic apparatus (psychoanalysis)
... implicates all functions and processes of the three instances of the second topographical model. In general it translates perceived information into movements.

Psychoanalysis
    ... is a science developed by Sigmund Freud, combining three different aspects: a research method of the unconscious meaning behind certain phenomena such as dreams, actions etc.; a psychotherapeutic treatment founded on the interpretation of the resistance, the transference and the unconscious wish; and all psychological and psychopathological theories emerging from psychoanalytic treatment and research.

Psychology
    ... is an empirical science studying mental processes, subjective experience and human behavior.

Real-time application (engineering)
    ... has to fulfill tasks within specified time constraints.

Reality testing (psychoanalysis)
    ... is an Ego function for differentiating between the inner presentation and the outer reality. Therefore it enables the differentiation between imagination and reality.

Redundancy (engineering)
    ... describes a method to increase, for example, the reliability of a system by duplicating critical functional units. Depending on the type of critical functional units one can distinguish between hardware-, information-, time-, and software redundancy.

Repression (psychoanalysis)
    ... is a type of defense mechanism that is responsible for transferring conscious and pre-conscious contents to the unconscious.

Safety (engineering)
    ... is the ability of an entity not to cause critical or catastrophic events. A device is considered to be safe if it is reliable enough to cause no harm.

Scenario (ARS)
    ... is a sequence of images.

Security (computer engineering)
    ... describes methods to avoid unauthorized access.

Sensor (engineering)
    ... is the counterpart to an actuator. A technical device that is able to capture a physical or chemical quantity as a measurable value.

Sensor fusion (engineering)
... describes the combination of sensor data that originates from different sensor types. Compared to single sensor measurements the quality of information should increase through the use of sensor fusion.

Simulation, social ~ (engineering)
... is a computer simulation used to research social phenomena like cooperation, competition, or social network dynamics. A subset is an Agent Based Social Simulation (ABSS) which deals with agent based modeling of the simulation.

Situation awareness (engineering)
... describes the process of perceiving the environment, the comprehension of its meaning and future impact.

Software (engineering)
... describes a collection of computer programs, procedures and documentation that perform tasks on a computer.

Software-radio (engineering)
Nearly all radio functions are implemented by software. The idea is to get the software as close to the antenna as is feasible and turn hardware problems into software problems.

Super-Ego (psychoanalysis)
One of the three instances defined in Freud's second topographical model. It contains restrictions, demands, and rewards.

State machine (engineering)
... is a functional model composed of a number of states, transitions between those states, and actions.

Symbol (engineering)
... is an object, picture, or other concrete representation of ideas, concepts, or other abstractions.

System (engineering)
A system consists of multiple units that are functionally or structurally connected. The term refers to a coherent, connected unit. A computer system, for example, can consist of a computer and its peripheral units like printer, screen, mouse and the like.

Template (engineering)
... is a predefined mask that can be used as an archetype or for comparison.

Test bench (engineering)
... is a virtual environment used to verify the correctness and soundness of a model or design. Therefore, input data are generated and the according output data provided by the model or design are validated against the expected output data.

Thinking (psychoanalysis)
The main task of thinking is to structure and organize primary-processes. This organizational task proceeds on the basis of culturally influenced logic and results in the conversion to the secondary processes.

Top-down design (engineering)
As opposed to *bottom-up design*, it refers to a design method that starts its design process with the problem to be solved or the task to be committed. The designer tries to identify the necessary functionality in order to overcome the problem. This functionality is then further subdivided until existing solutions can perform subtasks.

Ubiquitous computing (engineering)
... integrates information processing thoroughly into everyday objects and activities.

Unconscious, the ~ (psychoanalysis)
... describes all processes and functions of the human mental apparatus that the subject is not aware of. It represents the main part of the psychic processes.

Wish (psychoanalysis)
... drives humans to regain an already perceived, lustful situation to satisfy a need.

Word presentation (psychoanalysis)
... refers to the description of an object using a set of symbols, like verbal expression in case of humans. Word presentations are processed in the Ego and form the basis for thinking.

World representation (engineering)
... is a system (e.g. an autonomous agent) that has this structure to represent all data that it has available describing the environment it is located in.

Z3 (engineering)
... was the first functional programmable computer in the world completed in 1941 by Konrad Zuse (Germany).

Abbreviations

ACL	Agent Communication Languages
AI	Artificial Intelligence
ANN	Artificial Neural Network
ARS	Artificial Recognition System
ARS-PA	Artificial Recognition System-PsychoAnalysis
BA	Basic Assumption
BAD	Basic Assumption Dependency
BAFF	Basic Assumption Fight-Flight
BAP	Basic Assumption Pairing
BDI	Believe, Desire, Intention
BFG	Bubble Family Game
CI	Cognitive Intelligence
CIM	Computer Integrated Manufacturing
CogAff	Cognition and Affect
D	Depressive positions
EC	Extended Consciousness
ECT	Electroconvulsive Therapy
EMA	Emotion and Adaptation
ENF	First International Engineering & Neuropsychoanalysis Forum
FIPA	Foundation for Intelligent Physical Agents
HLL	High Level Language
IEEE	Institute of Electrical and Electronics Engineers, Inc.
IMF	Intrinsic Motive Formation
INDIN	International Conference on Industrial Informatics
KQML	Knowledge Query and Manipulation Language
MAS	Multi Agent System
MIT	Massachusetts Institute of Technology
NPSA	Neuropsychoanalysis
OCC	Ortony-Collins-Clore Model
OCD	Obsessive Compulsive Disorder
OOD	Object Oriented Design
PAG	Periaqueductal Gray
PC	Personal Computer
PC	Primary Consciousness
PI	Projective Identification
PLCs	Programmable Logic Controller
PROM	Programmable Read Only Memory
PS	Paranoid-schizoid Positions
PTSD	Post-traumatic Stress Disorder

RAM	Random Access Memory
ROM	Read Only Memory
SENSE	Smart Embedded Network of Sensing Entities
TMS	Transcranial Magnetic Stimulation
VLIW	Very Long Instruction Word

Index

436